Formation Control of Multiple Autonomous Vehicle Systems

Formation Control of Multiple Autonomous Vehicle Systems

Hugh H.T. Liu
University of Toronto
Toronto, Canada

Bo Zhu
University of Electronic Science and Technology of China
Chengdu, China

WILEY

This edition first published 2018

Registered Office
John Wiley & Sons Ltd, The Atrium, Southern Gate, Chichester, West Sussex, PO19 8SQ, UK

Editorial Offices
9600 Garsington Road, Oxford, OX4 2DQ, UK
The Atrium, Southern Gate, Chichester, West Sussex, PO19 8SQ, UK

For details of our global editorial offices, customer services, and more information about Wiley products visit us at www.wiley.com.

Wiley also publishes its books in a variety of electronic formats and by print-on-demand. Some content that appears in standard print versions of this book may not be available in other formats.

Library of Congress Cataloging-in-Publication Data:

Names: Liu, Hugh H.T., author. | Zhu, Bo (Mechanical engineer), author.
Title: Formation control of multiple autonomous vehicle systems / Hugh H.T. Liu, Bo Zhu.
Description: First edition. | Hoboken, NJ : Wiley, 2018. | Includes bibliographical references. |
Identifiers: LCCN 2018016968 (print) | LCCN 2018028265 (ebook) | ISBN 9781119263043 (Adobe PDF) | ISBN 9781119263050 (ePub) | ISBN 9781119263067 (hardcover)
Subjects: LCSH: Autonomous vehicles–Case studies. | Formation control (Machine theory)–Case studies. | Synchronization–Case studies. | Motion control devices–Case studies.
Classification: LCC TL152.8 (ebook) | LCC TL152.8 .L58 2018 (print) | DDC 629.04–dc23
LC record available at https://lccn.loc.gov/2018016968

Cover Design: Wiley
Cover Image: © Kirill_Savenko/Getty Images

Set in 10/12pt WarnockPro by SPi Global, Chennai, India

Printed and bound by CPI Group (UK) Ltd, Croydon, CR0 4YY

10 9 8 7 6 5 4 3 2 1

This book is dedicated to:

My wife, Hong, and my children, Alan and Connie
–Hugh H.T. Liu

My wife, Qian Mo, and my parents, Genyi Zhu and Gengxin Zhang
–Bo Zhu

Contents

Preface

Background

Autonomous vehicles, or more precisely, autonomous vehicle systems, have received much attention from the public, thanks to well publicized products and product prototypes. Self-driving cars, drones and humanoid robots, are popular in exhibitions and shows. Some have even seen successful commercial applications. The smart, self-navigating iRobot^{TM} is no longer a curiosity from the research laboratory. Instead, iRobot products are already on the shelves of mainstream retailers, offering consumers their services, from vacuuming to mopping. In research and development, autonomous vehicle systems (AVSs) are receiving close attention, not only because of the wide range of their potential applications, but also because of the wide variety of vehicle configurations and platforms: cars, robots, spacecraft, and unmanned aerial vehicles (UAVs), to name just a few.

Even though different vehicle systems may have different configurations, they do share some common characteristics. For example, they must have some motion mechanism to enable movement; they are equipped with multiple sensory components to measure and collect vehicular and environmental information; they have "brains" (computer processors) to make automatic decisions to generate or regulate their motion based on the collected information, giving them a certain level of autonomy in accomplishing their missions. The enabling technologies behind these vehicles or vehicle systems – for example, the guidance, control and navigation technologies, the sensors and sensing technologies, and the communication protocols – represent state-of-the-art capabilities and highlight the current challenges and limitations in the field. Their capabilities and challenges make AVSs an exciting, almost ideal field to work in. As a result, AVSs have exhibited rapid progress in terms of research and development, from design to deployment.

Naturally, if we treat each vehicle as a self-regulating system, a fleet of vehicles will form a network of systems, or a system of systems. Such a macro viewpoint definitely broadens the horizon of AVSs to a new level. One may recognize the potential when a network of AVSs is considered. Imagine a single autonomous mobile robot is dispatched to survey and map a particular terrain, or an autonomous UAV flies over an area on an aerial photography mission: a coordinated group of mobile robots or a fleet of UAVs in formation can increase the scope of coverage dramatically. In another scenario, an individual robot or UAV has limited payload capacity, yet such tasks may be carried out by a team of cooperative robots or UAVs. In space applications, satellites stay in formation, to offer a wider coverage in telecommunications. Examples go on and on. If we group multiple autonomous vehicle systems together, we can significantly enhance performance and capacity, and multiple vehicle systems in coordination accomplish missions or tasks that a single vehicle cannot handle. The benefits are obvious.

We introduce a general term, *formation in motion*, for the phenomenon of multiple AVSs moving in coordination. One critical technology behind successful formation in motion is the control strategy that commands, coordinates, and adjusts the vehicles autonomously. As such, the *formation control of multiple autonomous vehicle systems* is the topic of interest to which this book is dedicated.

Contents of the Book

First of all, let us clarify what "formation control" or "formation in motion" refers to in this book. The concept of formation in motion can be intuitively described as a group of vehicle systems in motion that stays in a fixed pattern (e.g. a geometric shape). We may extend the concept to allow these vehicles to follow a dynamic pattern (e.g. spinning in a geometric fashion). Accordingly, formation control refers to control actions that ensure the group of vehicles stays in fixed or dynamic formation during its movement. It is worth pointing out that formation control is relevant to several associated concepts, such as coordinated control and cooperative control. Coordination, by definition, refers to the the harmonious functioning of parts for effective results. Cooperative control is a more general term, to address control and communication mechanisms that work together for control of large-scale dynamic systems. It is obvious that formation control demonstrates distinctive features of shape-keeping and, by implication, of motion synchronization. In this book, we assume we are dealing with formation control unless otherwise specified.

Formation control of AVSs involves the dynamics and control aspects of the dynamic behaviour of multi-vehicle systems, the design of proper control techniques to regulate the formation motion of these vehicles, and the development of system-level decision-making strategies to increase the level of autonomy for the whole group of vehicles, enabling them to carry out their missions. The fundamental concepts of dynamics and control are the main focus presented in this book. Formation control involves communication protocols, network frameworks, and information technology. Covering these associated technologies goes beyond the scope of this book, but the relevant concepts will be introduced at suitable points in the text.

In terms of dynamics, attention is focused on multi-vehicle systems dynamics. The intention is to develop a uniform paradigm for describing vehicle systems' dynamic behaviour, addressing both individual vehicle motion and overall group movement. Interactions between vehicles are also covered. Considering various vehicle configuration possibilities and even heterogeneous vehicles, it is important to have a unified platform so as to provide a foundation for analysis.

Similarly, regarding control, the intention is to focus on formation control, in other words the control strategy and its implementation in each individual vehicle so that it remains in fixed or dynamic formation while in motion. We shall cover fundamental

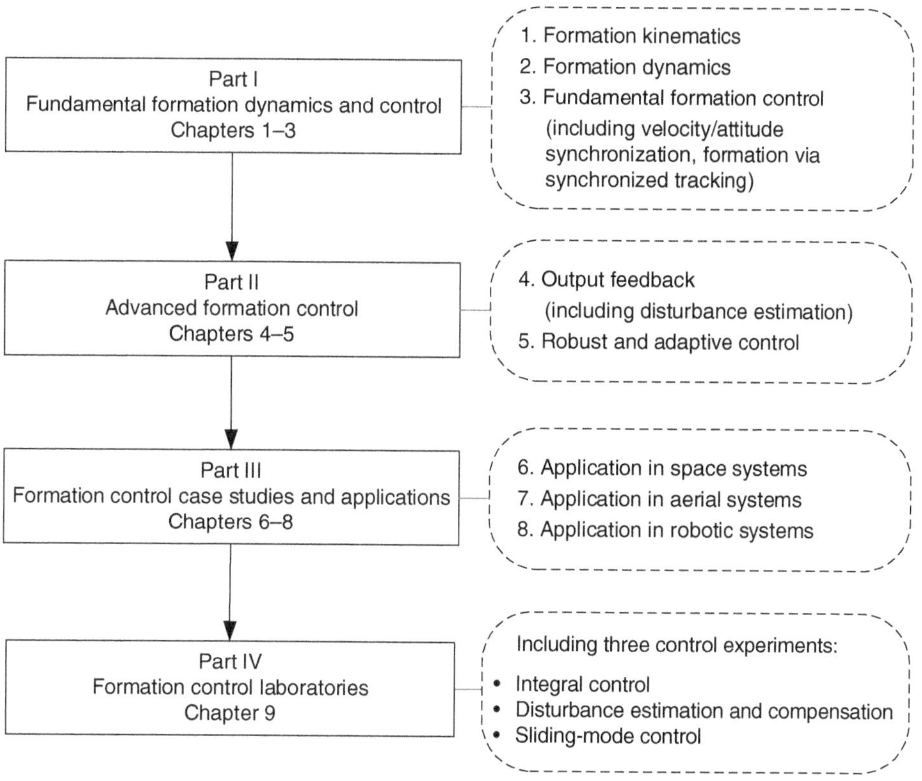

Figure 1 The roadmap of the book.

formation control concepts, as well as more advanced topics relating to special cases or applications.

In addition, several formation control application case studies are presented in this book. These cases cover descriptions, design, and control of different vehicle systems, including robotics, space applications, and UAVs. A dedicated section covers design of laboratory experiments. These exercises will give readers opportunities to further enhance their learning experience.

The roadmap of the rest of the book is shown in Figure 1.

After the Preface, the rest of the book is organized into four parts and nine chapters. Part I covers the fundamental topics, including vectorial formation dynamics (Chapter 2) and fundamental formation control (Chapter 3). Part II addresses advanced topics, including output feedback formation control (Chapter 4) and robust and adaptive formation control (Chapter 5). Part III focuses on formation control application case studies in a wide range of areas. Chapter 6 describes formation control for space systems, Chapter 7 presents formation control for aerial systems, Chapter 8 describes formation control for robotic systems. Part IV presents a unique laboratory, a system of three degree-of-freedom (DOF) desktop "helicopters", which is an excellent platform to investigate formation dynamics and control issues experimentally.

Curriculum

As mentioned before, in this book, we focus on specific formation behaviours and explore a unified dynamic behaviour description with various formation control strategies. The book is intended to be used as a textbook for a graduate-level course. It can certainly serve as a reference book or textbook for a senior undergraduate course. In North America, a standard graduate course for one semester typically consists of 2-hour lectures. A suggested course syllabus is provided in Table 1 as the recommended template for the course delivery.

Table 1 Suggested syllabus.

Week	Topic	Lab work
1	Chapter 1: Introduction	
2	Chapter 2: Formation Dynamics	
3	Chapter 2 (cont'd)	Lab: 3DOF modelling
4	Chapter 3: Fundamental Formation Control	
5	Chapter 3 (cont'd)	Lab: 3DOF Control
6	Chapter 4: Output Feedback	
7	Chapter 5: Robust and Adaptive Formation Control	Lab: 3DOF Advanced Control
8	Chapter 6: Space Systems	
9	Chapter 7: Aerial Systems	Lab: 3DOF UAV Control
10	Chapter 8: Robotic Systems	
11	Course Review	
12	Laboratory Project Demonstration	
13	Exam	

List of Tables

List of Figures

Acknowledgments

This book gives a self-contained presentation of our research work in control of multi-vehicle systems. Some materials have been adapted from a number of our publications. We acknowledge the IEEE, IET, AIAA, Elsevier, John Wiley & Sons, and Taylor & Francis for granting us the permission to reuse materials from those publications. We also gratefully acknowledge the support for our research by the Natural Sciences and Engineering Research Council of Canada (grant no. RGPIN-2017-06708), the National Nature Science Foundation of China (grant nos. 61773095 and 61304016), and the Fundamental Research Funds for the Central Universities (grant no. ZYGX2016J161).

In particular, we acknowledge the following publishers for granting us the permission to reuse materials from our publications copyrighted by these publishers in the book:

©2008 IEEE. Norman H.M. Li and Hugh H.T. Liu, Formation UAV flight control using virtual structure and motion synchronization, the 2008 American Control Conference, Seattle, Washington, pp. 1782–1787, June 11–13, 2008 (material found in Section 2.1.2)

©2013 IEEE. B. Zhu, W. Sun, and C. Meng, Position tracking of multi double-integrator dynamics by bounded distributed control without velocity measurements, the 2013 American Control Conference, Washington, DC, pp. 4033–4038, 2013. (material found in Section 4.4)

©2015 IEEE. C. Peng, B. Zhu, L. Yin, B. Yang, and C. Wang, Attitude synchronization of multiple 3-DOF helicopters without angular velocity measurements by bounded distributed control, the 34th Chinese Control Conference, Hangzhou, Zhejiang, pp. 7196–7201, 2015 (material found in Section 4.4)

©2014 IEEE. B. Zhu, Z. Li, Hugh H.T. Liu, and H. Gao, Robust second-order tracking of multi-vehicle systems without velocity measurements on directed communication topologies, the 2014 American Control Conference, Portland, Oregon, pp. 5414–5419, 2014 (material found in Sections 4.5 and 5.4)

©2015 IEEE. Z. Li, Hugh H.T. Liu, B. Zhu, and H. Gao, Robust second-order consensus tracking of multiple 3-DOF laboratory helicopters via output-feedback, *IEEE/ASME Transactions on Mechatronics*, vol. 20, no. 5, 2538–2549 (material found in Sections 5.5 and 9.5)

©2005 IET. Reproduced by permission of the Institution of Engineering & Technology J. Shan, Hugh H.T. Liu, and S. Nowotny, Synchronized trajectory tracking control of multiple 3-DOF experimental helicopters, *IEE Proceedings of Control Theory and Applications*, vol. 152, pp. 683–692, 2005.(material found in Sections 5.6 and 9.3)

©2005 AIAA. J. Shan and Hugh H.T. Liu, Close formation flight control with motion synchronization, *AIAA Journal of Guidance, Control, and Dynamics*, vol. 28, AIAA-13953-980, pp. 1316–1320, 2005 (material found in Chapter 7)

©2016 AIAA. Q. Zhang, Hugh H.T. Liu. Robust design of close formation flight control via uncertainty and disturbance estimator, the 2016 AIAA Guidance, Navigation, and Control Conference, San Diego, California, AIAA 2016-2102, pp. 1–15, 2016 (material found in Chapter 7)

©2007 ASME. Hugh H.T. Liu, J. Shan, and D. Sun, Adaptive synchronization control of multiple spacecraft formation flying, *Journal of Dynamic Systems, Measurement, and Control*, vol. 129, pp. 337–342, 2007. (material found in Section 5.7 and Chapter 6)

©2015 Elsevier. B. Zhu, Hugh H.T. Liu, and Z. Li, Robust distributed attitude synchronization of multiple three-DOF experimental helicopters, *Control Engineering Practice*, vol. 36, pp. 87–99, 2015 (material found in Sections 5.2 and 9.4)

©2016 Wiley. B. Zhu, C. Meng, and G. Hu, Robust consensus tracking of double integrator dynamics by bounded distributed control, *International Journal of Robust and Nonlinear Control*, vol. 26, no. 7, pp. 1489–1511, 2016 (material found in Sections 4.6 and 5.3)

Part I

Formation Control: Fundamental Concepts

1

Formation Kinematics

This chapter introduces the notation to be used in the book, as well as the subject of vectorial kinematics, which is frequently used to derive equations of motion.

1.1   Notation

$\rightarrow$	tends (or converges) to
$\Rightarrow$	implies
$\equiv$	identically equals (or equal)
$\overset{\Delta}{=}$	defined as
$\ll$	much smaller than
$\gg$	much greater than
$\forall$	for all
$\exists$	(if) there exists
$\in$	belongs to
$\notin$	does not belong to
$\subset$	a strict subset of
$\subseteq$	a subset of
$\cap$	intersection
$\cup$	union
$\emptyset$	empty set
$\mapsto$	maps to
Σ	summation
Π	left product
$\otimes$	Kronecker product
∞	positive infinity

Formation Control of Multiple Autonomous Vehicle Systems, First Edition. Hugh H.T. Liu and Bo Zhu.
© 2018 John Wiley & Sons Ltd. Published 2018 by John Wiley & Sons Ltd.

$\mathbb{R}$	set of real numbers
$\mathbb{R}^k$	set of $k \times 1$ real vectors
$\mathbb{R}^{m\times n}$	set of $m \times n$ real matrices
$\mathbb{C}$	set of complex numbers
$\mathbb{C}^k$	set of $k \times 1$ complex vectors
$\mathbb{C}^+$	set of complex numbers with positive real parts
$\mathbb{C}^-$	set of complex numbers with negative real parts
$B(x, \epsilon)$	open ball centered at x with radius ϵ
$\|z\|$	amplitude (or absolute value) of number z
$Re\,(z)$	real part of number z
$Im\,(z)$	imaginary part of number z
x^T	transpose of a real vector x
$\\|x\\|$	2-norm of a real vector x
$\\|x\\|_p$	p-norm of a real vector x
A^T	transpose of a real matrix A
$\\|A\\|$	induced 2-norm of a real matrix A
$\\|A\\|_p$	induced p-norm of a real matrix A
$\|A\|$ or $\det(A)$	the determinant of a square matrix A
$A > 0$	a positive matrix A
$A \geq 0$	a nonnegative matrix A
e^z	exponential of a real or complex number z
e^A	exponential of a real matrix A
$\rho(A)$	spectral radius of matrix A
$\lambda_i(A)$	the i-th eigenvalue of matrix A
$\lambda_{\max}(A)$	the maximum eigenvalue of a real symmetric matrix A
$\lambda_{\min}(A)$	the minimum eigenvalue of a real symmetric matrix A
$rank(A)$	rank of matrix A
$\mathrm{diag}\,(a_1, \cdots, a_k)$	a diagonal matrix with diagonal entries a_1 to a_k
$\mathrm{diag}\,(A_1, \cdots, A_k)$	a block diagonal matrix with diagonal blocks A_1 to A_k
max	maximum
min	minimum
sup	supremum, the least upper bound
inf	infimum, the greatest lower bound
sin	sine function

cos	cosine function
sign	signum function
tanh	tangent hyperbolic function
sec	tangent hyperbolic function
$\mathbf{1}_k$	$k \times 1$ column vector of all ones
$\mathbf{0}_k$	$k \times 1$ column vector of all zeros
I_m	$m \times m$ identity matrix
i	imaginary unit
$\mathbf{0}_{m\times n}$	$m \times n$ zero matrix

1.2 Vectorial Kinematics

The motion of an individual vehicle, is a six degree-of-freedom movement in space with respect to time. If we borrow the concept of rigid-body dynamic behaviour, such movement is often captured by a translational movement of a mass point (e.g. the centre of mass) and a rotational movement about an instantaneous axis through that point. Therefore, a distinctive description of translational or rotational dynamic behaviour is often developed through vectorial kinematics and dynamics.

1.2.1 Frame Rotation

It is essential to know how to deal with several reference frames and the transformation of the matrix representations of a vector (since the representation depends on the specific reference frame) from one frame to another. Only relative *rotation* (orientation change) between reference frames is important when considering representation of vectors. The relative translation does not affect the components of a vector since neither direction nor magnitude depends on the placement of the frame's origin. Translational motion between frames can be treated in the same way as Galilean transformation.

The physical description of motion of a mass point requires an origin to construct a vector. It is different from the general statement of independence of a vector from a reference frame origin.

Rotation Matrix

A vector $\underset{\rightarrow}{\boldsymbol{v}}$ has different expressions under two different frames $\mathcal{F}_a$ and $\mathcal{F}_b$:

$$\underset{\rightarrow}{\boldsymbol{v}} = \underset{\rightarrow a}{\boldsymbol{F}}^T \underline{\boldsymbol{v}}_a = \underset{\rightarrow b}{\boldsymbol{F}}^T \underline{\boldsymbol{v}}_b \tag{1.1}$$

where $\underline{\boldsymbol{v}}_a$ and $\underline{\boldsymbol{v}}_b$ are numerical expressions of vector $\underset{\rightarrow}{\boldsymbol{v}}$ under frames F_a and F_b respectively, sometimes referred to as numerical vectors. The vector-like $\underset{\rightarrow a}{\boldsymbol{F}}$ and $\underset{\rightarrow b}{\boldsymbol{F}}$ are vectorized representations of frame axes, $\underset{\rightarrow a}{\boldsymbol{F}} = [\underset{\rightarrow 1}{\boldsymbol{a}}\ \underset{\rightarrow 2}{\boldsymbol{a}}\ \underset{\rightarrow 3}{\boldsymbol{a}}]^T$, $\underset{\rightarrow b}{\boldsymbol{F}} = [\underset{\rightarrow 1}{\boldsymbol{b}}\ \underset{\rightarrow 2}{\boldsymbol{b}}\ \underset{\rightarrow 3}{\boldsymbol{b}}]^T$. We simply call these vectrices, a made-up name for axis vectors presented in matrix format. It is obvious that the relationship between two expressions lies in the relationship between these two frames.

Consider two reference frames $\mathcal{F}_a : \{\underset{\rightarrow}{\mathbf{a}}_1, \underset{\rightarrow}{\mathbf{a}}_2, \underset{\rightarrow}{\mathbf{a}}_3\}$ and $\mathcal{F}_b : \{\underset{\rightarrow}{\mathbf{b}}_1, \underset{\rightarrow}{\mathbf{b}}_2, \underset{\rightarrow}{\mathbf{b}}_3\}$. Rotating from $\mathcal{F}_a$ to $\mathcal{F}_b$ means that $\underset{\rightarrow}{\mathbf{F}}_a$ to $\underset{\rightarrow}{\mathbf{F}}_b$: $\underset{\rightarrow}{\mathbf{F}}_a \Rightarrow \underset{\rightarrow}{\mathbf{F}}_b$.

From (1.1) we have

$$\underline{v}_b = \underset{\rightarrow}{\mathbf{F}}_b \cdot \underset{\rightarrow}{v} = \underset{\rightarrow}{\mathbf{F}}_b \cdot \underset{\rightarrow}{\mathbf{F}}_a^T \underline{v}_a \triangleq \underline{C}_{ba}\underline{v}_a \tag{1.2}$$

where

$$\underset{\rightarrow}{\mathbf{F}}_b \cdot \underset{\rightarrow}{\mathbf{F}}_a^T = \begin{bmatrix} \underset{\rightarrow}{\mathbf{b}}_1 \cdot \underset{\rightarrow}{\mathbf{a}}_1 & \underset{\rightarrow}{\mathbf{b}}_1 \cdot \underset{\rightarrow}{\mathbf{a}}_2 & \underset{\rightarrow}{\mathbf{b}}_1 \cdot \underset{\rightarrow}{\mathbf{a}}_3 \\ \underset{\rightarrow}{\mathbf{b}}_2 \cdot \underset{\rightarrow}{\mathbf{a}}_1 & \underset{\rightarrow}{\mathbf{b}}_2 \cdot \underset{\rightarrow}{\mathbf{a}}_2 & \underset{\rightarrow}{\mathbf{b}}_2 \cdot \underset{\rightarrow}{\mathbf{a}}_3 \\ \underset{\rightarrow}{\mathbf{b}}_3 \cdot \underset{\rightarrow}{\mathbf{a}}_1 & \underset{\rightarrow}{\mathbf{b}}_3 \cdot \underset{\rightarrow}{\mathbf{a}}_2 & \underset{\rightarrow}{\mathbf{b}}_3 \cdot \underset{\rightarrow}{\mathbf{a}}_3 \end{bmatrix} \triangleq \begin{bmatrix} c_{11} & c_{12} & c_{13} \\ c_{21} & c_{22} & c_{23} \\ c_{31} & c_{32} & c_{13} \end{bmatrix} = \underline{C}_{ba}$$

The short expression then becomes

$$\underline{u}_b = \underline{C}_{ba}\underline{u}_a \tag{1.3}$$

$$\underset{\rightarrow}{\mathbf{F}}_b = \underline{C}_{ba}\underset{\rightarrow}{\mathbf{F}}_a \tag{1.4}$$

Switching the letters a and b,

$$\underline{C}_{ab} = \underset{\rightarrow}{\mathbf{F}}_a \cdot \underset{\rightarrow}{\mathbf{F}}_b^T = \begin{bmatrix} \underset{\rightarrow}{\mathbf{a}}_1 \cdot \underset{\rightarrow}{\mathbf{b}}_1 & \underset{\rightarrow}{\mathbf{a}}_1 \cdot \underset{\rightarrow}{\mathbf{b}}_2 & \underset{\rightarrow}{\mathbf{a}}_1 \cdot \underset{\rightarrow}{\mathbf{b}}_3 \\ \underset{\rightarrow}{\mathbf{a}}_2 \cdot \underset{\rightarrow}{\mathbf{b}}_1 & \underset{\rightarrow}{\mathbf{a}}_2 \cdot \underset{\rightarrow}{\mathbf{b}}_2 & \underset{\rightarrow}{\mathbf{a}}_2 \cdot \underset{\rightarrow}{\mathbf{b}}_3 \\ \underset{\rightarrow}{\mathbf{a}}_3 \cdot \underset{\rightarrow}{\mathbf{b}}_1 & \underset{\rightarrow}{\mathbf{a}}_3 \cdot \underset{\rightarrow}{\mathbf{b}}_2 & \underset{\rightarrow}{\mathbf{a}}_3 \cdot \underset{\rightarrow}{\mathbf{b}}_3 \end{bmatrix} \tag{1.5}$$

Orthonormality

It can be shown that matrices $\underline{C}_{ab}$ and $\underline{C}_{ba}$ are orthonormal when both frames of reference $\mathcal{F}_a$ and $\mathcal{F}_b$ are orthonormal; in other words they have orthonormal basis vectors.

$$\underline{C}_{ab}^T\underline{C}_{ab} = \underline{1} \tag{1.6}$$

$$\underline{C}_{ba}^T\underline{C}_{ba} = \underline{1} \tag{1.7}$$

$$\underline{C}_{ab}^T = \underline{C}_{ba} = \underline{C}_{ab}^{-1} \tag{1.8}$$

$$\det \underline{C}_{ba} = 1 \tag{1.9}$$

Principal Rotations

There are three principal (basic) rotations of our interest:

- **$\underset{\rightarrow}{\mathbf{F}}_a \Rightarrow \underset{\rightarrow}{\mathbf{F}}_b$ about $\underset{\rightarrow}{\mathbf{a}}_1$ or $\underset{\rightarrow}{x}$**

$$[\underset{\rightarrow}{\mathbf{b}}_1 \quad \underset{\rightarrow}{\mathbf{b}}_2 \quad \underset{\rightarrow}{\mathbf{b}}_3] = [\underset{\rightarrow}{\mathbf{a}}_1 \quad \underset{\rightarrow}{\mathbf{a}}_2 \quad \underset{\rightarrow}{\mathbf{a}}_3] \begin{bmatrix} 1 & 0 & 0 \\ 0 & c_\sigma & -s_\sigma \\ 0 & s_\sigma & c_\sigma \end{bmatrix}$$

$$\underset{\rightarrow}{\mathbf{F}}_b^T \triangleq \underset{\rightarrow}{\mathbf{F}}_a^T \underline{C}_1^T(\sigma) \tag{1.10}$$

where $s_\sigma = \sin\sigma, c_\sigma = \cos\sigma$.

- $\underset{\rightarrow}{F}_a \Rightarrow \underset{\rightarrow}{F}_b$ **about** $\underset{\rightarrow}{a}_2$ **or** $\underset{\rightarrow}{y}$

$$[\underset{\rightarrow}{b}_1 \quad \underset{\rightarrow}{b}_2 \quad \underset{\rightarrow}{b}_3] = [\underset{\rightarrow}{a}_1 \quad \underset{\rightarrow}{a}_2 \quad \underset{\rightarrow}{a}_3]\begin{bmatrix} c_\sigma & 0 & s_\sigma \\ 0 & 1 & 0 \\ -s_\sigma & 0 & c_\sigma \end{bmatrix}$$

$$\underset{\rightarrow}{F}_b^T \overset{\Delta}{=} \underset{\rightarrow}{F}_a^T \underline{C}_2^T(\sigma) \tag{1.11}$$

- $\underset{\rightarrow}{F}_a \Rightarrow \underset{\rightarrow}{F}_b$ **about** $\underset{\rightarrow}{a}_3$ **or** $\underset{\rightarrow}{z}$

$$[\underset{\rightarrow}{b}_1 \quad \underset{\rightarrow}{b}_2 \quad \underset{\rightarrow}{b}_3] = [\underset{\rightarrow}{a}_1 \quad \underset{\rightarrow}{a}_2 \quad \underset{\rightarrow}{a}_3]\begin{bmatrix} c_\sigma & -s_\sigma & 0 \\ s_\sigma & c_\sigma & 0 \\ 0 & 0 & 1 \end{bmatrix}$$

$$\underset{\rightarrow}{F}_b^T \overset{\Delta}{=} \underset{\rightarrow}{F}_a^T \underline{C}_3^T(\sigma) \tag{1.12}$$

1.2.2 The Motion of a Vector

The motion of a vector represents its rate of change with time, which is described as the *time derivative* of the vector. Consider a vector $\underset{\rightarrow}{u}$ and its expressions $\underline{u}_a$ and $\underline{u}_b$ with respect to the reference frame $\mathcal{F}_a : \{\underset{\rightarrow}{a}_1, \underset{\rightarrow}{a}_2, \underset{\rightarrow}{a}_3\}$ and $\mathcal{F}_b : \{\underset{\rightarrow}{b}_1, \underset{\rightarrow}{b}_2, \underset{\rightarrow}{b}_3\}$ respectively; that is,

$$\underset{\rightarrow}{u} = \underset{\rightarrow}{F}_a^T \underline{u}_a = \underset{\rightarrow}{a}_1 u_{a,1} + \underset{\rightarrow}{a}_2 u_{a,2} + \underset{\rightarrow}{a}_3 u_{a,3} \tag{1.13}$$

$$= \underset{\rightarrow}{F}_b^T \underline{u}_b = \underset{\rightarrow}{b}_1 u_{b,1} + \underset{\rightarrow}{a}_2 u_{b,2} + \underset{\rightarrow}{b}_3 u_{b,3}. \tag{1.14}$$

The time derivative is a vector itself, and also has different expressions under these two frames of reference:

$$\frac{d}{dt}\underset{\rightarrow}{u} = \left(\frac{d}{dt}\underset{\rightarrow}{F}_a\right)^T \underline{u}_a + \underset{\rightarrow}{F}_a^T \underline{\dot{u}}_a \tag{1.15}$$

$$\frac{d}{dt}\underset{\rightarrow}{u} = \left(\frac{d}{dt}\underset{\rightarrow}{F}_b\right)^T \underline{u}_b + \underset{\rightarrow}{F}_b^T \underline{\dot{u}}_b \tag{1.16}$$

$$\frac{d}{dt}\underset{\rightarrow}{u} \overset{\Delta}{=} \underset{\rightarrow}{v} = \underset{\rightarrow}{F}_a^T \underline{v}_a = \underset{\rightarrow}{F}_b^T \underline{v}_b \tag{1.17}$$

Obviously, the expression of time derivative vector $\underset{\rightarrow}{v}$ depends on the time derivative of the vectrix, or the change of rate of basis vectors of the associated frame of reference.

Absolute and Relative Time Derivatives

Assume $\mathcal{F}_a$ represents an inertial space (a Newtonian absolute space). To an observer in $\mathcal{F}_a$, the basis vectors of $\mathcal{F}_a$, or the vectrix $\underset{\rightarrow}{F}_a$, will remain unchanged (no orientation change, and no magnitude change of course, since they are unit vectors). In other words, the first term in (1.15) is zero. Since $\mathcal{F}_a$ is an inertial frame, we define the time derivative of a vector $\underset{\rightarrow}{u}$ in $\mathcal{F}_a$ as an *absolute time derivative*, denoted by a bullet • superscript:

$$\overset{\bullet}{\underset{\rightarrow}{u}} \overset{\Delta}{=} \frac{d}{dt}(\underset{\rightarrow}{u})|_{\mathcal{F}_a} = \underset{\rightarrow}{F}_a^T \underline{\dot{u}}_a = \underset{\rightarrow}{a}_1 \dot{u}_{a,1} + \underset{\rightarrow}{a}_2 \dot{u}_{a,2} + \underset{\rightarrow}{a}_3 \dot{u}_{a,3} \tag{1.18}$$

Assume $\mathcal{F}_b$ is a moving frame of reference relative to $\mathcal{F}_a$. Denote the moving (rotating) rate of change with time by $\underset{\rightarrow}{\boldsymbol{\omega}}_{ba}$. Similarly, to an observer in $\mathcal{F}_b$, the vectrix $\underset{\rightarrow}{\boldsymbol{F}}_b$ also remains unchanged. In other words, the first term in (1.16) is zero. Therefore, the time derivative of a vector $\underset{\rightarrow}{\boldsymbol{u}}$ in the moving frame $\mathcal{F}_b$ is defined as a *relative time derivative*, denoted by a circle ∘ superscript:

$$\overset{\circ}{\underset{\rightarrow}{\boldsymbol{u}}} \overset{\Delta}{=} \frac{d}{dt}(\underset{\rightarrow}{\boldsymbol{u}})|_{\mathcal{F}_b} = \underset{\rightarrow}{\boldsymbol{F}}_b^T \dot{\underline{u}}_b = \underset{\rightarrow}{\boldsymbol{b}}_1 \dot{u}_{b,1} + \underset{\rightarrow}{\boldsymbol{a}}_2 \dot{u}_{b,2} + \underset{\rightarrow}{\boldsymbol{b}}_3 \dot{u}_{b,3} \tag{1.19}$$

We here note that $\overset{\bullet}{\underset{\rightarrow}{\boldsymbol{u}}}$ and $\overset{\circ}{\underset{\rightarrow}{\boldsymbol{u}}}$ are two different vectors,

$$\underset{\rightarrow}{\boldsymbol{v}} = \overset{\bullet}{\underset{\rightarrow}{\boldsymbol{u}}} \qquad \underline{v}_a = \dot{\underline{u}}_a \tag{1.20}$$

$$\underset{\rightarrow}{\boldsymbol{v}} \neq \overset{\circ}{\underset{\rightarrow}{\boldsymbol{u}}} \qquad \underline{v}_b \neq \dot{\underline{u}}_b \tag{1.21}$$

However, they are closely related to each other.

The definitions of absolute and relative derivatives are special cases of the following general expressions for time derivatives in a frame of reference. To an observer in a frame of reference $\mathcal{F}_x$ ($x = a, b, c, \ldots$) the basis vectors of the frame, or the vectrix, remain unchanged. Hence,

$$\left(\frac{d}{dt}\underset{\rightarrow}{\boldsymbol{F}}_x\right)_x \overset{\Delta}{=} \frac{d}{dt}(\underset{\rightarrow}{\boldsymbol{F}}_x)|_{\mathcal{F}_x} = \underset{\rightarrow}{\boldsymbol{O}} \tag{1.22}$$

and the special cases are:

$$\overset{\bullet}{\underset{\rightarrow}{\boldsymbol{F}}}_a = \underset{\rightarrow}{\boldsymbol{O}} \qquad \overset{\circ}{\underset{\rightarrow}{\boldsymbol{F}}}_b = \underset{\rightarrow}{\boldsymbol{O}}$$

The time derivative of a vector $\underset{\rightarrow}{\boldsymbol{u}}$ in the frame of reference $\mathcal{F}_x$ becomes:

$$\left(\frac{d}{dt}\underset{\rightarrow}{\boldsymbol{u}}\right)_x = \left(\frac{d}{dt}\underset{\rightarrow}{\boldsymbol{F}}_x\right)_x^T \underline{u}_x + \underset{\rightarrow}{\boldsymbol{F}}_x^T \dot{\underline{u}}_x = \underset{\rightarrow}{\boldsymbol{F}}_x^T \dot{\underline{u}}_x \tag{1.23}$$

However, when dealing with multiple frames of reference in one frame $\mathcal{F}_x$, the basis vector of another frame $\mathcal{F}_y$ is no longer unchanged with time. Therefore, it leads to

$$\left(\frac{d}{dt}\underset{\rightarrow}{\boldsymbol{u}}\right)_x = \left(\frac{d}{dt}\underset{\rightarrow}{\boldsymbol{F}}_y\right)_x^T \underline{u}_y + \underset{\rightarrow}{\boldsymbol{F}}_y^T \dot{\underline{u}}_y = \left(\frac{d}{dt}\underset{\rightarrow}{\boldsymbol{F}}_y\right)_x^T \underline{u}_y + \left(\frac{d}{dt}\underset{\rightarrow}{\boldsymbol{u}}\right)_y \tag{1.24}$$

In other words, the motion of a vector in frame $\mathcal{F}_x$ consists of the motion of this vector in frame $\mathcal{F}_y$ and the motion of frame $\mathcal{F}_y$ relative to frame $\mathcal{F}_x$.

Returning to our previous special cases (absolute and relative derivatives), (1.15) and (1.16):

$$\left(\frac{d}{dt}\underset{\rightarrow}{\boldsymbol{u}}\right)_a = \overset{\bullet}{\underset{\rightarrow}{\boldsymbol{F}}}{}_a^T \underline{u}_a + \underset{\rightarrow}{\boldsymbol{F}}_a^T \dot{\underline{u}}_a = \underset{\rightarrow}{\boldsymbol{F}}_a^T \dot{\underline{u}}_a \tag{1.25}$$

$$\left(\frac{d}{dt}\underset{\rightarrow}{\boldsymbol{u}}\right)_a = \overset{\bullet}{\underset{\rightarrow}{\boldsymbol{F}}}{}_b^T \underline{u}_b + \underset{\rightarrow}{\boldsymbol{F}}_b^T \dot{\underline{u}}_b \tag{1.26}$$

$$\left(\frac{d}{dt}\underset{\rightarrow}{\boldsymbol{u}}\right)_b = \overset{\circ}{\underset{\rightarrow}{\boldsymbol{F}}}{}_a^T \underline{u}_a + \underset{\rightarrow}{\boldsymbol{F}}_a^T \dot{\underline{u}}_a \tag{1.27}$$

$$\left(\frac{d}{dt}\underset{\rightarrow}{u}\right)_b = \overset{\circ}{\underset{\rightarrow}{F}}{}_b^T \underline{u}_b + \underset{\rightarrow}{F}{}_b^T \dot{\underline{u}}_b = \underset{\rightarrow}{F}{}_b^T \dot{\underline{u}}_b \tag{1.28}$$

For the absolute derivative $(\frac{d}{dt}\underset{\rightarrow}{u})_a$, we often drop the subscript a.

The focus now is placed on the rate of change of frame $\mathcal{F}_b$ relative to another frame $\mathcal{F}_a$.

Relative Rotation

To begin with, we look at the rotation of $\mathcal{F}_b$ relative to $\mathcal{F}_a$ about a *fixed* axis $\underset{\rightarrow}{a}$,

$$\underset{\rightarrow}{\omega}{}_{ba} = \underset{\rightarrow}{a}\dot{\theta}(t) \tag{1.29}$$

One can conclude that, for a unit vector $\underset{\rightarrow}{b}$ such as the axes of $\mathcal{F}_b$,

$$\overset{\bullet}{\underset{\rightarrow}{b}} = \underset{\rightarrow}{\omega}{}_{ba} \times \underset{\rightarrow}{b} \tag{1.30}$$

where the symbol $\times$ between two vectors denotes the the cross product or vector product in three-dimensional space (see Section 4.5.3 in the book by Polyanin and Manzhirov [1] for a definition).

General Rotation

Generally speaking, the angular velocity not only changes with the magnitude $\dot{\theta}$, but also changes its rotational orientation (there is no fixed axis). In other words, we are looking for a general description for angular velocity (*general angular velocity*), one that it is not associated with $\underset{\rightarrow}{a}$ like the one we had before: we are dealing with the general rotation case. From the previous case, we would like to see a general description of angular velocity, such that for an unit vector $\underset{\rightarrow}{d}$, the following equation still holds:

$$\overset{\bullet}{\underset{\rightarrow}{d}} = \underset{\rightarrow}{\omega}{}_{ba} \times \underset{\rightarrow}{d} \tag{1.31}$$

On the one hand,

$$\overset{\bullet}{\underset{\rightarrow}{d}} = \frac{d}{dt}(\underset{\rightarrow}{F}{}_b^T \underline{d}_b) = \overset{\bullet}{\underset{\rightarrow}{F}}{}_b^T \underline{d}_b + \underset{\rightarrow}{F}{}_b^T \dot{\underline{d}}_b = \overset{\bullet}{\underset{\rightarrow}{F}}{}_b^T \underline{d}_b, \qquad \dot{\underline{d}}_b = \underline{0} \tag{1.32}$$

and on the other hand,

$$\overset{\bullet}{\underset{\rightarrow}{d}} = \underset{\rightarrow}{\omega}{}_{ba} \times \underset{\rightarrow}{d} = \underset{\rightarrow}{F}{}_b^T \underline{\omega}{}_b^{ba\times} \underline{d}_b \tag{1.33}$$

Therefore,

$$\underline{\omega}{}_b^{ba\times} \underline{d}_b = \underset{\rightarrow}{F}{}_b \cdot \overset{\bullet}{\underset{\rightarrow}{d}} = \underset{\rightarrow}{F}{}_b \cdot \overset{\bullet}{\underset{\rightarrow}{F}}{}_b^T \underline{d}_b \tag{1.34}$$

Consider the arbitrary unit vector $\underset{\rightarrow}{d}$. We must have

$$\underline{\omega}{}_b^{ba\times} = \underset{\rightarrow}{F}{}_b \cdot \overset{\bullet}{\underset{\rightarrow}{F}}{}_b^T \tag{1.35}$$

We define *general angular velocity* as:

$$\underrightarrow{\boldsymbol{\omega}}_{ba} \triangleq \underrightarrow{\boldsymbol{F}}_b^T \underline{\boldsymbol{\omega}}_b^{ba} \tag{1.36}$$

where the expression matrix $\underline{\boldsymbol{\omega}}_b^{ba}$ is given by

$$\underline{\boldsymbol{\omega}}_b^{ba\times} \triangleq \underrightarrow{\boldsymbol{F}}_b \cdot \dot{\underrightarrow{\boldsymbol{F}}}_b^T \tag{1.37}$$

By that definition, we have the conclusion of an expression for general rotation (1.31).

Proof: First, we show that $\underline{\boldsymbol{\omega}}_b^{ba\times}$ is *skew-symmetric*:

$$\underline{\boldsymbol{\omega}}_b^{ba\times} = \underrightarrow{\boldsymbol{F}}_b \cdot \dot{\underrightarrow{\boldsymbol{F}}}_b^T = \begin{bmatrix} \underrightarrow{\boldsymbol{b}}_1 \\ \underrightarrow{\boldsymbol{b}}_2 \\ \underrightarrow{\boldsymbol{b}}_3 \end{bmatrix} \cdot \begin{bmatrix} \dot{\underrightarrow{\boldsymbol{b}}}_1 & \dot{\underrightarrow{\boldsymbol{b}}}_2 & \dot{\underrightarrow{\boldsymbol{b}}}_3 \end{bmatrix}$$

$$= \begin{bmatrix} \underrightarrow{\boldsymbol{b}}_1 \cdot \dot{\underrightarrow{\boldsymbol{b}}}_1 & \underrightarrow{\boldsymbol{b}}_1 \cdot \dot{\underrightarrow{\boldsymbol{b}}}_2 & \underrightarrow{\boldsymbol{b}}_1 \cdot \dot{\underrightarrow{\boldsymbol{b}}}_3 \\ \underrightarrow{\boldsymbol{b}}_2 \cdot \dot{\underrightarrow{\boldsymbol{b}}}_1 & \underrightarrow{\boldsymbol{b}}_2 \cdot \dot{\underrightarrow{\boldsymbol{b}}}_2 & \underrightarrow{\boldsymbol{b}}_2 \cdot \dot{\underrightarrow{\boldsymbol{b}}}_3 \\ \underrightarrow{\boldsymbol{b}}_3 \cdot \dot{\underrightarrow{\boldsymbol{b}}}_1 & \underrightarrow{\boldsymbol{b}}_3 \cdot \dot{\underrightarrow{\boldsymbol{b}}}_2 & \underrightarrow{\boldsymbol{b}}_3 \cdot \dot{\underrightarrow{\boldsymbol{b}}}_3 \end{bmatrix}$$

From the orthonormal basis vector characteristics: $\underrightarrow{\boldsymbol{b}}_i \cdot \underrightarrow{\boldsymbol{b}}_i = 1, \underrightarrow{\boldsymbol{b}}_i \cdot \underrightarrow{\boldsymbol{b}}_j = 0$, we can draw the conclusion that $\underrightarrow{\boldsymbol{b}}_i \cdot \dot{\underrightarrow{\boldsymbol{b}}}_i = 0, \underrightarrow{\boldsymbol{b}}_i \cdot \dot{\underrightarrow{\boldsymbol{b}}}_j = -\dot{\underrightarrow{\boldsymbol{b}}}_i \cdot \underrightarrow{\boldsymbol{b}}_j$ and find the skew-symmetric matrix.

Secondly, we find the derivative of basis vector $\underrightarrow{\boldsymbol{b}}$, (or any arbitrary vector $\underrightarrow{\boldsymbol{u}}$ fixed with $\mathcal{F}_b$):

$$\underrightarrow{\boldsymbol{\omega}}_{ba} \times \underrightarrow{\boldsymbol{b}} = \underrightarrow{\boldsymbol{F}}_b^T \underline{\boldsymbol{\omega}}_b^{ba\times} \underline{\boldsymbol{b}}_b = \underbrace{\underrightarrow{\boldsymbol{F}}_b^T \underrightarrow{\boldsymbol{F}}_b}_{\text{identity dyadic}} \cdot \dot{\underrightarrow{\boldsymbol{F}}}_b^T \underline{\boldsymbol{b}}_b = \dot{\underrightarrow{\boldsymbol{F}}}_b^T \underline{\boldsymbol{b}}_b + \underrightarrow{\boldsymbol{F}}_b^T \underbrace{\dot{\underline{\boldsymbol{b}}}_b}_{\text{basis unit length}} = \dot{\underrightarrow{\boldsymbol{b}}}$$ □

To conclude, for the general rotation of a basis vector, we have

$$\dot{\underrightarrow{\boldsymbol{b}}} = \underrightarrow{\boldsymbol{\omega}}_{ba} \times \underrightarrow{\boldsymbol{b}} \quad \text{where} \quad \underline{\boldsymbol{\omega}}_b^{ba\times} \triangleq \underrightarrow{\boldsymbol{F}}_b \cdot \dot{\underrightarrow{\boldsymbol{F}}}_b^T$$

$$\dot{\underrightarrow{\boldsymbol{F}}}_b = \underrightarrow{\boldsymbol{\omega}}_{ba} \times \underrightarrow{\boldsymbol{F}}_b \quad \dot{\underrightarrow{\boldsymbol{F}}}_b^T = \underrightarrow{\boldsymbol{\omega}}_{ba} \times \underrightarrow{\boldsymbol{F}}_b^T$$

Matrix Expression

From definition (1.37) and relationship equation (1.4), we have

$$\begin{aligned} \underline{\boldsymbol{\omega}}_b^{ba\times} &= \underrightarrow{\boldsymbol{F}}_b \cdot \dot{\underrightarrow{\boldsymbol{F}}}_b^T \\ &= \underline{\boldsymbol{C}}_{ba} \underrightarrow{\boldsymbol{F}}_a \cdot (\underrightarrow{\boldsymbol{F}}_a^T \dot{\underline{\boldsymbol{C}}}_{ab} + \dot{\underrightarrow{\boldsymbol{F}}}_a^T \underline{\boldsymbol{C}}_{ab}) \\ &= \underline{\boldsymbol{C}}_{ba} \dot{\underline{\boldsymbol{C}}}_{ab} \\ &= -\dot{\underline{\boldsymbol{C}}}_{ba} \underline{\boldsymbol{C}}_{ba}^T \end{aligned}$$

From a different perspective, for a unit vector $\underset{\rightarrow}{\boldsymbol{d}}$

$$\begin{aligned}
\overset{\bullet}{\underset{\rightarrow}{\boldsymbol{d}}} &= \underset{\rightarrow a}{\boldsymbol{F}}^T \dot{\underline{\boldsymbol{d}}}_a = \underset{\rightarrow b}{\boldsymbol{F}}^T \underline{\boldsymbol{C}}_{ba} \frac{d}{dt}(\underline{\boldsymbol{C}}_{ab} \underline{\boldsymbol{d}}_b) \\
&= \underset{\rightarrow b}{\boldsymbol{F}}^T \underline{\boldsymbol{C}}_{ba} \underline{\boldsymbol{C}}_{ab} \dot{\underline{\boldsymbol{d}}}_b + \underset{\rightarrow b}{\boldsymbol{F}}^T \underline{\boldsymbol{C}}_{ba} \dot{\underline{\boldsymbol{C}}}_{ab} \underline{\boldsymbol{d}}_b, \quad \dot{\underline{\boldsymbol{d}}}_b = \underline{\boldsymbol{0}} \\
&= \underset{\rightarrow b}{\boldsymbol{F}}^T \underline{\boldsymbol{C}}_{ba} \dot{\underline{\boldsymbol{C}}}_{ab} \underline{\boldsymbol{d}}_b
\end{aligned}$$

On the other hand,

$$\overset{\bullet}{\underset{\rightarrow}{\boldsymbol{d}}} = \underset{\rightarrow ba}{\boldsymbol{\omega}} \times \underset{\rightarrow}{\boldsymbol{d}} = \underset{\rightarrow b}{\boldsymbol{F}}^T \underline{\boldsymbol{\omega}}_b^{ba\times} \underline{\boldsymbol{d}}_b$$

Therefore, $\underline{\boldsymbol{\omega}}_b^{ba\times} = \underline{\boldsymbol{C}}_{ba} \dot{\underline{\boldsymbol{C}}}_{ab}$

In summary,

$$\underline{\boldsymbol{\omega}}_b^{ba\times} = \underline{\boldsymbol{C}}_{ba} \dot{\underline{\boldsymbol{C}}}_{ab} \tag{1.38}$$

$$\dot{\underline{\boldsymbol{C}}}_{ba} + \underline{\boldsymbol{\omega}}_b^{ba\times} \underline{\boldsymbol{C}}_{ba} = \underline{\boldsymbol{0}} \tag{1.39}$$

1.2.3 The First Time Derivative of a Vector

Consider the time derivative of an arbitrary vector,

$$\overset{\bullet}{\underset{\rightarrow}{\boldsymbol{u}}} = \overset{\bullet}{\underset{\rightarrow b}{\boldsymbol{F}}}{}^T \underline{\boldsymbol{u}}_b + \underset{\rightarrow b}{\boldsymbol{F}}^T \dot{\underline{\boldsymbol{u}}}_b = \underset{\rightarrow ba}{\boldsymbol{\omega}} \times \underset{\rightarrow b}{\boldsymbol{F}}^T \underline{\boldsymbol{u}}_b + \underset{\rightarrow b}{\boldsymbol{F}}^T \dot{\underline{\boldsymbol{u}}}_b$$

Since

$$\overset{\circ}{\underset{\rightarrow}{\boldsymbol{u}}} = \underset{\rightarrow b}{\boldsymbol{F}}^T \dot{\underline{\boldsymbol{u}}}_b$$

we have:

$$\overset{\bullet}{\underset{\rightarrow}{\boldsymbol{u}}} = \overset{\circ}{\underset{\rightarrow}{\boldsymbol{u}}} + \underset{\rightarrow ba}{\boldsymbol{\omega}} \times \underset{\rightarrow}{\boldsymbol{u}} \tag{1.40}$$

Furthermore,

$$\begin{aligned}
\underset{\rightarrow a}{\boldsymbol{F}}^T \dot{\underline{\boldsymbol{u}}}_a &= \overset{\bullet}{\underset{\rightarrow b}{\boldsymbol{F}}}{}^T \underline{\boldsymbol{u}}_b + \underset{\rightarrow b}{\boldsymbol{F}}^T \dot{\underline{\boldsymbol{u}}}_b \\
&= \underset{\rightarrow b}{\boldsymbol{F}}^T \underline{\boldsymbol{\omega}}_b^{ba\times} \underline{\boldsymbol{u}}_b + \underset{\rightarrow b}{\boldsymbol{F}}^T \dot{\underline{\boldsymbol{u}}}_b \\
&= \underset{\rightarrow a}{\boldsymbol{F}}^T \underline{\boldsymbol{C}}_{ab} \left[\underline{\boldsymbol{\omega}}_b^{ba\times} \underline{\boldsymbol{u}}_b + \dot{\underline{\boldsymbol{u}}}_b \right]
\end{aligned}$$

In summary

$$\overset{\bullet}{\underset{\rightarrow}{\boldsymbol{u}}} = \overset{\circ}{\underset{\rightarrow}{\boldsymbol{u}}} + \underset{\rightarrow}{\boldsymbol{\omega}}^{ba} \times \underset{\rightarrow}{\boldsymbol{u}} \tag{1.41}$$

$$\dot{\underline{\boldsymbol{u}}}_a = \underline{\boldsymbol{C}}_{ab} \left[\underline{\boldsymbol{\omega}}_b^{ba\times} \underline{\boldsymbol{u}}_b + \dot{\underline{\boldsymbol{u}}}_b \right] \tag{1.42}$$

Pure Translation

We say $\mathcal{F}_b$ is in *pure translation* with respect to $\mathcal{F}_a$ if the rotation matrix $\underline{C}_{ba}$ is constant in time; in other words, if the orientation of the basis vectors of one frame relative to the basis vectors of the other frame remains fixed. In other words, there might be rotation from $\mathcal{F}_a$ to $\mathcal{F}_b$, but there is *no change of that rotation in time*. Then,

$$\overset{\bullet}{\underrightarrow{F}}_b = \frac{d}{dt}\left(\underline{C}_{ba}\underrightarrow{F}_a\right) = \dot{\underline{C}}_{ba}\underrightarrow{F}_a + \underline{C}_{ba}\overset{\bullet}{\underrightarrow{F}}_a = \underrightarrow{O} \tag{1.43}$$

leading to

$$\overset{\bullet}{\underrightarrow{u}} = \overset{\circ}{\underrightarrow{u}} \tag{1.44}$$

1.2.4 The Second Time Derivative of a Vector

We can treat the second derivative as the first derivative of vector $\overset{\bullet}{\underrightarrow{u}}$,

$$\begin{aligned}
\overset{\bullet\bullet}{\underrightarrow{u}} &= \overset{\bullet}{\overset{\bullet}{\underrightarrow{u}}} \\
&= \overset{\circ}{\overset{\bullet}{\underrightarrow{u}}} + \underrightarrow{\omega}_{ba} \times \overset{\bullet}{\underrightarrow{u}} \\
&= \frac{d}{dt}\Big|_{\mathcal{F}_b}(\overset{\circ}{\underrightarrow{u}} + \underrightarrow{\omega}_{ba} \times \underrightarrow{u}) + \underrightarrow{\omega}_{ba} \times (\overset{\circ}{\underrightarrow{u}} + \underrightarrow{\omega}_{ba} \times \underrightarrow{u}) \\
&= \overset{\circ\circ}{\underrightarrow{u}} + \overset{\circ}{\underrightarrow{\omega}}_{ba} \times \underrightarrow{u} + \underrightarrow{\omega}_{ba} \times \overset{\circ}{\underrightarrow{u}} + \underrightarrow{\omega}_{ba} \times \overset{\circ}{\underrightarrow{u}} + \underrightarrow{\omega}_{ba} \times (\underrightarrow{\omega}_{ba} \times \underrightarrow{u})
\end{aligned}$$

In summary

$$\begin{aligned}
\overset{\bullet\bullet}{\underrightarrow{u}} &= \overset{\circ\circ}{\underrightarrow{u}} + 2\underrightarrow{\omega}_{ba} \times \overset{\circ}{\underrightarrow{u}} + \overset{\circ}{\underrightarrow{\omega}}_{ba} \times \underrightarrow{u} + \underrightarrow{\omega}_{ba} \times (\underrightarrow{\omega}_{ba} \times \underrightarrow{u}) \\
\underline{\ddot{u}}_a &= \underline{C}_{ab}(\underline{\ddot{u}}_b + 2\underline{\omega}_b^{ba\times}\underline{\dot{u}}_b + \underline{\dot{\omega}}_b^{ba\times}\underline{u}_b + \underline{\omega}_b^{ba\times}\underline{\omega}_b^{ba\times}\underline{u}_b)
\end{aligned}$$

Another observation is:

$$\overset{\bullet}{\underrightarrow{\omega}}_{ba} = \overset{\circ}{\underrightarrow{\omega}}_{ba} + \underrightarrow{\omega}_{ba} \times \underrightarrow{\omega}_{ba} = \overset{\circ}{\underrightarrow{\omega}}_{ba} \tag{1.45}$$

1.2.5 Motion with Respect to Multiple Frames

Here, we will formally prove that

$$\underrightarrow{\omega}_{ca} = \underrightarrow{\omega}_{cb} + \underrightarrow{\omega}_{ba} \tag{1.46}$$

$$\underrightarrow{\omega}_{ab} + \underrightarrow{\omega}_{ba} = \underrightarrow{0} \tag{1.47}$$

Proof:

$$\begin{aligned}
\underline{\omega}_c^{ca\times} &= -\dot{\underline{C}}_{ca}\underline{C}_{ac} \\
&= -\frac{d}{dt}\left(\underline{C}_{cb}\underline{C}_{ba}\right)\underline{C}_{ab}\underline{C}_{bc}
\end{aligned}$$

$$= -\underbrace{\dot{\underline{C}}_{cb}\underline{C}_{ba}\underline{C}_{ab}}_{}\underline{C}_{bc} - \underline{C}_{cb}\dot{\underline{C}}_{ba}\underline{C}_{ab}\underline{C}_{bc}$$

$$\underline{1}$$

$$= -\dot{\underline{C}}_{cb}\underline{C}_{bc} + \underline{C}_{cb}(-\dot{\underline{C}}_{ba}\underline{C}_{ab})\underline{C}_{bc}$$

$$= \underline{\omega}_c^{cb\times} + \underline{C}_{cb}\underline{\omega}_b^{ba\times}\underline{C}_{bc}$$

$$= \underline{\omega}_c^{cb\times} + (\underline{C}_{cb}\underline{\omega}_b^{ba})^\times$$

Consider

$$\underline{C}_{cb}\underline{\omega}_b^{ba\times}\underline{C}_{bc} = (\underline{C}_{cb}\underline{\omega}_b^{ba})^\times$$

Then we can say

$$\underline{\omega}_c^{ca} = \underline{\omega}_c^{cb} + \underline{C}_{cb}\underline{\omega}_b^{ba}$$

which leads to the relationship equation

$$\underset{\rightarrow ca}{\omega} = \underset{\rightarrow cb}{\omega} + \underset{\rightarrow ba}{\omega}.$$

With the fact that $\underset{\rightarrow aa}{\omega} = \underset{\rightarrow}{0}$, we can readily derive from the above equation that $\underset{\rightarrow}{0} = \underset{\rightarrow ab}{\omega} + \underset{\rightarrow ba}{\omega}$. □

Time Derivatives

Assume a fixed (inertial, absolute) frame of reference $\mathcal{F}_a$ and two moving frames of reference $\mathcal{F}_b$ and $\mathcal{F}_c$, each rotating relative to $\mathcal{F}_a$ at rates of $\underset{\rightarrow ba}{\omega}$ and $\underset{\rightarrow ca}{\omega}$ respectively. Then we have

$$\left(\frac{d}{dt}\underset{\rightarrow}{u}\right)_a = \left(\frac{d}{dt}\underset{\rightarrow}{u}\right)_b + \underset{\rightarrow ba}{\omega} \times \underset{\rightarrow}{u} \tag{1.48}$$

$$\left(\frac{d}{dt}\underset{\rightarrow}{u}\right)_a = \left(\frac{d}{dt}\underset{\rightarrow}{u}\right)_c + \underset{\rightarrow ca}{\omega} \times \underset{\rightarrow}{u} \tag{1.49}$$

If we denote the absolute derivative by $\bullet$, the relative derivative in $\mathcal{F}_b$ by $\circ$, and the relative derivative in $\mathcal{F}_c$ by $*$, then the above equations become

$$\overset{\bullet}{\underset{\rightarrow}{u}} = \overset{\circ}{\underset{\rightarrow}{u}} + \underset{\rightarrow ba}{\omega} \times \underset{\rightarrow}{u}$$

$$\overset{\bullet}{\underset{\rightarrow}{u}} = \overset{*}{\underset{\rightarrow}{u}} + \underset{\rightarrow ca}{\omega} \times \underset{\rightarrow}{u}$$

Note that the absolute time derivative formulae looks the same as each other, but the rotating rates are different.

1.3 Euler Parameters and Unit Quaternion

Euler's theorem states that any rotation of an object in 3-D space leaves some axis fixed: this is the rotation axis. As a result, any rotation can be described by a unit vector $\underline{a}$ (satisfying $\underline{a}^T\underline{a} = 1$) in the direction of the rotational axis, and the angle of rotation, ϕ, about $\underline{a}$. The rotation matrix is represented by

$$\underline{C}(\underline{a}, \phi) = \cos\phi\underline{1} + (1 - \cos\phi)\underline{\underline{aa^T}} - \sin\phi\underline{a}^\times \tag{1.50}$$

where the set $(\underline{a}, \phi)$ is often called the Euler axis/angle variables.

To avoid a triangular calculation, these variables can be replaced by the so-called Euler parameters:

$$\eta = \cos\frac{\phi}{2} \tag{1.51}$$

$$\underline{\varepsilon} = \underline{a}\sin\frac{\phi}{2} \tag{1.52}$$

Note that $\underline{\varepsilon}^T\underline{\varepsilon} + \eta^2 = 1$. Then the rotation matrix becomes

$$\underline{C} = (2\eta^2 - 1)\underline{1} + 2\underline{\varepsilon}\underline{\varepsilon}^T - 2\eta\underline{\varepsilon}^\times \stackrel{\Delta}{=} \underline{C}(\underline{\varepsilon}, \eta) \tag{1.53}$$

Now, let us take a look at the rotation matrix corresponding to two consecutive rotations, represented by either their Euler axis/angle variables, or the Euler parameters,

$$\underline{C}(\underline{a}_3, \phi_3) = \underline{C}(\underline{a}_2, \phi_2)\underline{C}(\underline{a}_1, \phi_1) \tag{1.54}$$

$$\underline{C}(\underline{\varepsilon}_3, \eta_3) = \underline{C}(\underline{\varepsilon}_2, \eta_2)\underline{C}(\underline{\varepsilon}_1, \eta_1) \tag{1.55}$$

After some tedious matrix algebraic manipulation, this leads to the following relationship:

$$\begin{cases} \cos\frac{\phi_3}{2} = \cos\frac{\phi_2}{2}\cos\frac{\phi_1}{2} - \sin\frac{\phi_2}{2}\sin\frac{\phi_1}{2}\underline{a}_1^T\underline{a}_2 \\ \sin\frac{\phi_3}{2}\underline{a}_3 = \cos\frac{\phi_2}{2}\sin\frac{\phi_1}{2}\underline{a}_1 + \sin\frac{\phi_2}{2}\cos\frac{\phi_1}{2}\underline{a}_2 + \underline{a}_1^\times\underline{a}_2\sin\frac{\phi_2}{2}\sin\frac{\phi_1}{2} \end{cases} \tag{1.56}$$

or

$$\begin{cases} \eta_3 = \eta_2\eta_1 - \underline{\varepsilon}_1^T\underline{\varepsilon}_2 \\ \underline{\varepsilon}_3 = \eta_2\underline{\varepsilon}_1 + \eta_1\underline{\varepsilon}_2 + \underline{\varepsilon}_1^\times\underline{\varepsilon}_2 \end{cases} \tag{1.57}$$

In matrix format, we obtain the following

$$\begin{bmatrix} \underline{\varepsilon}_3 \\ \eta_3 \end{bmatrix} = \left[\begin{array}{c|c} \eta_2\underline{1} - \underline{\varepsilon}_2^\times & \underline{\varepsilon}_2 \\ -\underline{\varepsilon}_2^T & \eta_2 \end{array}\right]\begin{bmatrix} \underline{\varepsilon}_1 \\ \eta_1 \end{bmatrix} \stackrel{\Delta}{=} \underline{L}(\underline{\varepsilon}_2, \eta_2)\begin{bmatrix} \underline{\varepsilon}_1 \\ \eta_1 \end{bmatrix} \tag{1.58}$$

$$\begin{bmatrix} \underline{\varepsilon}_3 \\ \eta_3 \end{bmatrix} = \left[\begin{array}{c|c} \eta_1\underline{1} + \underline{\varepsilon}_1^\times & \underline{\varepsilon}_1 \\ -\underline{\varepsilon}_1^T & \eta_1 \end{array}\right]\begin{bmatrix} \underline{\varepsilon}_2 \\ \eta_2 \end{bmatrix} \stackrel{\Delta}{=} \underline{R}(\underline{\varepsilon}_1, \eta_1)\begin{bmatrix} \underline{\varepsilon}_2 \\ \eta_2 \end{bmatrix} \tag{1.59}$$

The matrix representation of $\{\underline{\varepsilon}, \eta\}$ is one of the expressions of the so-called *unit quaternion*, denoted by

$$\bar{\boldsymbol{q}} \stackrel{\Delta}{=} \begin{bmatrix} \underline{q} \\ q \end{bmatrix} \tag{1.60}$$

In the current case,

$$\bar{\boldsymbol{q}} = \begin{bmatrix} \underline{\varepsilon} \\ \eta \end{bmatrix}. \tag{1.61}$$

It is obvious that $|\bar{\boldsymbol{q}}|^2 = \underline{\varepsilon}^T\underline{\varepsilon} + \eta^2 = 1$. Therefore the Euler parameter is considered as a unit quaternion.

We define the quaternion multiplication as

$$\bar{\boldsymbol{q}} \otimes \bar{\boldsymbol{p}} \stackrel{\Delta}{=} \left[\begin{array}{c|c} q\underline{1} - \underline{q}^\times & \underline{q} \\ \hline -\underline{q}^T & q \end{array}\right]\bar{\boldsymbol{p}} = \underline{L}(\bar{\boldsymbol{q}})\bar{\boldsymbol{p}} \tag{1.62}$$

$$\bar{q} \otimes \bar{p} \overset{\Delta}{=} \left[\begin{array}{c|c} p\underline{1} + \underline{p}^\times & \underline{p} \\ \hline -\underline{p}^T & p \end{array} \right] \bar{q} = \underline{R}(\bar{p})\bar{q} \tag{1.63}$$

Its inverse (representing a reverse rotation of angle $-\phi$ about unit axis $\underline{a}$) is defined by

$$\bar{q}^{-1} \overset{\Delta}{=} \left[\begin{array}{c} -\underline{q} \\ q \end{array} \right] \tag{1.64}$$

and one can prove that

$$\bar{q} \otimes \bar{q}^{-1} = \left[\begin{array}{c} \underline{0} \\ 1 \end{array} \right] \overset{\Delta}{=} \bar{1} \tag{1.65}$$

Using these definitions, we can obtain the formulation of attitude error. Assume $(\underline{a}_d, \phi_d)$ and $(\underline{a}, \phi)$ represent the desired attitude and actual attitude, respectively. The error between the actual and the desired attitudes can be treated as a consecutive rotation from the desired attitude,

$$\underline{C}(\underline{a}, \phi) = \underline{C}(\underline{a}_e, \phi_e)\underline{C}(\underline{a}_d, \phi_d) \tag{1.66}$$

This leads to

$$\underline{C}(\underline{a}_e, \phi_e) = \underline{C}(\underline{a}, \phi)\underline{C}^{-1}(\underline{a}_d, \phi_d)$$

In other words,

$$\bar{q}_e = \bar{q} \otimes \bar{q}_d^{-1} = \left[\begin{array}{c|c} \eta_d \underline{1} - \underline{\varepsilon}_d^\times & -\underline{\varepsilon}_d \\ \hline \underline{\varepsilon}_d^T & \eta_d \end{array} \right] \left[\begin{array}{c} \underline{\varepsilon} \\ \eta \end{array} \right] = \underline{R}(\bar{q}_d^{-1})\bar{q} \tag{1.67}$$

This expression shows that the error attitude (quaternion) is now represented by the actual attitude and the desired attitude. Further, as $\bar{q} \to \bar{q}_d$, the steady error attitude $\bar{q}_e \to \bar{1}$.

Under the Euler parameters or unit quaternion, the angular rate vector $\overset{\rightarrow}{\omega}$ is described as

$$\underline{\omega} = \dot{\phi}\underline{a} - (1 - \cos \phi)\underline{a}^\times \dot{\underline{a}} + \sin \phi \dot{\underline{a}} \tag{1.68}$$

and

$$\dot{\bar{q}} = \left[\begin{array}{c} \dot{\underline{\varepsilon}} \\ \dot{\eta} \end{array} \right] = \frac{1}{2} \left[\begin{array}{c|c} -\underline{\omega}^\times & \underline{\omega} \\ \hline -\underline{\omega}^T & 0 \end{array} \right] \left[\begin{array}{c} \underline{\varepsilon} \\ \eta \end{array} \right] \overset{\Delta}{=} \frac{1}{2}\underline{\Omega}(\underline{\omega})\bar{q} \tag{1.69}$$

2

Formation Dynamics of Motion Systems

The concept of formation in motion can be intuitively described as a group of vehicle systems staying in a fixed pattern (e.g. a geometric shape) while in motion. We can extend the concept to allow these vehicles to follow a certain dynamic pattern (e.g. spinning in a geometric fashion). With this context in mind, in this chapter we shall formulate the dynamics of vehicle systems for establishing a fixed formation, but may extend this to allow for a time-varying, dynamic formation.

Technical descriptions of formation control can be categorized into three main approaches:

- the leader–follower approach
- the behavioural approach
- the virtual structure approach.

In this chapter, we will use similar categories to establish the dynamics formulations of the motion of multiple vehicle systems.

2.1 Virtual Structure

Imagine a group of multiple vehicles in a "fixed" formation, so that they move as if one entity. Intuitively, one would treat such a a group of vehicles as if they were a rigid body. In this way, the concept of a virtual structure is born.

Let us take a moment to consider the properties of a rigid body and its relationship with a group of vehicles in fixed formation. First of all, any infinitesimal mass point on the rigid body has a fixed position relative to the *centre of mass* (CM) of the body,[1] and these relative positions remain fixed for any time instant in motion. Secondly, each element on the body has a fixed orientation relative to the CM as well, and keeps an identical orientational rate of change in time – the same as the body's rotational rate. We assume a body-fixed frame is established such that the body angular rate is the rotational rate of the body-fixed frame with respect to the inertial frame. In summary, these rigid-body properties can be captured by the following mathematical descriptions:

$$\underset{\rightarrow cm}{r} = \frac{1}{m}\int_B \underset{\rightarrow}{r}\, dm \tag{2.1}$$

1 Or relative to any other fixed point on the body

Formation Control of Multiple Autonomous Vehicle Systems, First Edition. Hugh H.T. Liu and Bo Zhu.

$$\left(\frac{d}{dt}\underset{\rightarrow}{\rho}\right)_B = 0 \tag{2.2}$$

$$\underset{\rightarrow}{\omega} = \underset{\rightarrow}{\omega}_{BE} \tag{2.3}$$

Accordingly, we shall define a virtual structure in which each vehicle in a fixed formation is treated as one element of a virtual rigid body.

The centre formation centre of a virtual structure is defined as the centre of mass of the system of vehicles, when in their desired positions:

$$\underset{\rightarrow}{r}_F \stackrel{\Delta}{=} \frac{1}{m}\Sigma_{i=1}^{n} \underset{\rightarrow}{r}_i^d m_i, \quad m = \Sigma_{i=1}^{n} m_i \tag{2.4}$$

where $\underset{\rightarrow}{r}_i^d$ represents the ideal position vector of vehicle i with respect to the origin of an inertial frame $\mathcal{F}_E$. In other words, $\underset{\rightarrow}{r}_F = \underset{\rightarrow}{r}_{cm}^d$. If we are dealing with homogeneous vehicles, then obviously the formation centre is the same as the geometric centre of the formation.

Next, let us define the formation velocity and angular velocity of the virtual structure, denoted as velocity $\underset{\rightarrow}{v}_F$ and $\underset{\rightarrow}{\omega}_F$ respectively.

$$\underset{\rightarrow}{v}_F \stackrel{\Delta}{=} \underset{\rightarrow}{r}_F^{\bullet} = \left(\frac{d}{dt}\underset{\rightarrow}{r}_F\right)_E \tag{2.5}$$

In addition, we assume that the angular rate of the virtual structure is the same as the rotational rate of a formation reference frame $\mathcal{F}_F$ with respect to the inertial frame $\mathcal{F}_E$.

$$\underset{\rightarrow}{\omega}_F \equiv \underset{\rightarrow}{\omega}_{FE} \tag{2.6}$$

Like the relationship between the position vector and the velocity vector, the corresponding vector associated with the angular velocity $\underset{\rightarrow}{\omega}$ is the so-called attitude vector. The attitude vector is most often represented by a unit quaternion. It is defined by

$$\overline{q}_F \stackrel{\Delta}{=} \begin{bmatrix} \underline{q} \\ q \end{bmatrix} = \begin{bmatrix} \underline{\varepsilon} \\ \eta \end{bmatrix} = \begin{bmatrix} \underline{a} \cdot \sin(\phi/2) \\ \cos(\phi/2) \end{bmatrix} \tag{2.7}$$

where $\underline{a}$ is a unit vector representing the direction of rotation (axis of rotation) and ϕ is the rotating angle about $\underline{a}$.[2] To keep a unique unit quaternion representation, we further assume that $0 \le \phi \le \pi$ such that $q \ge 0$. It is obvious that $|\overline{q}|^2 = \underline{\varepsilon}^T\underline{\varepsilon} + \eta^2 = 1$, so the Euler parameter is considered a unit quaternion.

With the introduction of the concepts of formation position and velocity, attitude and angular rate, we now have the formation states of the virtual structure as follows:

$$\underset{\rightarrow}{x}_F(t) = \begin{Bmatrix} \underset{\rightarrow}{r}_F(t) \\ \underset{\rightarrow}{v}_F(t) \\ \overline{q}_F(t) \\ \underset{\rightarrow}{\omega}_F(t) \end{Bmatrix} \tag{2.8}$$

Note that the state vector is a function of time t.

2 In classical mechanics, this is often described as a Euler axis/angle representation.

In later discussions, we may encounter the idea of formation maneuvers; that is, when the formation patterns are expected to change, in both space and time, during the mission. This phenomenon may be represented as a series of desired discrete states of the virtual-structure formation. It is as if the continuous $\underset{\rightarrow}{x}_F(t)$ is represented by piecewise discrete states $\underset{\rightarrow}{x}_F(k)$ or $\underset{\rightarrow}{x}_F(t_k)$.

2.1.1 Formation Control Problem Statement

Once the formation virtual structure states are defined, formation control involves identifying the desired formation state for each individual vehicle $\underset{\rightarrow}{x}_i^d$ based on the formation virtual structure state $\underset{\rightarrow}{x}_F$ and designing suitable control laws to regulate each vehicle's state so that it tracks the desired one.

For each vehicle i, $(i = 1, \dots, n)$, we define its position vector $\underset{\rightarrow}{r}_i$. Then, its velocity vector

$$\underset{\rightarrow}{v}_i \stackrel{\Delta}{=} \overset{\bullet}{\underset{\rightarrow}{r}_i} = \left(\frac{d}{dt}\underset{\rightarrow}{r}_i\right)_E \tag{2.9}$$

For each vehicle i, the position and velocity vectors, together with the angular velocity, have the following relationships with the formation centre:

$$\underset{\rightarrow}{r}_i = \underset{\rightarrow}{r}_F + \underset{\rightarrow}{r}_{iF} \stackrel{\Delta}{=} \underset{\rightarrow}{r}_F + \underset{\rightarrow}{\rho}_i \tag{2.10}$$

$$\underset{\rightarrow}{v}_i = \underset{\rightarrow}{v}_F + \underset{\rightarrow}{v}_{iF} \stackrel{\Delta}{=} \underset{\rightarrow}{v}_F + \underset{\rightarrow}{\mu}_i \tag{2.11}$$

The angular velocity $\underset{\rightarrow}{\omega}_i$ is treated as the frame rotating rate (with respect to $\mathcal{F}_E$) when a specific vehicle-fixed frame $\mathcal{F}_i$ is defined,

$$\underset{\rightarrow}{\omega}_i = \underset{\rightarrow}{\omega}_F + \underset{\rightarrow}{\omega}_{iF} \tag{2.12}$$

We note that the time derivative of the relative position vector has different expressions in terms of respective moving frames,

$$\begin{aligned}\underset{\rightarrow}{\mu}_i \stackrel{\Delta}{=} \overset{\bullet}{\underset{\rightarrow}{\rho}_i} &= \left(\frac{d}{dt}\underset{\rightarrow}{\rho}_i\right)_E \\ &= \left(\frac{d}{dt}\underset{\rightarrow}{\rho}_i\right)_i + \underset{\rightarrow}{\omega}_i \times \underset{\rightarrow}{\rho}_i\end{aligned} \tag{2.13}$$

$$= \left(\frac{d}{dt}\underset{\rightarrow}{\rho}_i\right)_F + \underset{\rightarrow}{\omega}_F \times \underset{\rightarrow}{\rho}_i \stackrel{\Delta}{=} \overset{\circ}{\underset{\rightarrow}{\rho}_i} + \underset{\rightarrow}{\omega}_F \times \underset{\rightarrow}{\rho}_i \tag{2.14}$$

and we shall explore these different expressions whenever appropriate. In the context of a "relative" derivative, we often define $(\overset{\circ}{\cdot}) = \left(\frac{d}{dt}(\cdot)\right)_F$. Therefore, the key concept of a virtual structure, depending on the velocity and angular velocity for each vehicle, lies in the discussion of the relative information $\underset{\rightarrow}{\rho}_i$, $\underset{\rightarrow}{\mu}_i$, and $\underset{\rightarrow}{\omega}_{iF}$ respectively. First of all, when the vehicles are locked in the desired (ideal) fixed formation, we expect that any ith and

jth vehicle in formation will have identical ideal angular velocities. In other words, the relative rotating rate should be zero. This desired feature is represented As follows:

$$\underset{\rightarrow}{\omega}_i^d = \underset{\rightarrow}{\omega}_j^d = \underset{\rightarrow}{\omega}_F, \quad \underset{\rightarrow}{\omega}_{iF}^d = 0 \tag{2.15}$$

Assume $\mathcal{F}_i$ is the vehicle-fixed frame attached to the ith vehicle. In the ideal formation, it is obvious that all vehicles belong to the same "rigid-body" and frame $\mathcal{F}_F$ represents the body-fixed frame for the virtual structure. Even an individual frame $\mathcal{F}_i$ may not align identically to $\mathcal{F}_F$, but it should have a relatively fixed relationship with respect to $\mathcal{F}_F$ such that a relative angular rotation is allowed, but the angular rate will be zero.

Secondly, the relative position vector $\underset{\rightarrow}{\rho}_i$ is assumed to remain static in order to maintain formation. As such, it is expected that the change rate of that vector with respect to time is zero:

$$\underset{\rightarrow}{\mu}_i^d = \overset{\bullet}{\underset{\rightarrow}{\rho}}{}_i^d = \overset{\circ}{\underset{\rightarrow}{\rho}}{}_i^d + \underset{\rightarrow}{\omega}_F \times \underset{\rightarrow}{\rho}_i^d \tag{2.16}$$

and

$$\overset{\circ}{\underset{\rightarrow}{\rho}}{}_i^d \triangleq \left(\frac{d}{dt} \underset{\rightarrow}{\rho}_i^d \right)_F = 0 \tag{2.17}$$

Equations (2.15) and (2.17) are the key concepts of formation when applying the "rigid-body" concept to a virtual structure. Now, the formation control problem for the virtual structure can be described by the following statement:

> Given the desired reference (formation centre) state $\underset{\rightarrow}{x}_F(t)$ and the corresponding desired states for each individual vehicle $\underset{\rightarrow}{x}_i^d(t)$, design suitable control approaches to make individual vehicles track their desired states; that is, $\underset{\rightarrow}{x}_i(t) \rightarrow \underset{\rightarrow}{x}_i^d(t)$.

This statement can be further specified as:

$$\underset{\rightarrow}{x}_i \rightarrow \underset{\rightarrow}{x}_i^d \Leftrightarrow \begin{cases} \underset{\rightarrow}{r}_i \rightarrow \underset{\rightarrow}{r}_i^d, & \underset{\rightarrow}{\rho}_i \rightarrow \underset{\rightarrow}{\rho}_i^d & \\ \underset{\rightarrow}{v}_i \rightarrow \underset{\rightarrow}{v}_i^d, & \underset{\rightarrow}{\mu}_i \rightarrow \underset{\rightarrow}{\mu}_i^d = \underset{\rightarrow}{\omega}_F \times \underset{\rightarrow}{\rho}_i^d & \overset{\circ}{\underset{\rightarrow}{\rho}}{}_i \rightarrow 0 \\ \bar{q}_i \rightarrow \bar{q}_i^d & & \\ \underset{\rightarrow}{\omega}_i \rightarrow \underset{\rightarrow}{\omega}_i^d, & \underset{\rightarrow}{\omega}_i \rightarrow \underset{\rightarrow}{\omega}_F, & \underset{\rightarrow}{\omega}_{iF} \rightarrow 0 \end{cases} \tag{2.18}$$

Let us give expressions for the desired individual vehicle states in formation in terms of $\mathcal{F}_E$, which are the references for formation control. First, the desired vehicle position can be expressed as:

$$\left[\underline{r}_i^d\right]_E = \left[\underline{r}_F\right]_E + \underline{C}_{EF}\left[\underline{\rho}_i^d\right]_F, \quad \text{or} \quad \underline{r}_{i,E}^d = \underline{r}_{F,E} + \underline{C}_{EF}\underline{\rho}_{i,F}^d \tag{2.19}$$

Next, the desired vehicle velocity can be expressed as:

$$\left[\underline{v}_i^d\right]_E = \left[\underline{v}_F\right]_E + \underline{C}_{EF}\left(\left[\overset{\circ}{\underline{\rho}}{}_i^d\right]_F + \underline{\omega}_F^\times\left[\underline{\rho}_i^d\right]_F\right) = \left[\underline{v}_F\right]_E + \underline{C}_{EF}\left(\underline{\omega}_F^\times\left[\underline{\rho}_i^d\right]_F\right) \tag{2.20}$$

We will simplify this notation as $[\underline{\omega}_F]_E = \underline{\omega}_E$, $[\underline{\omega}_F]_F = \underline{\omega}_F$, as long as there is no confusion in the context.

From the formula $(\underline{C}_{ab}\underline{u}_b)^\times = \underline{C}_{ab}\underline{u}_b^\times\underline{C}_{ba}$, one gets that

$$\begin{aligned}\underline{C}_{EF}\underline{\omega}_F^\times\underline{\rho}_{i,F}^d &= \underline{C}_{EF}(\underline{C}_{FE}\underline{\omega}_E)^\times\underline{\rho}_{i,F}^d \\ &= \underline{C}_{EF}(\underline{C}_{FE}\underline{\omega}_E^\times C_{EF})\underline{\rho}_{i,F}^d \\ &= \underline{\omega}_E^\times(\underline{C}_{EF}\underline{\rho}_{i,F}^d)\end{aligned}$$

Therefore,

$$\left[\underline{v}_i^d\right]_E = \left[\underline{v}_F\right]_E + \underline{\omega}_E^\times\underline{C}_{EF}\left[\underline{\rho}_i^d\right]_F \tag{2.21}$$

It is obvious by now that

$$\left[\underline{\omega}_i^d\right]_E = \left[\underline{\omega}_F\right]_E + \left[\underline{\omega}_{iF}\right]_E = \left[\underline{\omega}_F\right]_E \tag{2.22}$$

The desired orientation of each individual vehicle is captured by the unit quaternion,

$$\bar{\boldsymbol{q}}_i^d = \bar{\boldsymbol{q}}_{iF}^d \otimes \bar{\boldsymbol{q}}_F \tag{2.23}$$

On the other hand, it is often useful to express angular velocity in terms of Euler angular rates:

$$\underline{\omega}_F = \begin{bmatrix} 1 & 0 & -s_\gamma \\ 0 & c_\mu & c_\gamma s_\mu \\ 0 & -s_\mu & c_\gamma c_\mu \end{bmatrix}\begin{bmatrix} \dot{\mu}_F \\ \dot{\gamma}_F \\ \dot{\sigma}_F \end{bmatrix} \triangleq \underline{S}_F\underline{\Theta}_F \tag{2.24}$$

where $\sigma_F, \gamma_F, \mu_F$ are, respectively, the heading angle, flight path angle, and bank angle of the virtual structure. The formation frame is chosen such that the direction of its x-axis points to the formation velocity. Therefore,

$$\left[\underline{v}_F\right]_F = \begin{bmatrix} V_F \\ 0 \\ 0 \end{bmatrix} \tag{2.25}$$

and

$$\left[\underline{v}_F\right]_E = \underline{C}_{EF}\left[\underline{v}_F\right]_F \tag{2.26}$$

Note the rotation matrix

$$\underline{C}_{FE} = \underline{C}_1(\mu_F)\underline{C}_2(\gamma_F)\underline{C}_3(\sigma_F). \tag{2.27}$$

The rates for the desired individual vehicle states are given as follows.

$$\begin{aligned}&\left[\underline{\dot{r}}_i^d\right]_E = \left[\underline{v}_i^d\right]_E \\ &\left[\underline{\dot{v}}_i^d\right]_E \triangleq \underline{a}_i^d = \left[\underline{\dot{v}}_F\right]_E + \underline{\dot{\omega}}_E^\times\underline{C}_{EF}\left[\underline{\rho}_i^d\right]_F + \underline{\omega}_E^\times\underline{\omega}_E^\times\underline{C}_{EF}\left[\underline{\rho}_i^d\right]_F \\ &\dot{\bar{\boldsymbol{q}}}_i^d = \bar{\boldsymbol{q}}_{iF}^d \otimes \dot{\bar{\boldsymbol{q}}}_F = \bar{\boldsymbol{q}}_{iF}^d \otimes \tfrac{1}{2}\underline{\Omega}(\underline{\omega}_F)\bar{\boldsymbol{q}}_F\end{aligned} \tag{2.28}$$

Similarly, one may obtain the desired individual vehicle states and their rates as expressed in the formation-centre frame $\mathcal{F}_F$.

2.1.2 Extended Formation Control Problem

The requirement for rigidity in formation can be "loosened to make the formation shape more flexible by allowing the place holders to expand or contract while still keeping fixed relative orientation" [2]. The translational expandsion/contraction factor is given by

$$\underline{\boldsymbol{\Lambda}} = \begin{bmatrix} \lambda_1 & & \\ & \lambda_2 & \\ & & \lambda_3 \end{bmatrix}, \qquad \dot{\underline{\boldsymbol{\Lambda}}} = \begin{bmatrix} \dot{\lambda}_1 & & \\ & \dot{\lambda}_2 & \\ & & \dot{\lambda}_3 \end{bmatrix} \tag{2.29}$$

and the relative position vector is given by $\underline{\boldsymbol{\Lambda}}[\underline{\rho}_i^d]_F$. Now, to take into account the relative movement, one may have the following expressions for the desired state of the individual vehicle:

$$\left[\underline{\boldsymbol{r}}_i^d\right]_E = [\underline{\boldsymbol{r}}_F]_E + \underline{\boldsymbol{C}}_{EF}\underline{\boldsymbol{\Lambda}}\left[\underline{\rho}_i^d\right]_F \tag{2.30}$$

$$\left[\underline{\boldsymbol{v}}_i^d\right]_E = [\underline{\boldsymbol{v}}_F]_E + \underline{\boldsymbol{C}}_{EF}\dot{\underline{\boldsymbol{\Lambda}}}\left[\underline{\rho}_i^d\right]_F + \underline{\boldsymbol{\omega}}_E^\times \underline{\boldsymbol{C}}_{EF}\underline{\boldsymbol{\Lambda}}\left[\underline{\rho}_i^d\right]_F \tag{2.31}$$

$$\underline{\boldsymbol{a}}_i^d = \left[\dot{\underline{\boldsymbol{v}}}_F\right]_E + \underline{\boldsymbol{C}}_{EF}\ddot{\underline{\boldsymbol{\Lambda}}}\left[\underline{\rho}_i^d\right]_F + 2\underline{\boldsymbol{\omega}}_E^\times \underline{\boldsymbol{C}}_{EF}\dot{\underline{\boldsymbol{\Lambda}}}\left[\underline{\rho}_i^d\right]_F + \dot{\underline{\boldsymbol{\omega}}}_E^\times \underline{\boldsymbol{C}}_{EF}\underline{\boldsymbol{\Lambda}}\left[\underline{\rho}_i^d\right]_F + \underline{\boldsymbol{\omega}}_E^\times \underline{\boldsymbol{\omega}}_E^\times \underline{\boldsymbol{C}}_{EF}\underline{\boldsymbol{\Lambda}}\left[\underline{\rho}_i^d\right]_F \tag{2.32}$$

In the virtual-structure approach, the entire formation is treated as a rigid body. The positions of the vehicles in the structure are usually defined in a frame with respect to a reference point in the structure. As a trajectory is given for the reference point, the desired position for each vehicle can be calculated as the virtual structure evolves over time. In our formation controller, we define the centroid of the desired formation as the reference point in the virtual structure.

Example 2.1 Consider two vehicles, UAV1 and UAV2, forming a virtual structure. The geometry in the two-dimensional xy-plane is shown in Figure 2.1. Again, $\mathcal{F}_E$ is an inertial frame and $\mathcal{F}_F$ is the formation reference frame located at the reference point, the centroid of the desired formation. The formation is described as a virtual rigid body with inertial position, expressed in $\mathcal{F}_E$, of $[\underline{\boldsymbol{r}}_F]_E$, velocity $[\underline{\boldsymbol{v}}_F]_E$, heading ψ_F, and angular velocity $\underline{\boldsymbol{\omega}}_E$. Each vehicle can be represented by its position $[\underline{\boldsymbol{r}}_i]_E$, velocity $[\underline{\boldsymbol{v}}_i]_E$, heading

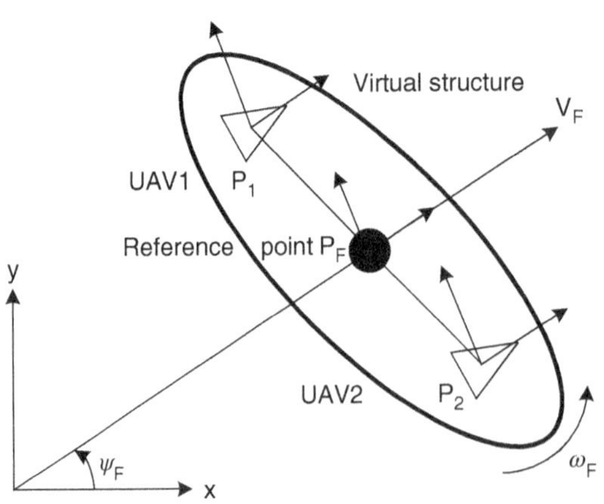

Figure 2.1 Coordinate frame of virtual structure.

ψ_i and angular velocity $[\underline{\omega}_i]_E$ with respect to the inertial frame or by $[\underline{\rho}_i]_F$, $[\underline{\mu}_i]_F$, ψ_{iF} and $[\underline{\omega}_{iF}]_F$ with respect to the formation reference frame.

The equations for the position and velocity dynamics for each vehicle ($i = 1, 2$) in the inertial frame are

$$\begin{aligned}\left[\underline{r}_i^d\right]_E(t) &= \left[\underline{r}_F\right]_E(t) + \underline{C}_{EF}(t)\left[\underline{\rho}_i^d\right]_F(t)\\ \left[\underline{v}_i^d\right]_E(t) &= \left[\underline{v}_F\right]_E(t) + \underline{\omega}_E^{\times}(t)\underline{C}_{EF}(t)\left[\underline{\rho}_i^d\right]_F(t)\end{aligned} \tag{2.33}$$

In the two-vehicle case, $[\underline{\rho}_2^d]_F(t) = -[\underline{\rho}_1^d]_F(t)$ if the reference point is set as the centroid of the entity. Therefore, defining $\underline{\Gamma}(t) = \underline{C}_{EF}(t)[\underline{\rho}_1^d]_F(t)$, the two parts of Equation (2.33) for UAV1 become

$$\begin{aligned}\left[\underline{r}_1^d\right]_E(t) &= \left[\underline{r}_F\right]_E(t) + \underline{\Gamma}(t)\\ \left[\underline{v}_1^d\right]_E(t) &= \left[\underline{v}_F\right]_E(t) + \underline{\omega}_E^{\times}(t)\underline{\Gamma}(t)\end{aligned} \tag{2.34}$$

and for UAV2, by substitutions of $[\underline{\rho}_2^d]_F(t)$ and $\Gamma(t)$,

$$\begin{aligned}\left[\underline{r}_2^d\right]_E(t) &= \left[\underline{r}_F\right]_E(t) - \underline{\Gamma}(t)\\ \left[\underline{v}_2^d\right]_E(t) &= \left[\underline{v}_F\right]_E(t) - \underline{\omega}_E^{\times}(t)\underline{\Gamma}(t)\end{aligned} \tag{2.35}$$

At the time instant $t = t_0$, where there is a non-zero magnitude of rotating rate, $\omega_E(t_0)$, the instantaneous attitude changes required for the aircraft in order to maintain the geometry (not the attitude of the vehicles alone) will be opposite, as shown in Figure 2.2, an example in which $\omega_E(t_0)$ is about the z-axis.

For the UAVs to go from P_i to P_i'

- UAV1's instantaneous heading change will be in the direction of $-\Delta\psi_F(t_0)$ with a magnitude larger than that of the heading change of the formation
- UAV2's heading change will be in the direction of $\Delta\psi_F(t_0)$ with a larger magnitude.

Therefore,

$$\begin{aligned}\Delta\psi_1^d(t_0) &= -(1+\kappa_1)\Delta\psi_F(t_0)\\ \Delta\psi_2^d(t_0) &= (1+\kappa_2)\Delta\psi_F(t_0)\end{aligned} \tag{2.36}$$

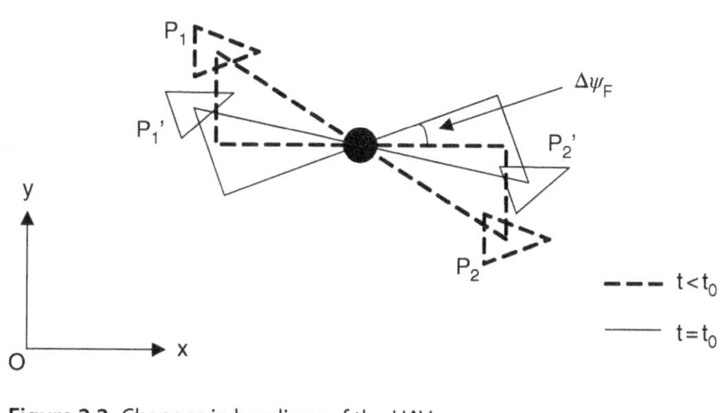

Figure 2.2 Changes in headings of the UAVs.

with $\kappa_i > 0$ as a constant. The same analogy can be applied to the roll ($\Delta\phi$) and pitch ($\Delta\theta$) attitude changes.

Based on the desired positions $[\underline{r}_i^d]_E$, velocities $[\underline{v}_i^d]_E$ and heading changes $\Delta\psi_i^d$ of the two vehicles, it can be observed that the relative values for each vehicle are opposite in magnitude with respect to a reference point that is the centroid of the virtual structure. Thus, it should be expected that the relative position errors, if there are any, between the actual vehicles' positions and the reference point in flight will be opposing in nature as well, when the same trajectory commands designated for the reference point are given to both vehicles. Therefore, a controller should be developed that eliminates these positional errors so as to maintain formation.

We here take the formation control of multiple aerial vehicles as an example. To eliminate the relative positional errors and to keep the vehicles in formation flight, a formation controller is needed. The objective of the formation controller is to maintain the formation geometry. Such a controller is typically implemented with a two-loop scheme in which the inner control loop allows tracking of the commanded velocity (v), heading (ψ) and altitude (h). In the outer control loop, the formation controller generates reference path commands for the inner controller. An autopilot that is capable of tracking velocity, altitude and heading is described in a paper by Ren and Beard [3], so we focus the development here on the outer loop controller. Details will be presented in Chapter 7.

From the desired and actual positions of the UAVs in trajectory during a formation flight, there can be longitudinal, lateral and vertical errors in the relative distances. An example in the xy-plane is shown in Figure 2.3.

Based on the reference trajectory commands $T_r = [v_r, \psi_r, h_r]^T$ for the reference point $p_r = [x_r, y_r, z_r]^T$ and the defined relative distances in the virtual structure, the desired position for each vehicle during the flight $p_i^d = [x_i^d, y_i^d, z_i^d]^T$ can be calculated. The actual position of each aircraft $p_i = [x_i, y_i, z_i]^T$ can be obtained from a GPS device located on the vehicle. Using the reference trajectory and the actual/desired positions for the vehicles as input, the formation controller generates modified trajectories for each UAV so as to maintain the geometry of the formation during the flight. We define the relative

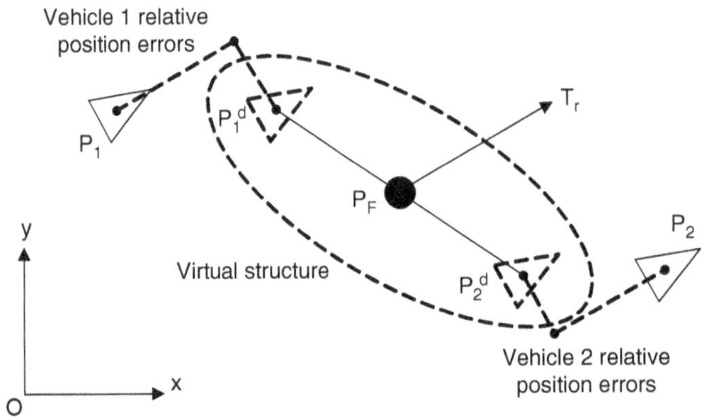

Figure 2.3 Relative errors of UAV positions.

position errors for the ith vehicle during the flight in the inertial frame as:

$$\begin{bmatrix} e_{xi} \\ e_{yi} \\ e_{zi} \end{bmatrix} = \begin{bmatrix} x_i^d - x_i \\ y_i^d - y_i \\ z_i^d - z_i \end{bmatrix} \tag{2.37}$$

To utilize the relative errors in (2.37), they need to be converted into errors in the formation frame using the rotation matrix $C_{FO}(t) = C_{OF}(t)^{-1}$. Therefore,

$$\begin{bmatrix} e_{xiF} \\ e_{yiF} \\ e_{ziF} \end{bmatrix} = C_{FO}(t) \begin{bmatrix} e_{xi} \\ e_{yi} \\ e_{zi} \end{bmatrix} \tag{2.38}$$

The modified trajectory commands for the inner loop controller are given by $T_{ci} = T_r + \Delta T_i$, where ΔT_i is calculated through a proportional-integral (PI) controller based on the relative positional errors in (2.38).

$$\begin{aligned} \Delta v_i(t) &= K_{px} e_{xiF}(t) + K_{ix} \int_0^t e_{xiF}(t)\, dt \\ \Delta \psi_i(t) &= K_{py} e_{yiF}(t) + K_{iy} \int_0^t e_{yiF}(t)\, dt \\ \Delta h_i(t) &= K_{pz} e_{ziF}(t) + K_{iz} \int_0^t e_{ziF}(t)\, dt \end{aligned} \tag{2.39}$$

$$\Delta T_i = \begin{bmatrix} \Delta v_i(t) \\ \Delta \psi_i(t) \\ \Delta h_i(t) \end{bmatrix} \tag{2.40}$$

The formation controller applies the corrected trajectory commands T_{ci} based on T_r and ΔT_i in (2.40) to the autopilots on the aircraft. These corrections account for the changes required to maintain the geometry of the formation. The architecture of the formation controller is illustrated in Figure 2.4. Two of these controllers are combined to form the overall structure in the two-vehicle system.

This book will introduce how to incorporate motion synchronization technology to achieve coordinated control of UAV. For example, the cross coupling concept developed by Shan and Liu [4] can be used to synchronize the relative position tracking motion of the aircraft. This approach uses synchronization errors, which incorporate error information from different agents in the system, to identify the performance of the synchronization.

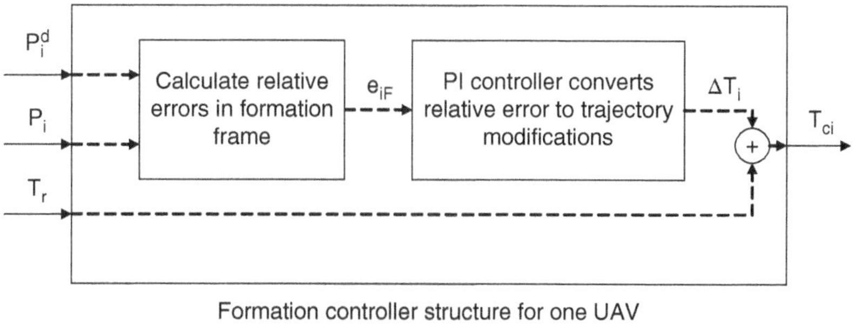

Figure 2.4 Controller for eliminating relative errors.

2.2 Behaviour-based Formation Dynamics

In parallel to the virtual structure, the other two well-known formation control approaches are behaviour-based and leader–follower. These share some common features. In this section, we will give a brief introduction of behaviour-based approach, highlighting its main concepts and principles. Perhaps one of the earliest published works in the area, and the one that gave it its name was by Balch and Arkin [5].

First, several formation geometries for a team of robots are defined, including line, column, diamond, and wedge. Secondly, formation is accomplished in two steps. Step 1 is a perceptual process, in which each robot's formation position is determined. This is followed by Step 2, a motion schema process to generate motor commands to direct the robot toward the correct location. Each robot computes its correct position in the formation based on the locations of the other robots.

Three techniques for formation position determination are identified.

1. In the unit-centre-referenced approach, each robot computes the unit centre independently, using the positions of all of the robots involved in the formation.
2. In the leader-referenced approach, each robot's formation position is determined in relation to the lead robot.
3. In the neighbor-referenced approach, each robot's position is maintained relative to another predetermined (neighbor) robot.

In addition, the orientation of the formation is defined by a direction from the unit centre to the next navigation waypoint, so the formation path for each robot is determined by the unit centre's waypoint position and velocity, as well as the position and orientation of each robot relative to the unit centre. The formation control strategies are derived by averaging several competing behaviours including goal seeking, collision avoidance and formation maintenance.

The behavioural approach is a decentralized one, and it can achieve flexibility and robustness. However, it is difficult to analyse mathematically. Note that each robot determines the positions of its peers by direct perception, by transmission of coordinates obtained from GPS or by dead reckoning.

In Balch and Arkin [5], the formation behaviors were implemented as motor schemas, within the autonomous robot architecture, and as steering and speed behaviors within the Unmanned Ground Vehicle (UGV) Demo II architecture. In both cases, the individual behaviors run as concurrent asynchronous processes, with each behavior representing a high-level behavioral intention of the agent. The motor-schema-based formation control approach includes several different schemas:

- move to goal
- avoid static obstacle
- avoid robot
- maintain formation.

The move-to-goal schema sets an adjustable gain value and (heading) direction toward a perceived goal to which the vehicle should be attracted. The avoid-static-obstacle schema (and also the avoid-robot one) repels the vehicle away from objects, with a variable gain and sphere of influence (distance). In the maintain-formation schema, a movement vector in the direction of the desired formation position and with adjustable magnitude is generated. This takes into account the so-called ballistic, controlled, or dead zones [5].

The motor schema's performance in terms of movement efficiency (e.g. turning or avoiding) can be measured using the metrics of path length ratio, average position error, and percentage of time out of formation.

2.3 Leader–Follower Formation Dynamics

The leader–follower approach in formation dynamics assigns the role of "leader" to one or more vehicles, while the rest are treated as "followers". The leader is assumed to be fully aware of the overall formation strategy and is given a desired trajectory that the formation will follow. The followers, on the other hand, will try to follow the command of the leader or to follow its motion.

The overall architecture is relatively simple. There is a clear command structure and the approach is easy to implement. However, it is an open-loop framework, and the lack of feedback from followers makes it a passive formation platform, which is sensitive to perturbations, and vulnerable to the quality of direction by the leader and particularly to the loss of leadership. In some papers in the literature, the leader is "a single point of failure for the formation", and the lack of explicit feedback from the followers may cause a failure to maintain formation if the follower's path is perturbed.

3

Fundamental Formation Control

In this chapter, we offer a comprehensive treatment of various formation control topics. Most topics covered relate to our own research work, with the addition of some other relevant studies for completeness. The topics involved are:

- velocity synchronization
- synchronized tracking
- formation control.

3.1 Unified Problem Description

In this section, we first give the definitions of several key concepts associated with formation control, considering a general multi-vehicle system (MVS) model. These definitions also apply to formation control when there are heterogeneous vehicles, but we will not dwell on this area. To help the reader understand the concepts involved, we take an MVS with dynamics modeled by N normalized double integrators as an illustrative example.

3.1.1 Some Key Definitions for Formation Control

Consider an MVS consisting of N vehicles with an interaction topology $\mathcal{G}$, and treat each vehicle as a node in $\mathcal{G}$. The dynamics of each vehicle are described by

$$\begin{cases} \dot{x}_i = f(x_i, u_i) \\ y_i = h(x_i) \end{cases} \tag{3.1}$$

where $i = 1, \cdots, N, x_i \in \mathbb{R}^n$ are the states, $u_i \in \mathbb{R}^m$ are the control inputs, $y_i \in \mathbb{R}^p$ are the outputs, and $f(x_i, u_i)$ and $h(x_i)$ are smooth functions with respect to their arguments.

A time-varying output formation is specified by a vector

$$F(t) = [F_1^T(t), \cdots, F_N^T(t)]^T \in \mathbb{R}^{pN} \tag{3.2}$$

where each $F_i(t)$ is piecewise continuously differentiable.

Definition 3.1 The MVS (3.1) is said to achieve a time-varying output formation (OF) specified by $F(t)$ if, for any given bounded initial states, there exists a vector $r(t) \in \mathbb{R}^p$ such that

$$y_i(t) - F_i(t) - r(t) = 0, i = 1, \cdots, N \tag{3.3}$$

Formation Control of Multiple Autonomous Vehicle Systems, First Edition. Hugh H.T. Liu and Bo Zhu.

where $r(t)$ is called a formation reference function, or the trajectory of a virtual leader. If objective (3.3) is satisfied with a time-invariant $F(t)$, then system (3.1) is said to achieve a time-invariant OF. If $F_i(t) = 0$ in Definition 3.1 for $i = 1, \cdots, N$, the above OF problem is reduced to the following output synchronization (OS) problem. That is, the OS problem can be regarded as a special case of an OF problem.

Definition 3.2 The MVS (3.1) is said to achieve mutual output synchronization if, for any given bounded initial states, there exists a vector $r(t) \in \mathbb{R}^p$ such that

$$y_i(t) - r(t) = 0, i = 1, \cdots, N \tag{3.4}$$

In particular, if $F_i(t) = 0$ and $h(x_i) = x_i$ for $i = 1, \cdots, N$, the OF problem is further reduced to a state synchronization (SS) problem. It is easy to verify that the above OS problem can be equivalently defined, as shown in Definition 3.3.

Definition 3.3 The MVS (3.1) is said to achieve mutual output synchronization if, for any given bounded initial states,

$$y_i(t) - y_j(t) = 0, i, j = 1, \cdots, N, j \neq i \tag{3.5}$$

Clearly, the existence of a vector $r(t) \in \mathbb{R}^p$ satisfying (3.4) is sufficient to guarantee (3.5). The converse is also true since, under (3.5), any $y_i(t)$ can be chosen as the reference function $r(t)$ to satisfy (3.4).

Definition 3.4 The MVS (3.1) is said to achieve synchronized output tracking (SOT) if, for a given time-varying reference $r_d(t)$,

$$y_i(t) - r_d(t) = 0, i = 1, \cdots, N \tag{3.6}$$

under any bounded initial states. In the SOT problem, the reference $r_d(t)$ denotes the desired output trajectory, which is often known to the designers or is artificially specified by them. In contrast, in the general OS problem, the function $r(t)$ may depend on the initial state of the MVS or the communication topology condition. Note that condition (3.6) is sufficient to guarantee (3.4) or (3.5). Therefore, the SOT problem may be regarded as a special case of the OS problem.

One of the main purposes of this book is to introduce approaches to developing u_i ($i = 1, \cdots, N$) for the MVS (3.1) that will meet these objectives in an asymptotic manner. In the following sections, we first give a simple example to illustrate these definitions and the requirements of asymptotic control. We then present several detailed designs for u_i to satisfy the requirements. The simplicity of these examples lies in the fact that the dynamics of each vehicle is modelled by a standard double integrator, which is a second-order linear system having no internal dynamics.

3.1.2 A Simple Illustrative Example

For simplicity, suppose that an MVS is modelled by N identical double integrators. These can be denoted in the form of (3.1) as:

$$\begin{aligned} f(x_i, u_i) &= \begin{bmatrix} 0 & 1 \\ 0 & 0 \end{bmatrix} x_i + \begin{bmatrix} 0 \\ 1 \end{bmatrix} u_i, \\ h(x_i) &= y_i = Cx_i. \end{aligned} \tag{3.7}$$

where the states $x_i \in \mathbb{R}^2$, the control input $u_i \in \mathbb{R}$, and the output $y_i \in \mathbb{R}$. Let $x_i = [r_i, v_i]^T$ in (3.7), where r_i denotes the linear or angular position, and v_i the linear or angular velocity, respectively. Then, MVS (3.1) with (3.7) is equivalent to

$$\begin{cases} \dot{r}_i = v_i, \\ \dot{v}_i = u_i. \end{cases} \tag{3.8}$$

$$y_i = C[r_i, v_i]' \tag{3.9}$$

which gives a simple mathematical description of Newton's second law, in which u_i is regarded as the input force or moment for vehicle i, and thus can be used to model simplified normalized translational or rotational dynamics.

For the MVS (3.8), if we specify

$$C = [0, 1], \tag{3.10}$$

the controlled output is the velocity $y_i = v_i$. On the other hand, if

$$C = [1, 0], \tag{3.11}$$

then the controlled output is the position $y_i = r_i$.

For the MVS (3.8)–(3.9), the mutual velocity synchronization at time T means that

$$v_i(T) - v_j(T) = 0, i, j = 1, \cdots, N, i \neq j. \tag{3.12}$$

In practical applications, velocity synchronization is often expected to be achieved in an asymptotic manner; that is,

$$\lim_{t\to\infty}(v_i(t) - v_j(t)) = 0, i, j = 1, \cdots, N, i \neq j. \tag{3.13}$$

Condition (3.13) is trivially satisfied if there exists a constant $\overline{v}$ such that for each $i = 1, \cdots, N$,

$$\lim_{t\to\infty}(v_i(t) - \overline{v}) = 0, \tag{3.14}$$

or if, for each $i = 1, \cdots, N$,

$$\lim_{t\to\infty}(v_i(t) - v_d(t)) = 0, \tag{3.15}$$

where $v_d(t)$ is referred to as the velocity reference or the desired velocity trajectory.

In the context of this book, $\overline{v}$ is an constant that is unknown for any vehicle, but depends on the initial state of the MVS and the interaction topology condition (this property will be examined in Section 3.4.1). In contrast, $v_d(t)$ may be a time-varying signal for the given reference; it is always assumed to be known for at least one vehicle to ensure that the associated synchronized trajectory tracking problem is solvable.

For MVS (3.8)–(3.9), the mutual angular-position synchronization at time T means that

$$r_i(T) - r_j(T) = 0, i, j = 1, \cdots, N, i \neq j. \tag{3.16}$$

Correspondingly, asymptotic angular-position synchronization means that

$$\lim_{t\to\infty}(r_i(t) - r_j(t)) = 0, i, j = 1, \cdots, N, i \neq j. \tag{3.17}$$

Condition (3.17) is trivially satisfied if there exists a constant $\bar{r}$ (which may be unknown) such that

$$\lim_{t\to\infty}(r_i(t) - \bar{r}) = 0, i = 1, \cdots, N, \tag{3.18}$$

or if

$$\lim_{t\to\infty}(r_i(t) - r_d(t)) = 0, i = 1, \cdots, N. \tag{3.19}$$

where $r_d(t)$ is a given angular-position reference or desired angular-position trajectory. In our discussions, $r_d(t)$ is known for at least one vehicle, and may be time varying.

3.2 Information Interaction Conditions

3.2.1 Algebraic Graph Theory

In this book, we are mainly concerned with the control problem of a group of vehicles. Generally, these vehicles interact with each other through a communication or sensing network or a combination of both. The information interaction conditions among the vehicles are important, because they may determine the mutual coupling level between vehicles and also the system stability. This is different from the stability of a traditional single-vehicle system.

Note that in mathematics and computer science, algebraic graphs are mathematical structures used to model pairwise relations between objects. Throughout this book, we will use algebraic graphs to model the information interaction between vehicles: each vehicle is regarded as a node of the algebraic graph associated with the MVS. In order to present the information interaction conditions clearly, we first set out some background about algebraic graphs.

A graph of order N is made up of N nodes and a certain number of edges that connect the nodes. It is denoted by $\mathcal{G}(\mathcal{I}, \mathcal{E})$, where $\mathcal{I} = \{1, \cdots, N\}$ is a finite nonempty node set, the edge formed by nodes i and j is denoted by (i, j), and $\mathcal{E} \subseteq \mathcal{I} \times \mathcal{I}$ is an edge set of pairs of nodes. In a undirected graph, the pairs of nodes are unordered, and the edge $(i, j) \in \mathcal{E}$ means that nodes i and j can obtain information from each other. In contrast, the pairs of nodes in a directed graph are ordered, and the edge $(i, j) \in \mathcal{E}$ means that node j can obtain information from node i, but not necessarily vice versa. Note that an undirected graph can be viewed as a special case of a directed graph, where an edge (i, j) in the undirected graph corresponds to the edges (i, j) and (j, i) in the directed graph. A path is a sequence of ordered edges of the form $(i, j), (j, k), \cdots$, where $(i, j), (j, k), \cdots \in \mathcal{E}$. If an edge $(i, j) \in \mathcal{E}$, then node i is said to be a neighbor of node j. The set of all neighbors of node i is denoted as $\mathcal{N}_i$.

A weighted graph associates a weight with every edge in the graph. In this book, all graphs are weighted. We use $\mathcal{A} = [a_{ij}] \in R^{N\times N}$ to denote the weighted adjacency matrix associated with a graph $\mathcal{G}$ of order N. For an undirected graph $\mathcal{G}$, if $(i, j) \in \mathcal{E}$, $a_{ij} = a_{ji} > 0$, and otherwise $a_{ij} = a_{ji} = 0$; $a_{ii} = 0$ for $i \in \mathcal{I}$. Thus $\mathcal{A}$ is a symmetric matrix for an undirected $\mathcal{G}$. The Laplacian matrix of $\mathcal{G}$, denoted by $\mathcal{L}_N[l_{ij}] \in R^{N\times N}$, is defined as: $l_{ij} = -a_{ij}$ for $i, j \in \mathcal{I}$ and $j \neq i$, and $l_{ii} = \sum_{j=1}^{N} a_{ij}$ for $i \in \mathcal{I}$. According to this definition, the Laplacian matrix of an undirected graph is symmetric. An undirected graph is connected if there is an undirected path between every pair of distinct nodes, and is fully connected if there is an edge between every pair of distinct nodes. In undirected graphs, a tree is a graph in which every pair of nodes is connected by exactly one undirected path.

For a directed graph $\mathcal{G} = (\mathcal{I}, \mathcal{E})$, an edge $(j, i) \in \mathcal{E}$ means that the ith vehicle can obtain information from the jth vehicle. Node j is called a neighbor of node i if $(j, i) \in \mathcal{E}$. If $\mathcal{G}$ has a sequence of edges $(i, i_1), (i_1, i_2), \cdots, (i_l, j)$ with distinct nodes i_k $(k = 1, 2, \cdots, l)$, the set $\{(i, i_1), (i_1, i_2), \cdots, (i_l, j)\}$ is called a path from node i to node j or, equivalently, node j is said to be reachable from node i, or node i has a directed path to node j. If there is a node which has a direct path to each of other nodes, this node is called a root node and $\mathcal{G}$ is said to have a directed spanning tree. The weighted adjacency matrix of $\mathcal{G}$ is a nonnegative matrix $\mathcal{A}_N = [a_{ij}] \in R^{N\times N}$ with $a_{ii} = 0$, $a_{ij} > 0$ if and only if $(j, i) \in \mathcal{E}$, and if $(j, i) \notin \mathcal{E}$, $a_{ij} = 0$. The definition of the Laplacian matrix of a directed graph $\mathcal{G}$ is the same as that of an undirected graph $\mathcal{G}$.

In a directed graph, a cycle is a directed path that starts and ends at the same node. A directed graph is strongly connected if there is a directed path from every node to every other node. A directed graph is complete if there is an edge from every node to every other node. A directed tree is a directed graph in which every node has exactly one parent except for the root node (which has no parent and has directed paths to all other nodes). A directed tree has no cycle because every edge is oriented away from the root. The in-degree and out-degree of node i are defined as $\sum_{j=1}^{N} a_{ij}$ and $\sum_{j=1}^{N} a_{ji}$, respectively. For both undirected and directed graphs, a node i is balanced if $\sum_{j=1}^{N} a_{ij} = \sum_{j=1}^{N} a_{ji}$. A graph is balanced if all the involved nodes are balanced. An undirected graph is balanced with symmetric $\mathcal{A}_N$.

For both undirected and directed graphs, 0 is an eigenvalue of L with an associated eigenvector $\mathbf{1}_N$ because $\mathcal{L}_N$ has zero row sums. Note that $\mathcal{L}_N$ is diagonally dominant and has nonnegative diagonal entries. According to Gershgorin's disc theorem (see Theorem 6.1.1 in Horn and Johnson, [6]), for an undirected graph, all nonzero eigenvalues of $\mathcal{L}_N$ are positive (because $\mathcal{L}_N$ is symmetric positive semidefinite), whereas all nonzero eigenvalues of $\mathcal{L}_N$ associated with a directed graph have positive real parts.

As for the case with a leader, in addition to matrices $\mathcal{A}_N$ and $\mathcal{L}_N$ for the N vehicles, we use a diagonal matrix $\mathcal{B}_N = diag(b_1, b_2, \cdots, b_N)$ to describe the information interaction condition among the N vehicles and the leader, which is defined as: If vehicle i can access the leader's information (including the velocity, position and acceleration), the weight $b_i > 0$ is a positive real number; and otherwise, $b_i = 0$. In the synchronized tracking problem, we view the leader as the node 0 and denote by $\overline{\mathcal{G}}$ the graph describing the information interaction among all the $N + 1$ nodes. Note that the graph $\mathcal{G}$ associated with the N vehicles can be regarded as a subgraph of the graph $\overline{\mathcal{G}}$ with an additional virtual node 0 (serving as the leader and having no neighbors). In this book, matrix $\mathcal{L}_N + \mathcal{B}_N$ is termed the information-exchange matrix for the synchronized tracking problem or the leader–follower formation control problem.

In this book, without special explanation, we generally assume that the interaction topology is time-invariant and thus the weights a_{ij}, $i, j \in \mathcal{I}$, and b_i are constants (i.e. time invariant).

3.2.2 Conditions for the Case without a Leader

For the information interaction models for the N vehicles, we mainly consider a communication topology satisfying one of the following conditions:

- **Condition 1:** Graph $\mathcal{G}$ has a directed spanning tree, ensuring that 0 is a simple eigenvalue of the associated Laplacian matrix $\mathcal{L}_N$.
- **Condition 2:** Graph $\mathcal{G}$ is undirected and connected, indicating that the associated Laplacian matrix $\mathcal{L}_N$ is symmetric, with 0 as a simple eigenvalue.

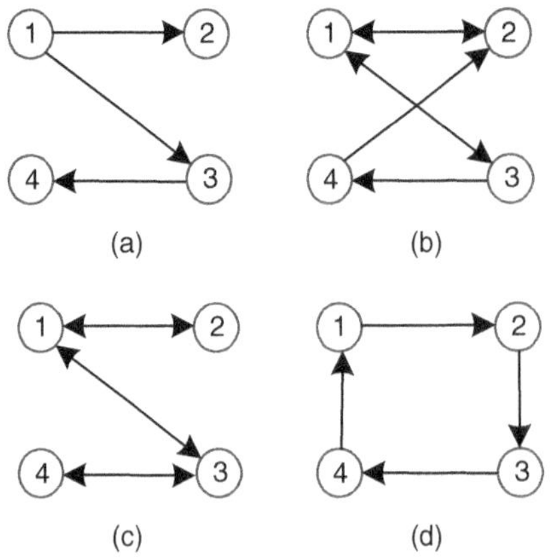

Figure 3.1 Interaction graphs without a leader.

To better understand Conditions 1 and 2, we here take as examples the four interaction graphs shown in Figure 3.1. We suppose that all the nonzero weights of the adjacency matrices associated with the four graphs are equal to 1.

- The graph shown in Figure 3.1a satisfies Condition 1, with vehicle 1 (or node 1) as the root node, having direct paths to all other vehicles. However, it is not strongly connected, since vehicle 2 (or vehicle 4) does not have a directed path to any other vehicle. It is not balanced either, since none of vehicles 1, 2 and 4 is a balanced node. In addition,

$$\mathcal{A}_4 = \begin{bmatrix} 0 & 0 & 0 & 0 \\ 1 & 0 & 0 & 0 \\ 1 & 0 & 0 & 0 \\ 0 & 0 & 1 & 0 \end{bmatrix}, \quad \mathcal{L}_4 = \begin{bmatrix} 0 & 0 & 0 & 0 \\ -1 & 1 & 0 & 0 \\ -1 & 0 & 1 & 0 \\ 0 & 0 & -1 & 1 \end{bmatrix}. \tag{3.20}$$

- The graph shown in Figure 3.1b satisfies Condition 1 and is strongly connected, since each vehicle has a directed path to all other vehicles. However, the graph is not balanced because vehicle 2, with in-degree 2 and out-degree 1, is not a balanced node. In addition,

$$\mathcal{A}_4 = \begin{bmatrix} 0 & 1 & 1 & 0 \\ 1 & 0 & 0 & 1 \\ 1 & 0 & 0 & 0 \\ 0 & 0 & 1 & 0 \end{bmatrix}, \quad \mathcal{L}_4 = \begin{bmatrix} 2 & -1 & -1 & 0 \\ -1 & 2 & 0 & -1 \\ -1 & 0 & 1 & 0 \\ 0 & 0 & -1 & 1 \end{bmatrix}. \tag{3.21}$$

- The graph shown in Figure 3.1c satisfies Condition 2, and is balanced. Further, we have

$$\mathcal{A}_4 = \begin{bmatrix} 0 & 1 & 1 & 0 \\ 1 & 0 & 0 & 0 \\ 1 & 0 & 0 & 1 \\ 0 & 0 & 1 & 0 \end{bmatrix}, \quad \mathcal{L}_4 = \begin{bmatrix} 2 & -1 & -1 & 0 \\ -1 & 1 & 0 & 0 \\ -1 & 0 & 2 & -1 \\ 0 & 0 & -1 & 1 \end{bmatrix}. \tag{3.22}$$

- The graph shown in Figure 3.1d is a cycle, strongly connected and balanced, and satisfies Condition 1. Either the in-degree or the out-degree of each vehicle is equal to 1.

Further, we have

$$\mathcal{A}_4 = \begin{bmatrix} 0 & 0 & 0 & 1 \\ 1 & 0 & 0 & 0 \\ 0 & 1 & 0 & 0 \\ 0 & 0 & 1 & 0 \end{bmatrix}, \quad \mathcal{L}_4 = \begin{bmatrix} 1 & 0 & 0 & -1 \\ -1 & 1 & 0 & 0 \\ 0 & -1 & 1 & 0 \\ 0 & 0 & -1 & 1 \end{bmatrix}. \tag{3.23}$$

3.2.3 Conditions for the Case with a Leader

For the case with a leader, we are mainly concerned with the following two conditions:

- **Condition 3:** The leader has directed paths to all vehicles or, equivalently, $\overline{\mathcal{G}}$ has a directed spanning tree with the leader as the root node.
- **Condition 4:** The subgraph $\mathcal{G}$ is connected and undirected (Condition 2 is satisfied), and the leader has a directed path to one vehicle (the leader is a neighbor of one vehicle).

Under Condition 4, Condition 3 is trivially satisfied. However, Condition 3 does not require graph $\mathcal{G}$ to be undirected or connected. Therefore, Condition 4 is sufficient but not necessary to render Condition 3 satisfied, i.e., Condition 3 is weaker than Condition 4. In addition, we give the following explanations to clarify the difference between Conditions 3 and 4.

- By definition, a directed graph is strongly connected if there is a directed path from every node to every other node, and the existence of a directed spanning tree is weaker than the condition of being strongly connected. Under Condition 3, the graph $\mathcal{G}$ may be directed and do not have a spanning tree.
- Under Condition 4, there is a path between any pair of vehicles, since $\mathcal{G}$ is connected and undirected. Thus the additional condition that the leader has a directed path to one vehicle ensures that the leader has directed paths to all vehicles.

To better understand Conditions 3 and 4, we here analyse the four interaction graphs shown in Figure 3.2. These will be frequently considered in the control development approach presented in this book. For simplicity, suppose that all the nonzero entries of the adjacency matrices associated with the four graphs are equal to 1.

- The graph $\overline{\mathcal{G}}$ shown in Figure 3.2a satisfies Condition 3 but not Condition 4, because graph $\mathcal{G}$ is directed and not connected. It can also be seen that the leader is a neighbor of only node 1, and that neither vehicle 2 nor vehicle 4 has a directed path to any other vehicle. Graph $\mathcal{G}$ is not balanced since none of nodes 1, 2 and 4 is a balanced node. In addition, $\mathcal{B}_4 = diag(1, 0, 0, 0)$, $\mathcal{A}_4$ and $\mathcal{L}_4$ are the same as those given by (3.20).
- The graph shown in Figure 3.2b satisfies both Conditions 3 and 4. Graph $\mathcal{G}$ is not only strongly connected but also balanced (all the nodes in $\mathcal{G}$, including node 1, are balanced). However, we also note that node 1 in graph $\overline{\mathcal{G}}$ is not balanced, with in-degree 4 and out-degree 3. In addition, $\mathcal{B}_4 = diag(1, 0, 0, 0)$ and

$$\mathcal{A}_4 = \begin{bmatrix} 0 & 1 & 1 & 1 \\ 1 & 0 & 1 & 0 \\ 1 & 1 & 0 & 1 \\ 1 & 0 & 1 & 0 \end{bmatrix}, \quad \mathcal{L}_4 = \begin{bmatrix} 3 & -1 & -1 & -1 \\ -1 & 2 & -1 & 0 \\ -1 & -1 & 3 & -1 \\ -1 & 0 & -1 & 2 \end{bmatrix}. \tag{3.24}$$

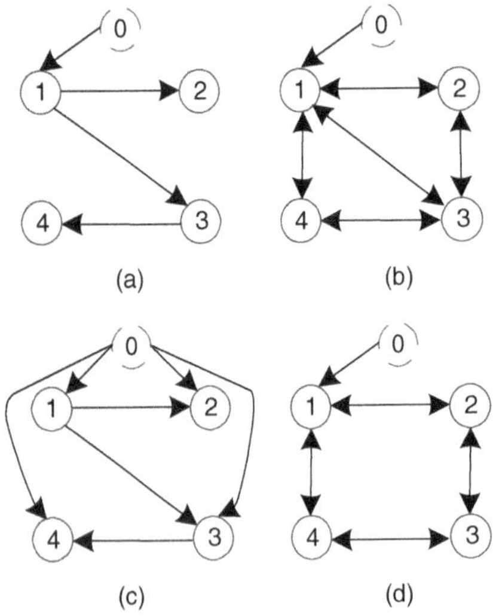

Figure 3.2 Interaction graphs with node 0 as the leader.

- The graph shown in Figure 3.2c satisfies Condition 3 but not Condition 4, because graph $\mathcal{G}$ is directed and not connected. It is also seen that the leader is a neighbor of each vehicle, which is different to the case shown in Figure 3.2a. The graph $\mathcal{G}$ is not balanced since none of vehicles 1, 2 and 4 is a balanced node. In addition, $\mathcal{B}_4 = diag(1, 1, 1, 1)$, $\mathcal{A}_4$ and $\mathcal{L}_4$ are the same as those given by (3.20).
- The graph shown in Figure 3.2d satisfies both Conditions 3 and 4. Moreover, the leader is a neighbor of only node 1, the same as the cases shown in Figures 3.2a and 3.2b. Graph $\mathcal{G}$ is balanced, since both the in-degree and out-degree of each involved node are 2. In addition, $\mathcal{B}_4 = diag(1, 0, 0, 0)$,

$$\mathcal{A}_4 = \begin{bmatrix} 0 & 1 & 0 & 1 \\ 1 & 0 & 1 & 0 \\ 0 & 1 & 0 & 1 \\ 1 & 0 & 1 & 0 \end{bmatrix}, \quad \mathcal{L}_4 = \begin{bmatrix} 2 & -1 & 0 & -1 \\ -1 & 2 & -1 & 0 \\ 0 & -1 & 2 & -1 \\ -1 & 0 & -1 & 2 \end{bmatrix}. \tag{3.25}$$

3.3 Synchronization Errors

For MVSs (3.8), we define the position synchronization error and velocity synchronization error between the ith and jth vehicles as follows:

$$\tilde{v}_{ij} = v_i - v_j, \tag{3.26}$$

$$\tilde{r}_{ij} = r_i - r_j, \tag{3.27}$$

with $i, j \in \mathcal{I}, i \neq j$. These are called the standard synchronization errors in this book.

As stated in the last section, the objective of synchronization control is to design a controller that can achieve asymptotic convergence of synchronization error. By the definitions in (3.26) and (3.27), we know that for each vehicle in the MVS (consisting of N vehicles), there are, in total, $N-1$ velocity synchronization errors (or $N-1$ position synchronization errors) to be asymptotically stabilized. For the MVS, there are a total of $(N-1)N$ velocity (or position) synchronization errors. When N is small (for instance,

$N = 2$ or $N = 3$), we can construct the synchronization controller using the variables defined by (3.26) and (3.27), and then evaluate the control performance by analysing the closed-loop dynamics of these variables. However, when $N > 4$, $(N-1)N > 12$ and, due to system complexity, it is not practical to use all these variables for control design and analysis.

To deal with the complexity problem, an obvious idea is to seek a single error variable for each vehicle, and then convert the synchronization control problem of MVS to a stabilization problem of each such error variable. This section introduces three different error variables that are used in the design of synchronization controllers throughout this book. Each of them is a local synchronization error (LSE) variable, meaning that only measurements of the vehicle itself and its neighbors are used.

3.3.1 Local Synchronization Error: Type I

For velocity synchronization, the Type I LSE is defined as

$$\xi_{vi} = \sum_{j\in N_i} a_{ij}(v_i - v_j), i \in \mathcal{I}, \tag{3.28}$$

which is a linear combination of the synchronization errors defined in (3.26), with coefficients $a_{ij}, j \in N_i$.

For position synchronization, the Type I LSE is defined as

$$\xi_{ri} = \sum_{j\in N_i} a_{ij}(r_i - r_j), i \in \mathcal{I}, \tag{3.29}$$

which is a linear combination of the synchronization errors defined in (3.27) with coefficients $a_{ij}, j \in N_i$. From the definition of $\mathcal{L}_N$, we can readily see that the definitions (3.28) and (3.29) are equivalent to

$$\begin{bmatrix} \xi_{v1} \\ \vdots \\ \xi_{vN} \end{bmatrix} = \mathcal{L}_N \begin{bmatrix} v_1 \\ \vdots \\ v_N \end{bmatrix}, \tag{3.30}$$

and

$$\begin{bmatrix} \xi_{r1} \\ \vdots \\ \xi_{rN} \end{bmatrix} = \mathcal{L}_N \begin{bmatrix} r_1 \\ \vdots \\ r_N \end{bmatrix}, \tag{3.31}$$

respectively.

Equations (3.30) and (3.31) show the linear relationship between Type I LSEs and the standard synchronization errors defined by (3.26) and (3.27). In addition, from (3.30) and (3.31), we can readily derive the following important property of Type I LSEs.

Lemma 3.1 Under Condition 1 or 2, the velocity synchronization objective (3.13) is achieved if and only if

$$\lim_{t\to\infty} \xi_{vi}(t) = 0, i \in \mathcal{I}, \tag{3.32}$$

and the position synchronization objective (3.17) is achieved if and only if

$$\lim_{t\to\infty} \xi_{ri}(t) = 0, i \in \mathcal{I}. \tag{3.33}$$

Proof: This lemma can be easily proven by the relationships (3.30) and (3.31) along with the fact that under Condition 1 or 2, $rank(\mathcal{L}_N) = N - 1$ and 0 is a simple eigenvalue of $\mathcal{L}_N$ associated with eigenvector $\mathbf{1}_N$. □

3.3.2 Local Synchronization Error: Type II

Given a velocity reference v_d and a position reference r_d, we define velocity tracking error $\tilde{v}_i$ and position tracking error $\tilde{r}_i$, velocity tracking error vector $\tilde{\mathbf{v}}$ and position tracking error vector, as follows.

$$\tilde{v}_i = : v_i - v_d \in R, i \in \mathcal{I}, \tag{3.34}$$

$$\tilde{r}_i = : r_i - r_d \in R, i \in \mathcal{I}, \tag{3.35}$$

$$\tilde{\mathbf{v}} = : [\tilde{v}_1, \cdots, \tilde{v}_N]^T \in R^N, \tag{3.36}$$

$$\tilde{\mathbf{r}} = : [\tilde{r}_1, \cdots, \tilde{r}_N]^T \in R^N. \tag{3.37}$$

As for synchronized tracking control tasks, the asymptotic convergence of both synchronization errors and tracking errors are required. Using the tracking errors defined by (3.34) and (3.35), we define the Type II LSEs as

$$\varepsilon_{vi}(t) = \sum_{j \in N_i} a_{ij}\tilde{v}_{ij} = \sum_{j \in N_i} a_{ij}(\tilde{v}_i - \tilde{v}_j), i \in \mathcal{I}, \tag{3.38}$$

$$\varepsilon_{ri}(t) = \sum_{j \in N_i} a_{ij}\tilde{r}_{ij} = \sum_{j \in N_i} a_{ij}(\tilde{r}_i - \tilde{r}_j), i \in \mathcal{I}. \tag{3.39}$$

These are also called generalized synchronization errors (GSEs) [7, 8]. The a_{ij} are the entries in the weighted adjacency matrix $\mathcal{A}_N$.

Note that the Type II LSE is a linear combination of either the standard synchronization errors defined by (3.26) and (3.27), or the tracking errors defined by (3.34) and (3.35). Moreover, by the definition of $\mathcal{L}_N$, we can readily see that the definitions (3.38) and (3.39) are equivalent to

$$\begin{bmatrix} \varepsilon_{v1} \\ \vdots \\ \varepsilon_{vN} \end{bmatrix} = \mathcal{L}_N \begin{bmatrix} \tilde{v}_1 \\ \vdots \\ \tilde{v}_N \end{bmatrix}, \tag{3.40}$$

and

$$\begin{bmatrix} \varepsilon_{r1} \\ \vdots \\ \varepsilon_{rN} \end{bmatrix} = \mathcal{L}_N \begin{bmatrix} \tilde{r}_1 \\ \vdots \\ \tilde{r}_N \end{bmatrix}, \tag{3.41}$$

respectively.

Equations (3.40) and (3.41) show the linear relationship between Type II LSEs and the tracking errors defined by (3.34) and (3.35). In addition, from (3.40) and (3.41), we can easily obtain the following property of Type II LSEs.

Lemma 3.2 Under Condition 1 or 2, the velocity synchronization objective (3.13) is achieved if and only if

$$\lim_{t \to \infty} \varepsilon_{vi}(t) = 0, i \in \mathcal{I}, \tag{3.42}$$

and the position synchronization objective (3.17) is achieved if and only if

$$\lim_{t\to\infty} \varepsilon_{ri}(t) = 0, i \in \mathcal{I}. \tag{3.43}$$

Proof: This lemma can be proven using the same arguments as used in the proof of Lemma 3.1, along with the fact that $\tilde{v}_1 = \cdots = \tilde{v}_N$ and $\tilde{r}_1 = \cdots = \tilde{r}_N$ are equivalent to $v_1 = \cdots = v_N$ and $v_1 = \cdots = v_N$, respectively. □

We here remind the reader that the above lemma is concerned with the convergence of mutual synchronization errors $\tilde{v}_{ij}$ and $\tilde{r}_{ij}$, rather than the tracking errors $\tilde{v}_i$ and $\tilde{r}_i$. Moreover, throughout the book, three specific realizations of the Type II LSEs are mainly considered. For the first realization,

$$\begin{aligned}
\varepsilon_{vi} &= \tilde{v}_{i(i+1)} = \tilde{v}_i - \tilde{v}_{i+1}, \\
\varepsilon_{ri} &= \tilde{r}_{i(i+1)} = \tilde{r}_i - \tilde{r}_{i+1}, \\
\varepsilon_{vN} &= \tilde{v}_{N1} = \tilde{v}_N - \tilde{v}_1, \\
\varepsilon_{rN} &= \tilde{r}_{N1} = \tilde{r}_N - \tilde{r}_1,
\end{aligned}$$

where $i = 1, \cdots, N-1$. Note that for this realization, only one $\tilde{r}_{ij}$ is used to construct ε_{ri} with $a_{ij} = 1$ for $j \in N_i$. This realization corresponds to

$$\mathcal{A}_N = \begin{bmatrix} 0 & 1 & & & \\ & 0 & 1 & & \\ & \ddots & \ddots & \ddots & \\ & & & 0 & 1 \\ 1 & & & & 0 \end{bmatrix}, \quad \mathcal{L}_N = \begin{bmatrix} 1 & -1 & & & \\ & 1 & -1 & & \\ & \ddots & \ddots & \ddots & \\ & & & 1 & -1 \\ -1 & & & & 1 \end{bmatrix}. \tag{3.44}$$

For the second realization,

$$\begin{aligned}
\varepsilon_{v1} &= \tilde{v}_{12} + \tilde{v}_{1N} = 2v_1 - v_2 - v_N, \\
\varepsilon_{r1} &= \tilde{r}_{12} + \tilde{r}_{1N} = 2r_1 - r_2 - r_N, \\
\varepsilon_{vi} &= \tilde{v}_{i(i+1)} + \tilde{v}_{i(i-1)} = 2v_i - v_{i+1} - v_{i-1}, \\
\varepsilon_{ri} &= \tilde{r}_{i(i+1)} + \tilde{r}_{i(i-1)} = 2r_i - r_{i+1} - r_{i-1}, \\
\varepsilon_{vN} &= \tilde{v}_{N1} + \tilde{v}_{N(N-1)} = 2v_N - v_1 - v_{N-1}, \\
\varepsilon_{rN} &= \tilde{r}_{N1} + \tilde{r}_{N(N-1)} = 2r_N - r_1 - r_{N-1},
\end{aligned}$$

where $i = 2, \ldots, N-1$.

Note that for the second realization, two $\tilde{r}_{ij}$ are used to construct ε_{ri}, and $a_{ij} = 1$ for $j \in N_i$. This realization corresponds to

$$\mathcal{A}_N = \begin{bmatrix} 0 & 1 & & & 1 \\ 1 & 0 & 1 & & \\ & \ddots & \ddots & \ddots & \\ & & 1 & 0 & 1 \\ 1 & & & 1 & 0 \end{bmatrix}, \quad \mathcal{L}_N = \begin{bmatrix} 2 & -1 & & & -1 \\ -1 & 2 & -1 & & \\ & \ddots & \ddots & \ddots & \\ & & -1 & 2 & -1 \\ -1 & & & -1 & 2 \end{bmatrix}. \tag{3.45}$$

For the third realization,

$$\begin{aligned}
\varepsilon_{v1} &= \tilde{v}_{1N} + \tilde{v}_{12} + \tilde{v}_{13} = 3v_1 - v_N - v_2 - v_3,\\
\varepsilon_{r1} &= \tilde{r}_{1N} + \tilde{r}_{12} + \tilde{r}_{13} = 3r_1 - r_N - r_2 - r_3,\\
\varepsilon_{vi} &= \tilde{v}_{i(i-1)} + \tilde{v}_{i(i+1)} + \tilde{v}_{i(i+2)} = 3v_i - v_{i-1} - v_{i+1} - v_{i+2},\\
\varepsilon_{ri} &= \tilde{r}_{i(i-1)} + \tilde{r}_{i(i+1)} + \tilde{r}_{i(i+2)} = 3r_i - r_{i-1} - r_{i+1} - r_{i+2},\\
\varepsilon_{v(N-1)} &= \tilde{v}_{(N-1)(N-2)} + \tilde{v}_{(N-1)N} + \tilde{v}_{(N-1)1} = 3v_{N-1} - v_{N-2} - v_N - v_1,\\
\varepsilon_{r(N-1)} &= \tilde{r}_{(N-1)(N-2)} + \tilde{r}_{(N-1)N} + \tilde{r}_{(N-1)1} = 3r_{N-1} - r_{N-2} - r_N - r_1,\\
\varepsilon_{vN} &= \tilde{v}_{N(N-1)} + \tilde{v}_{N1} + \tilde{v}_{N2} = 3v_N - v_{N-1} - v_1 - v_2,\\
\varepsilon_{rN} &= \tilde{r}_{N(N-1)} + \tilde{r}_{N1} + \tilde{r}_{N2} = 3r_N - r_{N-1} - r_1 - r_2,
\end{aligned}$$

where $i = 2, \cdots, N-2$.

Note that for the third realization, three $\tilde{r}_{ij}$ are used to construct ε_{ri} with $a_{ij} = 1$ for $j \in N_i$. This realization corresponds to

$$\mathcal{A}_N = \begin{bmatrix} 0 & 1 & 1 & & 1 \\ 1 & 0 & 1 & 1 & \\ & \ddots & \ddots & \ddots & \\ 1 & & 1 & 0 & 1 \\ 1 & 1 & & 1 & 0 \end{bmatrix}, \quad \mathcal{L}_N = \begin{bmatrix} 3 & -1 & -1 & & -1 \\ -1 & 3 & -1 & -1 & \\ & \ddots & \ddots & \ddots & \\ -1 & & -1 & 3 & -1 \\ -1 & -1 & & -1 & 3 \end{bmatrix}. \tag{3.46}$$

It is easy to verify that the graphs associated with realizations 1 and 3 are directed and satisfy Condition 1, whereas the graph associated with realizations 2 is undirected and satisfies Condition 2.

3.3.3 Local Synchronization Error: Type III

As shown by Lemmas 3.1 and 3.2, Type I and Type II LSEs have the common property that their convergence is sufficient and necessary to ensure the convergence of the standard synchronization errors defined by (3.26) and (3.27). As for synchronized tracking control, we require not only the convergence of synchronization errors but also the convergence of tracking errors. In order to make the design of a synchronized tracking controller easier, an obvious idea is to seek a single error variable for each vehicle, to unify the task of the convergence of standard synchronization errors and tracking errors. To this end, we here introduce Type III LSEs as follows:

$$e_{vi} = b_i\tilde{v}_i + \sum_{j\in N_i} a_{ij}\tilde{v}_{ij}, i \in \mathcal{I}, \tag{3.47}$$

$$e_{ri} = b_i\tilde{r}_i + \sum_{j\in N_i} a_{ij}\tilde{r}_{ij}, i \in \mathcal{I}. \tag{3.48}$$

There are also known as local neighborhood synchronization errors (LNSEs) [9, 10]. The synchronization errors $\tilde{v}_{ij}$ and $\tilde{r}_{ij}$ are defined by (3.26) and (3.27), and tracking errors $\tilde{v}_i$ and $\tilde{r}_i$ are defined by (3.34) and (3.35).

For the sake of simplicity of the MVS analysis, we define the following vectors associated with scale LSEs (3.47) and (3.48):

$$\mathbf{e}_v = [e_{v1}, \cdots, e_{vN}]^T \in R^N, \tag{3.49}$$

$$\mathbf{e}_r = [e_{r1}, \cdots, e_{rN}]^T \in R^N. \tag{3.50}$$

Then, the relationships (3.47) and (3.48) are equivalent to

$$\mathbf{e}_v = (\mathcal{L}_N + \mathcal{B}_N)\tilde{\mathbf{v}}, \tag{3.51}$$

$$\mathbf{e}_r = (\mathcal{L}_N + \mathcal{B}_N)\tilde{\mathbf{r}}, \tag{3.52}$$

where the included tracking error vectors $\tilde{\mathbf{v}}$ and $\tilde{\mathbf{r}}$ are defined by (3.36)–(3.37).

Equations (3.51) and (3.52) describe the linear relationships between the synchronization error vectors $\mathbf{e}_v$, $\mathbf{e}_r$ and tracking error vectors $\tilde{\mathbf{v}}$, $\tilde{\mathbf{r}}$. By the definitions of $\tilde{\mathbf{v}}$ and $\tilde{\mathbf{r}}$, we can easily see that if $\tilde{\mathbf{v}}$ and $\tilde{\mathbf{r}}$ converge to zero, so will the the standard synchronization errors $\tilde{v}_{ij}$ and $\tilde{r}_{ij}$, where $i, j \in \mathcal{I}$ and $i \neq j$. Furthermore, if $(\mathcal{L}_N + \mathcal{B}_N)$ is non-singular, the convergence of $\tilde{\mathbf{v}}$ and $\tilde{\mathbf{r}}$ can be achieved by asymptotically stabilizing $\mathbf{e}_v$ and $\mathbf{e}_r$. Therefore, the velocity (or position) synchronization tracking control problem can be reduced to the stabilization problem of $\mathbf{e}_v$ (or $\mathbf{e}_r$). The following lemma summarizes this property of this type of LSE.

Lemma 3.3 Under Condition 3 or 4,

$$\tilde{\mathbf{v}} = (\mathcal{L}_N + \mathcal{B}_N)^{-1}\mathbf{e}_v, \tag{3.53}$$

$$\tilde{\mathbf{r}} = (\mathcal{L}_N + \mathcal{B}_N)^{-1}\mathbf{e}_r, \tag{3.54}$$

and thus the velocity synchronized tracking objective (3.15) is achieved if and only if

$$\lim_{t\to\infty} e_{vi}(t) = 0, i \in \mathcal{I}. \tag{3.55}$$

The position synchronized tracking objective (3.19) is achieved if and only if

$$\lim_{t\to\infty} e_{ri}(t) = 0, i \in \mathcal{I}. \tag{3.56}$$

Proof: This lemma can be proven using the fact that under Condition 3 or 4, matrix $(\mathcal{L}_N + \mathcal{B}_N)$ indeed has full rank (as shown in Lemma 1.6 in the book by Ren and Cao [11]). □

In (3.47), the first component, $b_i\tilde{v}_i$, accounts for the velocity dissimilarity between node i and the leader, and the second component, $\sum_{j\in N_i} a_{ij}\tilde{v}_{ij}$, penalizes the velocity dissimilarity between neighboring nodes. Therefore, e_{vi} links the velocity synchronization error and tracking error into one control variable. Similarly, e_{ri} links the position synchronization error and tracking error into one control variable. Comparing (3.47) and (3.48) with (3.38) and (3.39), we can see that Type III LSEs include the component $b_i\tilde{v}_i$ or $b_i\tilde{r}_i$, whereas Type II LSEs do not. Despite the fact that b_i may be zero for some $i \in \mathcal{I}$, there exists at least one $b_i > 0$ to ensure that at least one vehicle can access the leader's information and therefore that the tracking problem is solvable. This slight difference is important in the sense that it ensures $\mathcal{L}_N + \mathcal{B}_N$ has full rank under Condition 3 or 4. In contrast, by the definition of $\mathcal{L}_N$, the maximum rank of $\mathcal{L}_N$ is $N - 1$ (under Condition 1 or 2).

From the definitions, we can see that for each vehicle, the computation of the three types of LSE only requires local measurements from the invidivual vehicle and its neighbors. The design of a network-level synchronization controller for a networked MVSs can therefore be converted into the design of a node-level controller that will drive the LSEs on each vehicle to zero. This greatly reduces the complexity of the design of synchronized tracking control and the approach can be easily adapted to other similar applications. Furthermore, by a slight modification of the definitions of e_{ri}, a leader–follower formation controller can be designed by applying synchronized tracking control approaches. This will be shown in Section 3.6.

In Sections 3.4–3.6, we will consider the problems of mutual velocity synchronization, mutual angular-position synchronization and formation control, and present control solutions for each. Finally, the differences and connections between these solutions will be discussed.

3.4 Velocity Synchronization Control

This section is to introduce the design of several control laws for MVS (3.8)–(3.10), which ensure that the objectives (3.14) and (3.15) are achieved.

3.4.1 Velocity Synchronization without a Leader

Recall the following simple control law for MVSs (3.8)–(3.10) [12, 13]:

$$u_i = -\xi_{vi}, i \in \mathcal{I}, \tag{3.57}$$

where ξ_{vi} is the LSE of Type I for each i. The control law (3.57) is distributed in the sense that only ξ_{vi}, as a local variable, and is used for the control of vehicle i.

The following two theorems show the relationship between the steady-state velocities (equilibrium velocity) and the initial velocities of vehicles, which correspond to Conditions 1 and 2, respectively. Hereafter, we use $v_i(0)$, $i = 1, \cdots, N$, to denote the initial velocities of the N vehicles.

Theorem 3.4 Under Condition 1, the control law (3.57) for systems (3.8)–(3.10) achieves the velocity synchronization objective (3.13), and, moreover, $v_i(t) \to \overline{v} = \sum_{i=1}^{N} \rho_i v_i(0)$ as $t \to \infty$, where $\rho = [\mu_1, \cdots, \mu_N]^T \geq 0$ and $\mathbf{1}_N^T \rho = 1$ and $\mathcal{L}_N^T \rho = 0_N$. (This is presented as Theorem 2.8 in Ren and Cao [13].)

Theorem 3.5 Under Condition 2, the control law (3.57) for systems (3.8)–(3.10) achieves the velocity synchronization objective (3.13), and, moreover, $v_i(t) \to \overline{v} = \frac{1}{N} \sum_{i=1}^{N} v_i(0)$ as $t \to \infty$. (This is presented as Corollary 2.9 in Ren and Cao [13].)

Both Theorems 3.4 and 3.5 show that the common velocity equilibrium is a linear combination of the N initial velocities. In particular, the common velocity equilibrium under Condition 2 is the average of the N initial velocities. This is therefore also referred to as average synchronization or average consensus [12, 13]. Since Condition 1 is weaker than Condition 2, the result of Theorem 3.4 is more general than the result of Theorem 3.5.

In what follows, we consider the SOT problem of MVSs (3.8)–(3.10), where the velocities of the N vehicles are required to track a common reference.

3.4.2 Velocity Synchronization with a Leader

For MVS (3.8)–(3.10), consider the control law

$$u_i = \dot{v}_d - k_v e_{vi}, i \in \mathcal{I}, \tag{3.58}$$

used to achieve the synchronized tracking objective (3.15), where the Type III LSEs are used and the constant gain $k_v > 0$. The resulting closed-loop equations are

$$\dot{\tilde{v}}_i = -k_v e_{vi}, i \in \mathcal{I}. \tag{3.59}$$

With relationships (3.36), (3.49) and (3.51), the N equations in (3.59) can be written in compact form as

$$\dot{\tilde{\mathbf{v}}} = -k_v(\mathcal{L}_N + \mathcal{B}_N)\tilde{\mathbf{v}}, \tag{3.60}$$

or, equivalently,

$$\dot{\mathbf{e}}_v = -k_v(\mathcal{L}_N + \mathcal{B}_N)\mathbf{e}_v, \tag{3.61}$$

provided $(\mathcal{L}_N + \mathcal{B}_N)$ is non-singular.

It is clear that $\tilde{\mathbf{v}}$ converges asymptotically to zero if and only if every eigenvalue of $\mathcal{L}_N + \mathcal{B}_N$ has a positive real part; that is, $-(\mathcal{L}_N + \mathcal{B}_N)$ is a Hurwitz matrix. Furthermore, if this is the case, the convergence velocity depends on gain k_v as well as the eigenvalues of $\mathcal{L}_N + \mathcal{B}_N$.

Generally, the location of the eigenvalues of $\mathcal{L}_N + \mathcal{B}_N$ depends on $\mathcal{L}_N$ as well as B_N. To derive a condition that ensures that $-(\mathcal{L}_N + \mathcal{B}_N)$ is Hurwitz, we may impose conditions on $\mathcal{L}_N$ and B_N, respectively. However, this may yield a conservative result since $\mathcal{L}_N$ and $\mathcal{B}_N$ may complement each other to render $-(\mathcal{L}_N + \mathcal{B}_N)$ Hurwitz. Throughout this book, we regard the desired trajectory as being the one generated by the leader, and attempt to seek a condition on graph $\overline{G}$, which are associated with all the $N+1$ vehicles (the leader and N vehicles). Under either Condition 3 or 4, $-(\mathcal{L}_N + \mathcal{B}_N)$ is Hurwitz. This property is seen from the following lemma.

Lemma 3.6 The matrix $-(\mathcal{L}_N + \mathcal{B}_N)$ is a Hurwitz matrix if and only if Condition 3 is satisfied. If this is the case with a directed subgraph G, all eigenvalues of $\mathcal{L}_N + \mathcal{B}_N$ have positive real parts; if, in particular, Condition 4 is satisfied, $\mathcal{L}_N + \mathcal{B}_N$ is symmetric positive definite and all the eigenvalues of $-(\mathcal{L}_N + \mathcal{B}_N)$ are negative real numbers. (This is presented as Lemma 1.6 in Ren and Cao [11].)

Theorem 3.7 Under Condition 3 or 4, MVS (3.8)–(3.10) with (3.58) achieves the velocity synchronized tracking objective (3.15).

This control law (3.58) is not distributed and restrictive, in the sense that the desired acceleration, $\dot{v}_d$, is required for each vehicle. To avoid such a limitation, we may consider the following control law (similar to the control law discussed by Liu *et al.* [14]):

$$u_i = k_i \left[b_i \dot{v}_d + \sum_{j \in N_i} a_{ij} u_j - k_v e_{vi} \right], i \in \mathcal{I}, \tag{3.62}$$

where the gain k_i is defined by

$$k_i = \frac{1}{b_i + \sum\limits_{j \in N_i} a_{ij}}, \tag{3.63}$$

depending on weights b_i and a_{ij}.

Under Condition 3, each node has at least one neighbor (the leader may be the neighbor of some vehicles, as shown in Figure 3.2), and thus $b_i + \sum\limits_{j \in N_i} a_{ij} > 0$ for each $i = 1, \cdots, N$, indicating that the control law (3.62) is well defined. In particular, the node-level control law given by (3.62) can be written in the compact form:

$$(\mathcal{L}_N + \mathcal{B}_N)\begin{bmatrix} u_1 \\ \vdots \\ u_N \end{bmatrix} = \begin{bmatrix} b_1\dot{v}_d - k_v e_{v1} \\ \vdots \\ b_N\dot{v}_d - k_v e_{vN} \end{bmatrix}, \tag{3.64}$$

which, under Condition 3, is equivalent to

$$\begin{bmatrix} u_1 \\ \vdots \\ u_N \end{bmatrix} = (\mathcal{L}_N + \mathcal{B}_N)^{-1}\begin{bmatrix} b_1\dot{v}_d - k_v e_{v1} \\ \vdots \\ b_N\dot{v}_d - k_v e_{vN} \end{bmatrix}. \tag{3.65}$$

A simple calculation shows that the closed-loop velocity trajectories driven by control law (3.62) satisfy

$$\dot{\tilde{\mathbf{v}}} = -k_v \tilde{\mathbf{v}}, \tag{3.66}$$

or, equivalently,

$$\dot{\mathbf{e}}_v = -k_v \mathbf{e}_v, \tag{3.67}$$

which is exponentially stable due to $k_v > 0$.

Theorem 3.8 Under Condition 3, MVS (3.8)–(3.10) with (3.62) achieves the velocity synchronized tracking objective (3.15).

Note that in (3.62) the control signals of neighboring vehicles, u_j with $j \in N_i$, are used to compute u_i. This may cause an implementation loop issue in practical applications. To deal with this problem, we may implement control law (3.62) in an approximation manner: use u_j obtained during the previous sampling period, $u_j(t - \tau)$, to compute the current $u_i(t)$, where τ denotes the (fixed) sampling step. In other words, we implement the following approximation of (3.62):

$$u_i(t) = k_i \left[b_i \dot{v}_d(t) + \sum_{j \in N_i} a_{ij} u_j(t - \tau) - k_v e_{vi} \right]. \tag{3.68}$$

As for the effect of the time delay τ on system performance, the following three statements are true for a connected undirected graph G:

- When $v_d(t)$ is a constant, asymptotic synchronization is guaranteed for any $\tau > 0$ by (3.68): the synchronization errors still converge asymptotically to zero.
- If $v_d(t)$ is time-varying, the synchronization errors are ultimately bounded and their ultimate bounds depend on the value of τ as well as the Lipschitz constant of $v_d(t)$.

- The convergence speed of synchronization errors is continuously dependent on τ, indicating that the difference in synchronization speed resulting from the introduction of τ is slight when τ is small enough.

See, for instance, Liu *et al.* for a proof [14].

3.5 Angular-position Synchronization Control

The objective of this section is to design a control law for systems (3.8)–(3.10) to achieve (3.18) and (3.19), respectively.

3.5.1 Synchronization without a Position Reference

The result in Equation (3.57) may be applied to construct a distributed control law for asymptotic angular-position synchronization. To this end, consider the coordinate transformation

$$\begin{bmatrix} r_i \\ s_i \end{bmatrix} = \begin{bmatrix} 1 & 0 \\ k & 1 \end{bmatrix} \begin{bmatrix} r_i \\ v_i \end{bmatrix}, i \in \mathcal{I}, \tag{3.69}$$

where the design parameter $k > 0$, ensuring the transformation is invertible. According to the transformation,

$$s_i = v_i + kr_i, \tag{3.70}$$

or, equivalently,

$$\dot{r}_i = v_i = -kr_i + s_i. \tag{3.71}$$

From (3.8) and (3.70), we obtain

$$\dot{s}_i = \dot{v}_i + kv_i = u_i + kv_i. \tag{3.72}$$

Applying the input replacement

$$u_i = \mu_i - kv_i \tag{3.73}$$

to (3.72) gives

$$\dot{s}_i = \mu_i, \tag{3.74}$$

which has the same form as the forced velocity equation, $\dot{v}_i = u_i$, in (3.8). Thus the control law for velocity synchronization can be straightforwardly applied to (3.74).

Combining (3.71) and (3.74) gives

$$\begin{cases} \dot{r}_i = -kr_i + s_i \\ \dot{s}_i = \mu_i \end{cases} i \in \mathcal{I}, \tag{3.75}$$

These statements describe the controlled vehicle dynamics in the new coordinate $(r_i, s_i)^T$.

Applying the control law (3.57) to (3.75), yields

$$\mu_i = -\xi_{si}, i \in \mathcal{I}, \tag{3.76}$$

where ξ_{si} are the Type I LSEs, defined similarly to ξ_{vi} in (3.28), i.e., $\xi_{si} = \sum_{j \in N_i} a_{ij}(s_i - s_j)$, $i \in \mathcal{I}$.

Substituting (3.70) and (3.76) into (3.73) gives

$$u_i = -kv_i + \sum_{j\in N_i} a_{ij}[k(r_j - r_i) + (v_j - v_i)], i \in \mathcal{I}. \tag{3.77}$$

Note that the linear damping, $-kv_i$, is included in (3.77).

We then have the following results.

Theorem 3.9 Under Condition 1, MVS (3.8)–(3.10) with the control law (3.77) achieves the asymptotic angular-position synchronization objective (3.17), and, moreover, $r_i(t) \to \bar{r} = \frac{1}{k}\sum_{i=1}^{N} \rho_i(kr_i(0) + v_i(0))$, $v_i(t) \to 0$ as $t \to \infty$, where $\rho = [\mu_1, \cdots, \mu_N]^T \geq 0$, $\mathbf{1}_N^T\rho = 1$ and $\mathcal{L}_N^T\rho = 0$.

Proof: By applying the results of Theorem 3.1 to (3.74) with (3.76), we obtain that, under Condition 1, $\lim_{t\to\infty}(s_j(t) - s_i(t)) = 0$ for any $i,j = 1,\cdots,N$ and $i \neq j$, and $\lim_{t\to\infty} s_i(t) = \sum_{i=1}^{N} \rho_i s_i(0)$, $i = 1,\cdots,N$. Since $s_j(t) - s_i(t) = k(r_j(t) - r_i(t)) + v_j(t) - v_i(t)$ by definition (3.70), the condition of $\lim_{t\to\infty}(s_j(t) - s_i(t)) = 0$ ensures that

$$\begin{aligned} &\lim_{t\to\infty}[k(r_j(t) - r_i(t)) + v_j(t) - v_i(t)] = 0 \\ &\Leftrightarrow \lim_{t\to\infty}[\dot{r}_j(t) - \dot{r}_i(t) + k(r_j(t) - r_i(t))] = 0 \\ &\Rightarrow \lim_{t\to\infty}(r_i(t) - r_j(t)) = 0 \\ &\Rightarrow \lim_{t\to\infty}(v_i(t) - v_j(t)) = 0 \end{aligned} \tag{3.78}$$

Thus, both angular-position synchronization and angular-velocity synchronization are achieved. From $\lim_{t\to\infty} s_i(t) = \sum_{i=1}^{N} \rho_i s_i(0)$, we obtain

$$\begin{aligned} &\lim_{t\to\infty}[v_i(t) + kr_i(t)] = \sum_{i=1}^{N} \rho_i s_i(0) = \sum_{i=1}^{N} \rho_i(kr_i(0) + v_i(0)) \\ &\Leftrightarrow \lim_{t\to\infty}[\dot{r}_i(t) + kr_i(t)] = \sum_{i=1}^{N} \rho_i(kr_i(0) + v_i(0)) \\ &\Rightarrow \lim_{t\to\infty} r_i(t) = \frac{1}{k}\sum_{i=1}^{N} \rho_i(kr_i(0) + v_i(0)) \\ &\Rightarrow \lim_{t\to\infty} v_i(t) = 0. \end{aligned} \tag{3.79}$$

This ends the proof of Theorem 3.9. □

Theorem 3.10 Under Condition 2, MVS (3.8)–(3.10) with the control law (3.77) achieves the asymptotic angular-position synchronization objective (3.17) and moreover, $r_i(t) \to \bar{r} = \frac{1}{kN}\sum_{i=1}^{N}(kr_i(0) + v_i(0))$, $v_i(t) \to 0$ as $t \to \infty$.

Proof: Theorem 3.10 can be proved along the same lines as for Theorem 3.9 and using the average synchronization result $\lim_{t\to\infty} s_i(t) = \frac{1}{N}\sum_{i=1}^{N} s_i(0)$. The detailed proof is omitted here for simplicity. □

The above results show that under the control law (3.77), the angular-position trajectories of the N vehicles synchronize asymptotically to a common position equilibrium, which is dependent on the initial angular positions and angular velocities. The angular-velocity trajectory of each vehicle synchronizes to zero asymptotically.

From the above control design, we see that the problem of angular-position synchronization without a reference can be converted to the synchronization problem of the N auxiliary variables s_i, $i = 1, \cdots, N$. The main purpose of introducing the auxiliary variable s_i is to reduce the system's relative order to 1 so that the design result for single integrator dynamics can be applied to construct control laws for higher relative-order systems. In fact, Equation (3.74) shows that the relative order between control u_i and the auxiliary variable s_i is 1, and, in contrast, the relative order between u_i and r_i is 2, as shown by (3.8). In the following section, this idea will be further applied to design a control law for (3.8) to achieve the objective of angular-position synchronized tracking.

3.5.2 Synchronization to a Position Reference

We use $v_d = \dot{r}_d$ and $\ddot{r}_d$ to denote the desired angular-velocity trajectory and angular-acceleration trajectory, respectively. Noting that s_i satisfies (3.70), we define

$$s_d = v_d + kr_d = \dot{r}_d + kr_d, \tag{3.80}$$

as the common desired trajectory for the controlled variables s_i ($i \in \mathcal{I}$). We define the tracking errors as

$$\tilde{s}_i \;=\; : s_i - s_d \in R, \tag{3.81}$$

$$\tilde{\mathbf{s}} \;=\; : [\tilde{s}_1, \cdots, \tilde{s}_N]^T \in R^N. \tag{3.82}$$

To achieve $s_i \to s_d$ as $t \to \infty$, we apply (3.58) to MVS (3.75) and obtain

$$\mu_i = \dot{s}_d - k_v b_i (s_i - s_d) - \sum_{j \in N_i} k_v a_{ij} (s_i - s_j), i = 1, \cdots, N. \tag{3.83}$$

With (3.83), Equations (3.73) become

$$u_i = \dot{s}_d - k_v b_i (s_i - s_d) - \sum_{j \in N_i} k_v a_{ij} (s_i - s_j) - kv_i, i = 1, \cdots, N. \tag{3.84}$$

Substituting (3.34), (3.70), (3.80) and (3.35) into (3.84) yields

$$u_i = \ddot{r}_d - k\tilde{v}_i - k_v e_{vi} - k_v k e_{ri}, \tag{3.85}$$

where the Type III LSEs are explicitly used.

In what follows, two simpler variations of (3.85) are discussed to help the reader understand the control law. If vehicle i has only access to the leader (i.e., $b_i > 0$ and N_i is an empty set), the control law (3.85) is reduced to

$$u_i = \ddot{r}_d - k\tilde{v}_i - k_v b_i (k\tilde{r}_i + \tilde{v}_i), \tag{3.86}$$

which has the standard proportional-derivative (PD) control form, with an acceleration compensation term $\ddot{r}_d$.

If the leader has a fixed position, vehicle i does not access the leader's information and has only one neighbor vehicle k (i.e., $\ddot{r}_d = \dot{r}_d = 0$, $b_i = 0$, $a_{ik} > 0$), the control law (3.85) is simply reduced to

$$u_i = -kv_i - k_v a_{ik}[k(r_i - r_k) + v_i - v_k], \tag{3.87}$$

which includes v_i as the damping term, $r_i - r_k$ as the relative position error, and $v_i - v_k$ as the relative velocity error.

The closed-loop dynamics consisting of (3.75) and (3.83) are

$$\begin{cases} \dot{\tilde{\mathbf{s}}} = -k_v(\mathcal{L}_N + \mathcal{B}_N)\tilde{\mathbf{s}} \\ \dot{\tilde{r}}_i = -k\tilde{r}_i + \tilde{s}_i, i \in \mathcal{I}, \end{cases} \tag{3.88}$$

As for the stability of system (3.88), we have the following result.

Theorem 3.11 Under Condition 3, MVS (3.8)–(3.10) driven by (3.85) achieves asymptotic synchronization in tracking the desired state $(r_d(t), \dot{r}_d(t))^T$, i.e., $\lim_{t\to\infty}\tilde{r}_i(t) = 0$, $\lim_{t\to\infty}\tilde{v}_i(t) = 0$ for all $i \in \mathcal{I}$.

If we use the result (3.62) to design control laws for (3.74), we will obtain

$$u_i = k_i \left[b_i\ddot{r}_d + \sum_{j\in N_i} a_{ij}u_j - kb_i\tilde{v}_i - k_v e_{vi} - k_v k e_{ri} \right], i \in \mathcal{I}, \tag{3.89}$$

which shows that the desired acceleration, $\ddot{r}_d$, is only required for the vehicles with the leader as a neighbor.

Under Condition 3, the explicit form of control law (3.89) is

$$\begin{bmatrix} u_1 \\ \vdots \\ u_N \end{bmatrix} = (\mathcal{L}_N + \mathcal{B}_N)^{-1} \begin{bmatrix} b_1\ddot{r}_d - kb_1\tilde{v}_1 - k_v e_{v1} - k_v k e_{r1} \\ \vdots \\ b_N\ddot{r}_d - kb_N\tilde{v}_N - k_v e_{vN} - k_v k e_{rN} \end{bmatrix}. \tag{3.90}$$

which, however, is not a distributed control law. Then, we have the following result.

Theorem 3.12 Under Condition 3, MVS (3.8)–(3.10) with (3.89) achieves asymptotic synchronization in tracking the desired state $(r_d(t), \dot{r}_d(t))^T$, i.e., $\lim_{t\to\infty}\tilde{r}_i(t) = 0$ and $\lim_{t\to\infty}\tilde{v}_i(t) = 0$ for all $i = 1, \cdots, N$.

To avoid a possible implementation loop issue associated with (3.89), we modify it as

$$u_i(t) = k_i \left[b_i\ddot{r}_d(t) + \sum_{j\in N_i} a_{ij}u_j(t-\tau) - kb_i\tilde{v}_i - k_v e_{vi} - k_v k e_{ri} \right], \tag{3.91}$$

where τ denotes a small sampling-step delay.

3.6 Formation via Synchronized Tracking

In the synchronized tracking problem in Section 3.5, the output trajectories of the N vehicles were driven to a common reference output trajectory. In this section, we try to design control laws to make the position trajectories of the N vehicles have a desired deviation from the reference position trajectory. The reference position trajectory may be considered as the virtual center of the formation, or as the trajectory of a virtual leader. For the N vehicles, all the desired deviations may be time varying and different from each other. Despite these facts, there are close connections between synchronized motion and formation motion. In particular, we will show in the following that the formation control problem can be converted to a (linear) position synchronized tracking problem.

We use $\delta_i(t)$, $i = 1, \cdots, N$, to denote the desired position deviation of vehicle i from the leader's trajectory $r_d(t)$. The formation is achieved at some time T when

$$r_i(T) - \delta_i(T) - r_d(T) = 0, i = 1, \cdots, N. \tag{3.92}$$

The formation is asymptotically achieved when

$$\lim_{t\to\infty}(r_i(t) - \delta_i(t) - r_d(t)) = 0, i = 1, \cdots, N. \tag{3.93}$$

By introducing some auxiliary variables, the formation control objective (3.93) can be converted to a synchronization control objective. To show this fact, we introduce auxiliary variables η_i, $i = 1, \cdots, N$, as

$$\eta_i(t) = r_i(t) - \delta_i(t), i = 1, \cdots, N. \tag{3.94}$$

Then, (3.93) is equivalent to

$$\lim_{t\to\infty}(\eta_i(t) - r_d(t)) = 0, i = 1, \cdots, N, \tag{3.95}$$

which can be regarded as a specific SOT objective, as defined in (3.19), with $\eta_i(t)$ as the controlled variables to track $r_d(t)$.

From (3.94),

$$\dot{\eta}_i(t) = v_i(t) - \dot{\delta}_i(t), i = 1, \cdots, N, \tag{3.96}$$

$$\ddot{\eta}_i(t) = u_i(t) - \ddot{\delta}_i(t), i = 1, \cdots, N, \tag{3.97}$$

where, according to Equations (3.8), $u_i(t)$ is included to replace $\dot{v}_i(t)$.

Considering the input replacement

$$u_i(t) = \overline{u}_i(t) + \ddot{\delta}_i(t), i = 1, \cdots, N, \tag{3.98}$$

Equations 3.98 then become

$$\ddot{\eta}_i(t) = \overline{u}_i(t), i = 1, \cdots, N. \tag{3.99}$$

Up to now, the design of formation control has been converted to a design of synchronization control $\overline{u}_i(t)$ for $\eta_i(t)$. Note that the equation of $\eta_i(t)$ has been converted into the form of double integrators, as in (3.99). Synchronization control can be designed using the results proposed in previous sections. In what follows, we will develop two control laws for $\overline{u}_i(t)$, by applying the results (3.85) and (3.89), respectively.

It is important to note that the design of formation control can also be converted to the design of synchronization control by introducing a virtual desired trajectory, $r_d^i(t) = r_d(t) - \delta_i(t)$, for vehicle i. Then, the objective (3.93) is equivalent to

$$\lim_{t\to\infty}(r_i(t) - r_d^i(t)) = 0, i = 1, \cdots, N. \tag{3.100}$$

This corresponds to the control problem with N different leaders, and there is one leader for each vehicle to track. However, discussions about this kind of control issue are beyond the scope of this book. Interested readers are referred to the literature for details [15].

3.6.1 Formation Control Solution 1

We apply the control law (3.85) to MVS (3.99). This is implemented by replacing $r_i(t)$, $\tilde{r}_i(t)$, $v_i(t)$, and $\tilde{v}_i(t)$ in (3.85) with $\eta_i(t)$, $\eta_i(t) - r_d(t)$, $\dot{\eta}_i(t)$, and $\dot{\eta}_i(t) - \dot{r}_d(t)$, respectively, and yields the expression for $\overline{u}_i(t)$ as

$$\begin{aligned}\overline{u}_i &= \ddot{r}_d - k(\dot{\eta}_i - \dot{r}_d) - k_v b_i[k(\eta_i - r_d) + \dot{\eta}_i - \dot{r}_d] \\ &\quad - \sum_{j\in N_i} k_v a_{ij}[k(\eta_i - \eta_j) + \dot{\eta}_i - \dot{\eta}_j]. \end{aligned} \tag{3.101}$$

Correspondingly, u_i in (3.98) satisfies

$$\begin{aligned}u_i &= \ddot{\delta}_i + \ddot{r}_d - k(\dot{\eta}_i - \dot{r}_d) - k_v b_i[k(\eta_i - r_d) + \dot{\eta}_i - \dot{r}_d] \\ &\quad - \sum_{j\in N_i} k_v a_{ij}[k(\eta_i - \eta_j) + \dot{\eta}_i - \dot{\eta}_j]. \end{aligned} \tag{3.102}$$

Note that the signal $\ddot{\delta}_i + \ddot{r}_d$ is required for the control of each vehicle.

Substituting (3.94) and (3.98) into (3.102) gives the control law

$$\begin{aligned}u_i &= \ddot{\delta}_i + \ddot{r}_d - k(v_i - \dot{\delta}_i - \dot{r}_d) - k_v b_i[k(r_i - \delta_i - r_d) + v_i - \dot{\delta}_i - \dot{r}_d] \\ &\quad - \sum_{j\in N_i} k_v a_{ij}[k(r_i - \delta_i - r_j + \delta_j) + v_i - \dot{\delta}_i - v_j + \dot{\delta}_j], \\ &= \ddot{\delta}_i + \ddot{r}_d - k(\tilde{v}_i - \dot{\delta}_i) - k_v b_i[k(\tilde{r}_i - \delta_i) + \tilde{v}_i - \dot{\delta}_i] \\ &\quad - \sum_{j\in N_i} k_v a_{ij}[k(\tilde{r}_{ij} - \delta_i + \delta_j) + \tilde{v}_{ij} - \dot{\delta}_i + \dot{\delta}_j]. \end{aligned} \tag{3.103}$$

Theorem 3.13 Under Condition 3, MVS (3.8)–(3.10) with (3.103) achieves the formation control objective (3.93).

To achieve asymptotically a time-invariant formation with constants δ_i, we propose the following formation control law:

$$\begin{aligned}u_i &= \ddot{r}_d - k(v_i - \dot{r}_d) - k_v b_i[k(r_i - \delta_i - r_d) + v_i - \dot{r}_d] \\ &\quad - \sum_{j\in N_i} k_v a_{ij}[k(r_i - \delta_i - r_j + \delta_j) + v_i - v_j], \\ &= \ddot{r}_d - k\tilde{v}_i - k_v b_i[k(\tilde{r}_i - \delta_i) + \tilde{v}_i] \\ &\quad - \sum_{j\in N_i} k_v a_{ij}[k(\tilde{r}_{ij} - \delta_i + \delta_j) + \tilde{v}_{ij}], \end{aligned} \tag{3.104}$$

which is obtained by letting $\dot{\delta}_i = 0$ and $\ddot{\delta}_i = 0$ $(i = 1, \cdots, N)$ in (3.103).

3.6.2 Formation Control Solution 2

We apply the control law (3.89) to MVS (3.99). This is done by replacing $r_i(t)$, $\tilde{r}_i(t)$, $v_i(t)$, $\tilde{v}_i(t)$ and $u_j(t)$ in (3.89) with $\eta_i(t)$, $\eta_i(t) - r_d(t)$, $\dot{\eta}_i(t)$, $\dot{\eta}_i(t) - \dot{r}_d(t)$ and $\overline{u}_j$, respectively, and ultimately yields the expressions for $\overline{u}_i(t)$ as

$$\overline{u}_i = k_i \Bigg[b_i \ddot{r}_d + \sum_{j \in N_i} a_{ij} \overline{u}_j - k_v b_i (\dot{\eta}_i - \dot{r}_d + k(\eta_i - r_d)) - k b_i (\dot{\eta}_i - \dot{r}_d) - \sum_{j \in N_i} k_v a_{ij} (\dot{\eta}_i - \dot{\eta}_j + k(\eta_i - \eta_j)) \Bigg] . \tag{3.105}$$

Correspondingly, u_i in (3.98) with (3.105) satisfies

$$u_i = k_i \Bigg[b_i (\ddot{\delta}_i + \ddot{r}_d) + \sum_{j \in N_i} a_{ij} (\ddot{\delta}_i - \ddot{\delta}_j) + \sum_{j \in N_i} a_{ij} u_j - k_v b_i (\dot{\eta}_i - \dot{r}_d + k(\eta_i - r_d)) - k b_i (\dot{\eta}_i - \dot{r}_d) - \sum_{j \in N_i} k_v a_{ij} (\dot{\eta}_i - \dot{\eta}_j + k(\eta_i - \eta_j)) \Bigg] . \tag{3.106}$$

Note that the signal $\ddot{\delta}_i + \ddot{r}_d$ is only needed for the control laws of vehicles i with $b_i > 0$.

Substituting equations (3.94) and (3.98) into (3.106), gives

$$u_i = k_i \Bigg[b_i (\ddot{\delta}_i + \ddot{r}_d) + \sum_{j \in N_i} a_{ij} (\ddot{\delta}_i - \ddot{\delta}_j) + \sum_{j \in N_i} a_{ij} u_j - k_v b_i (\tilde{v}_i - \dot{\delta}_i + k(\tilde{r}_i - \delta_i)) - k b_i (\tilde{v}_i - \dot{\delta}_i) - \sum_{j \in N_i} k_v a_{ij} (\tilde{v}_{ij} - \dot{\delta}_i + \dot{\delta}_j + k(\tilde{r}_{ij} - \delta_i + \delta_j)) \Bigg] . \tag{3.107}$$

Theorem 3.14 Under Condition 3, MVS (3.8)–(3.10) with (3.107) achieves the formation control objective (3.93).

To avoid a possible implementation loop issue associated with (3.107), we modify it as:

$$u_i = k_i \Bigg[b_i (\ddot{\delta}_i + \ddot{r}_d) + \sum_{j \in N_i} a_{ij} (\ddot{\delta}_i - \ddot{\delta}_j) + \sum_{j \in N_i} a_{ij} u_j(t - \tau) - k_v b_i (\tilde{v}_i - \dot{\delta}_i + k(\tilde{r}_i - \delta_i)) - k b_i (\tilde{v}_i - \dot{\delta}_i) - \sum_{j \in N_i} k_v a_{ij} (\tilde{v}_{ij} - \dot{\delta}_i + \dot{\delta}_j + k(\tilde{r}_{ij} - \delta_i + \delta_j)) \Bigg] , \tag{3.108}$$

where τ denotes a small sampling step. Note that the only difference between (3.107) and (3.108) is that the delayed control signal $u_j(t - \tau)$, rather than the current control signal $u_j(t)$, is used in (3.108).

To achieve asymptotically a time-invariant formation with constants δ_i $(i = 1, \cdots, N)$, we propose the following formation control law:

$$u_i = k_i \left[b_i \ddot{r}_d + \sum_{j \in N_i} a_{ij} u_j - k_v b_i (\tilde{v}_i + k(\tilde{r}_i - \delta_i)) - k b_i \tilde{v}_i \right. \\ \left. - \sum_{j \in N_i} k_v a_{ij} (\tilde{v}_{ij} + k(\tilde{r}_{ij} - \delta_i + \delta_j)) \right], \quad (3.109)$$

which is obtained by letting $\dot{\delta}_i = 0$ and $\ddot{\delta}_i = 0$ $(i = 1, \cdots, N)$ in (3.107).

3.7 Simulations

3.7.1 Verification of Theorem 3.12

We verify the results of Theorem 3.12 by considering two scenarios with a directed graph $\mathcal{G}$ and an undirected graph $\mathcal{G}$, respectively.

For each scenario, we suppose that:

- all the nonzero entries of the associated matrices $\mathcal{A}_4$ and $\mathcal{B}_4$ are equal to 1
- the control gains $k_v = 0.5$ and $k = 0.5$
- the initial controls $u_i(0) = 0$ $(i = 1, 2, 3, 4)$
- the initial states of the four vehicles are:

$$(r_1(0), v_1(0)) = (5, 6)$$
$$(r_2(0), v_2(0)) = (10, 2)$$
$$(r_3(0), v_3(0)) = (-5, -5)$$
$$(r_4(0), v_4(0)) = (-10, -10)$$

- the desired trajectories are:

$$r_d(t) = \sin(t)$$
$$\dot{r}_d(t) = \cos(t)$$
$$\ddot{r}_d(t) = -\sin(t).$$

Matlab R2014a/simulation solver ode4 (Runge-Kutta) with a fixed-step size of 0.01 s is used in the simulation.

Scenario 3.1 ***The interaction graph for the four vehicles is directed***

The interaction graph is shown in Figure 3.2a (satisfying Condition 3 with node 0 (the leader) as the root node). Thus,

$$\mathcal{L}_4 + \mathcal{B}_4 = \begin{bmatrix} 1 & 0 & 0 & 0 \\ -1 & 1 & 0 & 0 \\ -1 & 0 & 1 & 0 \\ 0 & 0 & -1 & 1 \end{bmatrix}, \quad (3.110)$$

which is clearly invertible.

The simulation results are presented in Figures 3.3 and 3.4. It is seen from Figure 3.3 that the position tracking errors of the four vehicles converge to zero, and the settling time (with a 5% error band) for each vehicle is about 26 s. Figure 3.4 shows that the velocity tracking errors of the four vehicles converge to zero, and that the settling time (with a 5% error band) for each vehicle is about 22 s.

Scenario 3.2 ***The interaction graph for the four vehicles is undirected***

The interaction graph is shown in Figure 3.2b (satisfying Condition 4). Thus,

$$\mathcal{L}_4 + \mathcal{B}_4 = \begin{bmatrix} 4 & -1 & -1 & -1 \\ -1 & 2 & -1 & 0 \\ -1 & -1 & 3 & -1 \\ -1 & 0 & -1 & 2 \end{bmatrix}, \tag{3.111}$$

which is symmetric and invertible.

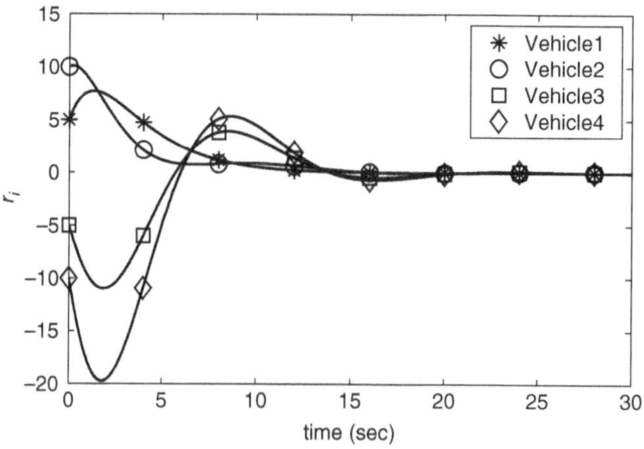

Figure 3.3 Position error responses for Scenario 1.

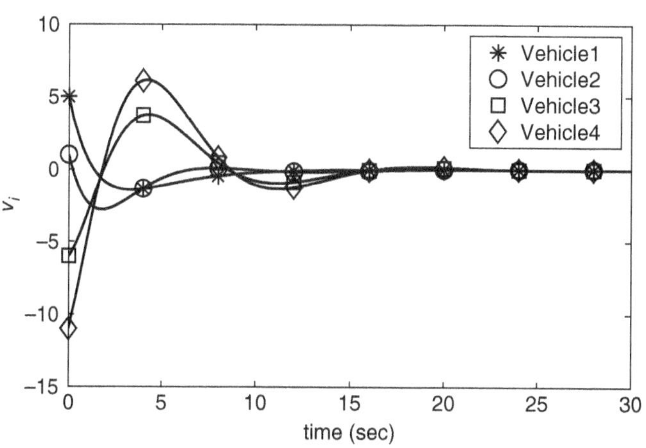

Figure 3.4 Velocity error responses for Scenario 1.

The simulation results are presented in Figures 3.5 and 3.6. It can be seen from Figure 3.5 that the position tracking errors of the four vehicles converge to zero, and that the settling time (with a 5% error band) for each vehicle is about 26 s. It can be seen from Figure 3.6 that the velocity tracking errors of the four vehicles converge to zero, and that the settling time (with a 5% error band) for each vehicle is about 22 s.

3.7.2 Verification of Theorem 3.13

We verify the results of Theorem 3.13 by considering two scenarios with time-invariant desired position deviations (δ_i, $i = 1, \cdots, 4$) and switching desired position deviations, respectively. These correspond to a time-invariant formation and a time-variant formation in 2-D space, respectively.

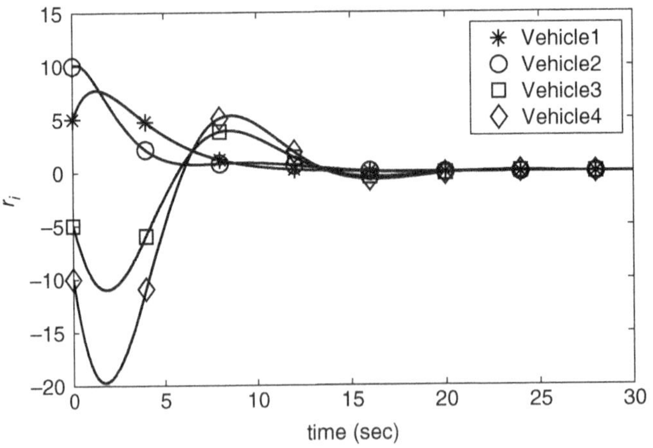

Figure 3.5 Position error responses for Scenario 2.

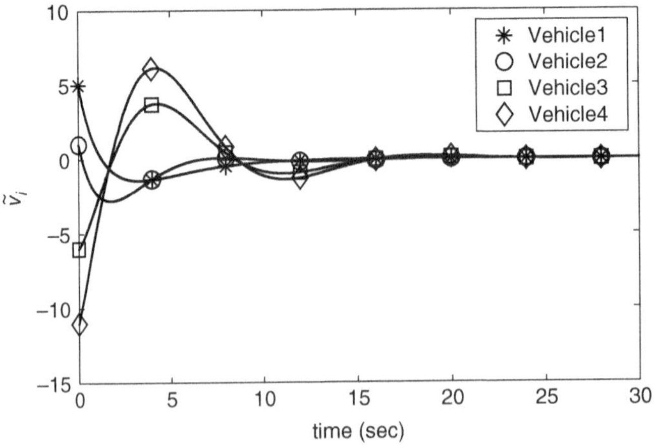

Figure 3.6 Velocity error responses for Scenario 2.

For each scenario, all the non-zero entries of matrices $\mathcal{A}_4$ and $\mathcal{B}_4$ are equal to 1 and the interaction graph is as shown in Figure 3.2c (satisfying Condition 3). Thus,

$$\mathcal{L}_4 + \mathcal{B}_4 = \begin{bmatrix} 4 & -1 & -1 & -1 \\ -1 & 2 & -1 & 0 \\ -1 & -1 & 3 & -1 \\ -1 & 0 & -1 & 2 \end{bmatrix}, \tag{3.112}$$

which is symmetric and invertible.

The parameters are as follows:

- the control gains are $k_v = 0.5$ and $k = 0.5$
- the initial controls $u_i(0) = (0,0)^T$ $(i = 1,2,3,4)$
- the initial states of the vehicles are:

$$\begin{aligned}
(r_1^x(0), r_1^y(0), v_1^x(0), v_1^y(0)) &= (3, 10, 6, 8) \\
(r_2^x(0), r_2^y(0), v_2^x(0), v_2^y(0)) &= (5, 8, 2, 4) \\
(r_3^x(0), r_3^y(0), v_3^x(0), v_3^y(0)) &= (0, 5, -5, -3) \\
(r_4^x(0), r_4^y(0), v_4^x(0), v_4^y(0)) &= (-2, 12, -10, -8)
\end{aligned}$$

- the desired trajectories are:

$$\begin{aligned}
r_d^x(t) &= -0.4t \\
\dot{r}_d^x(t) &= -0.4 \\
\ddot{r}_d^x(t) &= 0 \\
r_d^y(t) &= 10 \\
\dot{r}_d^y(t) &= 0 \\
\ddot{r}_d^y(t) &= 0.
\end{aligned}$$

Matlab R2014a/simulation solver ode3 (Bogacki–Shampine) with a fixed-step size of 0.01 s is used in the simulation.

Scenario 3.3 *Time-invariant desired position deviations*

The following deviations are required: $\delta_1 = (5,5)^T$, $\delta_2 = (-5,5)^T$, $\delta_3 = (-5,-5)^T$, $\delta_4 = (5,-5)^T$. The simulation results are presented in Figures 3.7–3.9. It is seen from Figures 3.7 and 3.8 that both the x-axis and y-axis position tracking errors of the four vehicles converge to zero. It is seen from Figure 3.9 that the 2-D space formation is achieved after about 10 s.

Scenario 3.4 *Switching desired position deviations*

The following deviations are required:

- for $t \in (0, 40)$

$$\begin{aligned}
\delta_1 &= (5,5)^T \\
\delta_2 &= (-5,5)^T \\
\delta_3 &= (-5,-5)^T \\
\delta_4 &= (5,-5)^T
\end{aligned}$$

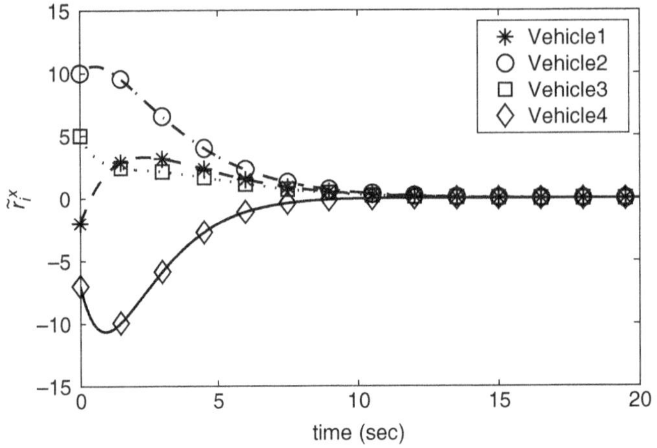

Figure 3.7 *x*-axis position error responses for Scenario 3.

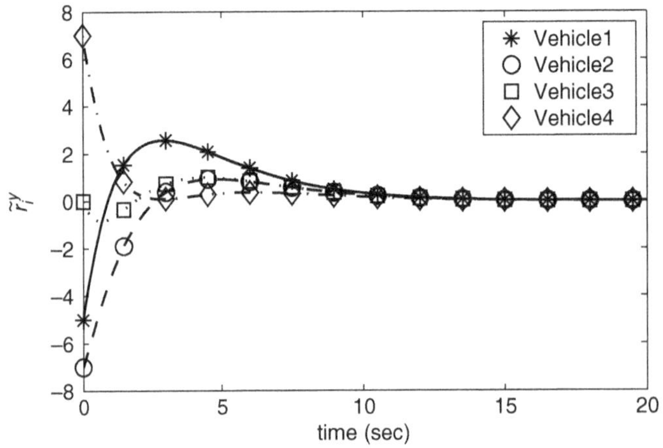

Figure 3.8 *y*-axis position error responses for Scenario 3.

- for $t \in [40, \infty)$:

$$\delta_1 = (10, 10)^T$$
$$\delta_2 = (-10, 10)^T$$
$$\delta_3 = (-10, -10)^T$$
$$\delta_4 = (10, -10)^T.$$

The simulation results are presented in Figures 3.10–3.12. It can be seen from Figures 3.10 and 3.11 that both the x-axis and y-axis position tracking errors of each vehicle converge to zero. It can be seen from Figure 3.12 that the first 2-D formation pattern (starting at $t = 0$ s) is achieved in about 10 s, and that the second 2-D formation pattern (starting at $t = 40$ s) is achieved at about $t = 50$ s, in other words, 10 s after the change in desired position deviations.

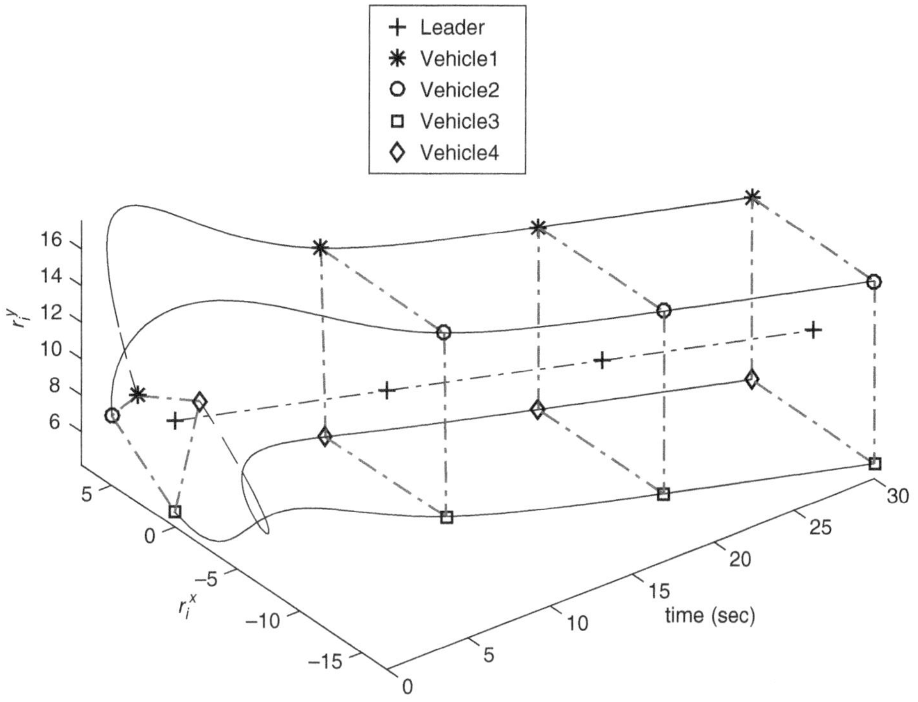

Figure 3.9 2-D formation trajectories with respect to time for Scenario 3.

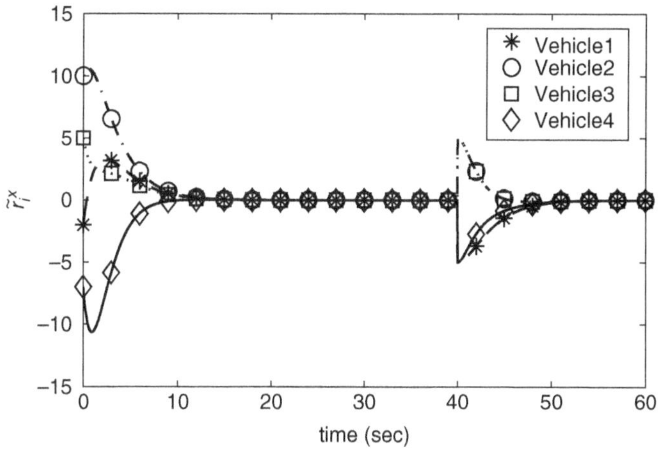

Figure 3.10 x-axis position error responses for Scenario 4.

3.7.3 Verification of Theorem 3.14

Consider a scenario with the interaction graph shown in Figure 3.2c. The desired position deviations are time-varying and given by:

$$\delta_1 = (5\cos(t), 5\sin(t))^T, \dot{\delta}_1 = (-5\sin(t), 5\cos(t))^T, \ddot{\delta}_1 = (-5\cos(t), -5\sin(t))^T$$

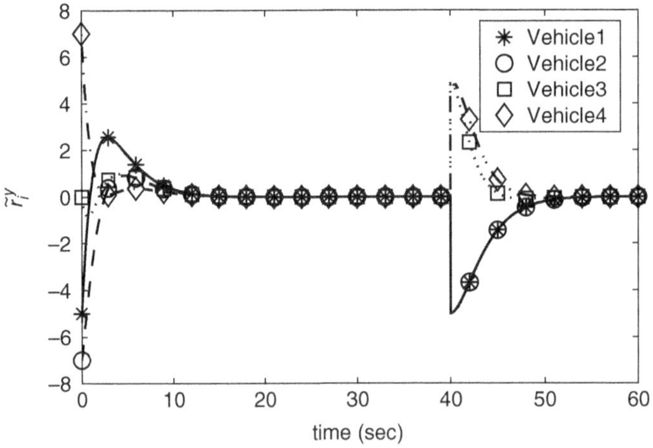

Figure 3.11 *y*-axis position error responses for Scenario 4.

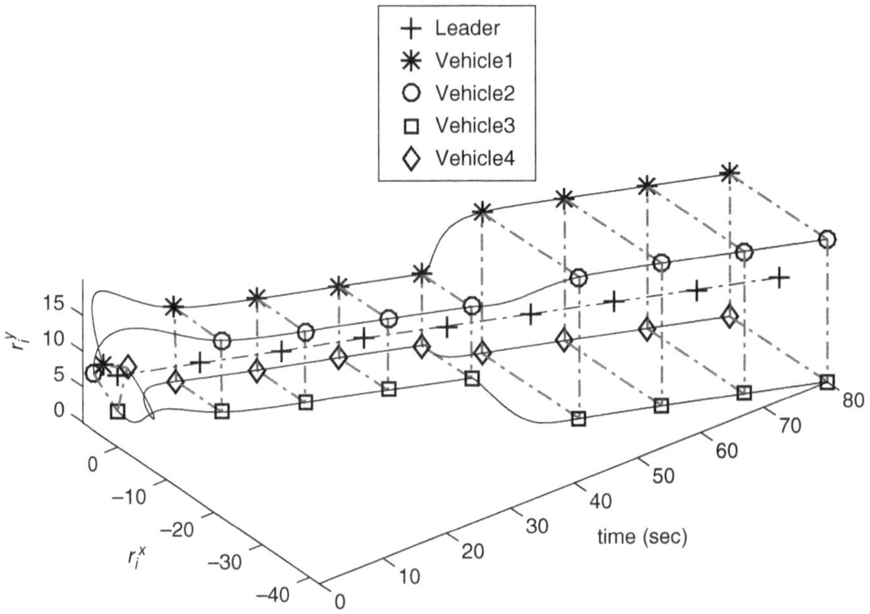

Figure 3.12 2-D formation trajectories with respect to time for Scenario 4.

$$\delta_2 = (-5\sin(t), 5\cos(t))^T, \dot{\delta}_2 = (-5\cos(t), -5\sin(t))^T, \ddot{\delta}_2 = (5\sin(t), -5\cos(t))^T$$
$$\delta_3 = (-5\cos(t), -5\sin(t))^T, \dot{\delta}_3 = (5\sin(t), -5\cos(t))^T, \ddot{\delta}_3 = (5\cos(t), 5\sin(t))^T$$
$$\delta_4 = (5\sin(t), -5\cos(t))^T, \dot{\delta}_4 = (5\cos(t), 5\sin(t))^T, \ddot{\delta}_4 = (-5\sin(t), 5\cos(t))^T,$$

which differ from those considered in Scenarios 3 and 4. With respect to the settings:

- the control gains k_v and k
- the initial controls $u_i(0)$ $(i = 1, 2, 3, 4)$
- the initial states of the vehicles $(r_i^x(0), r_i^y(0), v_i^x(0), v_i^y(0))$ $(i = 1, 2, 3, 4)$

- the reference trajectories $r_d^x(t)$, $\dot{r}_d^x(t)$, $\ddot{r}_d^x(t)$, $r_d^y(t)$, $\dot{r}_d^y(t)$, $\ddot{r}_d^y(t)$
- the Matlab R2014a/simulation solver

are the same as those specified for Scenarios 3 and 4. To avoid an algebraic-loop issue, we consider a one-step delay in neighbor control inputs, u_j ($j \in N_i$), for computing the control law (3.107); in other words, the control law (3.108) with $\tau = 0.01$ s is implemented.

The simulation results are presented in Figures 3.13–3.15. It can be seen from Figures 3.13 and 3.14 that both the x-axis and y-axis position tracking errors of each vehicle converge to zero in about 10 s. It can be seen from Figure 3.15 that the 2-D rectangular formation pattern is achieved after about 10 s, and that the actual position deviations of the four vehicles from the desired position trajectories are time-varying; they rotate around the virtual leader with the same angular velocity of 1.

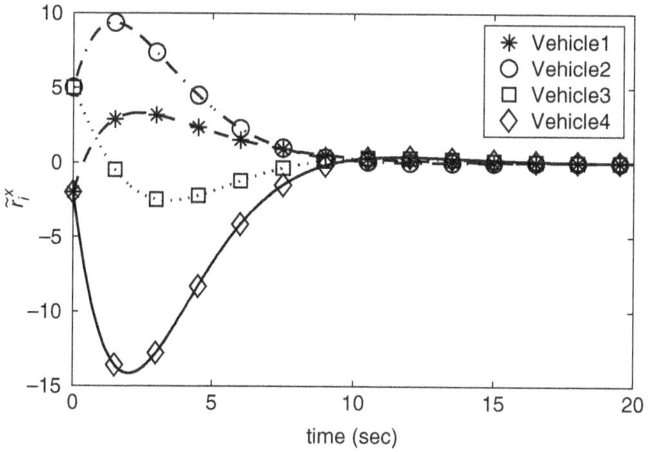

Figure 3.13 x-axis position error responses for Scenario 5.

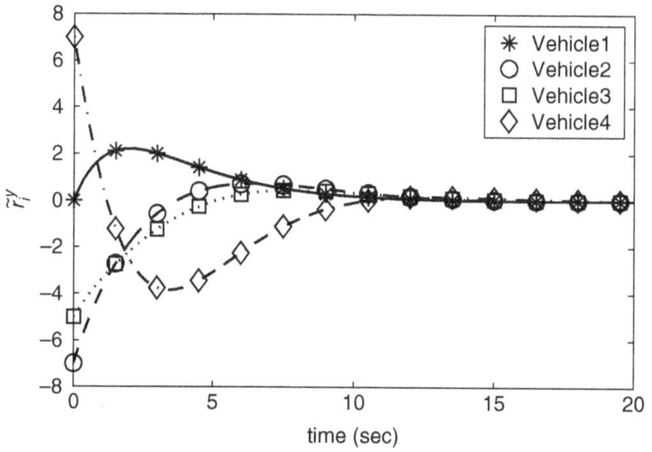

Figure 3.14 y-axis position error responses for Scenario 5.

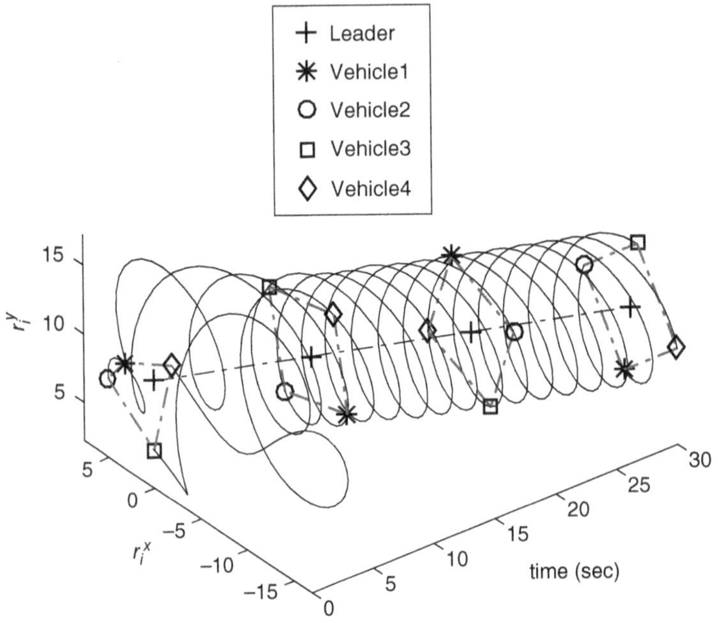

Figure 3.15 2-D formation trajectories with respect to time for Scenario 5.

3.8 Summary

In this chapter, we have presented the definitions of formation control, synchronization control, and synchronized trajectory tracking, considering a general MVS model. By taking the second-order MVS (3.8) as a simple illustrative example, we have introduced the controllers for the following three motion control problems:

- velocity synchronization control, with or without a given velocity reference;
- angular-position synchronization control, with or without a given position reference;
- formation control via synchronized trajectory tracking.

Note that the angular-position synchronization controller is obtained by applying the result developed for velocity synchronization as well as the relative-order reducing technique, and that the formation controller is obtained by applying the result developed for position synchronized tracking. In other words, the control approaches for the three different problems are closely related to each other.

We note that for each problem, this chapter has introduced at least one state-feedback control solution. Some other advanced control topics associated with the three problems (such as output-feedback control, disturbance estimation and compensation, robust control and so on) will be discussed in the forthcoming chapters.

Bibliography

1 A. D. Polyanin and A. V. Manzhirov, *Handbook of Mathematics for Engineers and Scientists.* CRC Press, 2006.
2 W. Ren and R. W. Beard, "Decentralized scheme for spacecraft formation flying via the virtual structure approach," *Journal of Guidance, Control, and Dynamics*, vol. 27, no. 1, pp. 73–82, 2004.
3 W. Ren and R. Beard, "Trajectory tracking for unmanned air vehicles with velocity and heading rate constraints," *IEEE Transactions on Control Systems and Technology*, vol. 12, no. 5, pp. 706–716, 2004.
4 J. Shan and H.-T. Liu, "Close-formation flight control with motion synchronization," *Journal of Guidance, Control, and Dynamics*, vol. 28, no. 6, pp. 1316–1320, 2001.
5 T. Balch and R. C. Arkin, "Behavior-based formation control for multirobot teams," *IEEE Transactions on Robotics and Automation*, vol. 14, no. 6, pp. 926–939, 1998.
6 R. A. Horn and C. R. Johnson, *Matrix Analysis.* Cambridge University Press, 1985.
7 A. Rodriguez-Angeles and H. Nijmeijer, "Mutual synchronization of robots via estimated state feedback: a cooperative approach," *IEEE Transactions on Control Systems Technology*, vol. 12, no. 4, pp. 542–554, 2004.
8 D. Sun, *Synchronization and Control of Multiagent Systems*, vol. 41. CRC Press, 2010.
9 S. Khoo, L. Xie, and Z. Man, "Robust finite-time consensus tracking algorithm for multirobot systems," *IEEE/ASME Transactions on Mechatronics*, vol. 14, no. 2, pp. 219–228, 2009.
10 A. Das and F. L. Lewis, "Distributed adaptive control for synchronization of unknown nonlinear networked systems," *Automatica*, vol. 46, no. 12, pp. 2014–2021, 2010.
11 W. Ren and Y. Cao, *Distributed Coordination of Multi-agent Networks.* Springer-Verlag, London, 2011.
12 R. Olfati-Saber and R. M. Murray, "Consensus problems in networks of agents with switching topology and time-delays," *IEEE Transactions on Automatic Control*, vol. 49, no. 9, pp. 1520–1533, 2004.
13 W. Ren and R. W. Beard, *Distributed Consensus in Multi-vehicle Cooperative Control.* Springer, 2008.

Formation Control of Multiple Autonomous Vehicle Systems, First Edition. Hugh H.T. Liu and Bo Zhu.

14 S. Liu, L. Xie, and F. L. Lewis, “Synchronization of multi-agent systems with delayed control input information from neighbors,” *Automatica*, vol. 47, no. 10, pp. 2152–2164, 2011.

15 Z. Li, W. Ren, X. Liu, and M. Fu, “Distributed containment control of multi-agent systems with general linear dynamics in the presence of multiple leaders,” *International Journal of Robust and Nonlinear Control*, vol. 23, no. 5, pp. 534–547, 2013.

Part II

Formation Control: Advanced Topics

4

Output-feedback Solutions to Formation Control

4.1 Introduction

For an angular-position synchronization control problem or a formation control problem, we may view the (linear or angular) position signals as the output of the considered MVS. In this chapter, we introduce a comprehensive treatment of angular-position synchronization control and formation control without velocity signals. These control problems are two typical output feedback control problems associated with MVSs.

4.2 Problem Statement

Recall the following MVS model:

$$\begin{cases} \dot{r}_i = v_i, \dot{v}_i = u_i \\ y_i = r_i. \end{cases} \tag{4.1}$$

where r_i denotes the linear (or angular) position of vehicle i and serves as the output of the ith vehicle system, v_i is the corresponding linear (or angular) velocity, and u_i is the control input. For simplicity, we here assume that the dimensions of all the involved variables is 1.

In this section, we mainly introduce the design of synchronization control for MVSs (4.1) under the constraint that only position signals r_i, $i \in \mathcal{I}$ are accessible for control design. In other words, the velocity signals v_i, $i = 1, \cdots, N$, v_d are not accessible for control design. This control problem can be regarded as an output-feedback control problem of a higher-order (2Nth-order) linear system. Since the MVS is completely observable, an intuitive idea to deal with this problem is based on the well-known Luenberger state observer (LSO) theory for observable linear systems [1]. The application of this theory generally consists of two steps:

- Design a full-order state observer for each vehicle to obtain an asymptotic estimate of the actual state.
- Replace the actual state variables of vehicles used in state-feedback controllers with their estimates (the estimates of neighboring vehicles' states are also often needed).

The separation principle ensures that this LSO-based approach works well for a linear system. A detailed design is omitted here for simplicity, and the readers are referred to the literature for design examples [2]. Instead, we will detail here the application of

Formation Control of Multiple Autonomous Vehicle Systems, First Edition. Hugh H.T. Liu and Bo Zhu.

an alternative approach, called the passivity approach, which is based on the design of a simple first-order linear filter. We will further show that by introducing this filter into the system, the asymptotic estimates of vehicle velocities are not required. Similar applications of such an approach can be found in Section 4.3 of the book by Ren and Beard [3] and the paper by Lawton *et al.* [4].

4.3 Linear Output-feedback Control

The control strategy (3.85) for synchronized position tracking requires that the relative velocities between neighbors, and/or the velocity tracking error, are known. However, in the context of this chapter, neither the relative velocity or the velocity tracking error is accessible for control design. To resolve a similar issue, Lawton *et al.* developed a decentralized approach using a passivity technique to inject relative damping into the system, where the relative damping is produced by a first-order linear filter [4].

Motivated by their approach, and using the Type III LSEs, e_{ri}, defined by (3.48), we construct the following dynamic output-feedback control law for system (4.1):

$$\begin{cases} \dot{\zeta}_i = -k_\zeta \zeta_i - e_{ri}, \\ u_i = \ddot{r}_d - k_e e_{ri} + \dot{\zeta}_i = \ddot{r}_d - (k_e + 1)e_{ri} - k_\zeta \zeta_i, i = 1, \cdots, N, \end{cases} \tag{4.2}$$

where $\zeta_i \in R$, with an arbitrary initial value $\zeta_i(0)$, is the state of the introduced first-order auxiliary equation included in (4.2), and $k_e \in R$ and $k_\zeta \in R$ are strictly positive control gains.

It is seen from (4.2) that the desired acceleration signal $\ddot{r}_d$ is required to be accessible to each vehicle. However, when the desired velocity, $\dot{r}_d$, is a constant, $\ddot{r}_d \equiv 0$ and the requirement is naturally removed.

Combining (4.1) and (4.2), yields the closed-loop system

$$\begin{cases} \dot{\zeta}_i = -k_\zeta \zeta_i - e_{ri}, \\ \ddot{\tilde{r}}_i = -(k_e + 1)e_{ri} - k_\zeta \zeta_i, i = 1, \cdots, N. \end{cases} \tag{4.3}$$

The two equations included in (4.3) are mutually coupled, because e_{ri} appears in the first equation and ζ_i in the second. Moreover, the state variables used in the two sides of (4.3) are not identical. Specifically, the left-hand side adopts variable $\tilde{r}_i$, while the right-hand side adopts variable e_{ri}. For these reasons, it is not realistic to analyse the stability of the node-level system with the Routh criterion or matrix theory. Applying the linear system approaches to network-level closed-loop dynamics is not easy either, because the system order, $3N$, is high and not known (the number of vehicles involved, N, is not fixed in advance).

Therefore, we here do not dwell on the application of linear system approaches to MVSs (4.3). Instead, we apply the Lyapunov function approach and LaSalle's invariance principle to obtain the following result.

Theorem 4.1 Consider MVS (4.3) under Condition 4. We have the following:

- e_{ri}, e_{vi}, $\tilde{r}_i$, $\tilde{v}_i$, ζ_i and $\dot{\zeta}_i$ are globally bounded,
- $\lim_{t\to\infty}\zeta_i(t) = 0$, $\lim_{t\to\infty}\dot{\zeta}_i(t) = 0$, $\lim_{t\to\infty}e_{ri}(t) = 0$, $\lim_{t\to\infty}e_{vi}(t) = 0$, $\lim_{t\to\infty}\tilde{r}_i(t) = 0$, and $\lim_{t\to\infty}\tilde{v}_i(t) = 0$, $i \in \mathcal{I}$.

We prove each of these statements separately as follows.

First statement proof: Consider the following Lyapunov function candidate

$$V = V_1 + V_2 + V_3 + V_4, \tag{4.4}$$

where

$$
\begin{aligned}
V_1 &= \frac{1}{2}\sum_{i=1}^{N}(v_i - \dot{r}_d - \zeta_i)^2,\\
V_2 &= \frac{k_e}{2}\sum_{i=1}^{N}\zeta_i^2,\\
V_3 &= \frac{k_e}{2}\sum_{i=1}^{N}b_i(r_i - r_d)^2,\\
V_4 &= \frac{k_e}{4}\sum_{i=1}^{N}\sum_{j=1}^{N}a_{ij}(r_i - r_j)^2.
\end{aligned} \tag{4.5}
$$

Note that under Condition 4, there exists at least one vehicle with $b_i > 0$, and thus $V_3 > 0$ for any $(r_i - r_d) \neq 0$. The time derivatives of V_i, $i = 1, 2, 3, 4$, along the trajectories of closed-loop dynamics (4.3) are

$$\dot{V}_1 = -k_e\sum_{i=1}^{N}(v_i - \dot{r}_d - \zeta_i)e_{ri}, \tag{4.6}$$

$$\dot{V}_2 = -k_e k_\zeta\sum_{i=1}^{N}\zeta_i^2 - k_e\sum_{i=1}^{N}\zeta_i e_{ri}, \tag{4.7}$$

$$\dot{V}_3 = k_e\sum_{i=1}^{N}b_i(r_i - r_d)(v_i - \dot{r}_d), \tag{4.8}$$

$$\dot{V}_4 = \frac{k_e}{2}\sum_{i=1}^{N}\sum_{j=1}^{N}a_{ij}(r_i - r_j)(v_i - v_j). \tag{4.9}$$

Under Condition 4, the interaction graph $\mathcal{G}$ is undirected, and thus $a_{ji} = a_{ij}$ for all $i, j \in \mathcal{I}$. By applying Lemma 4.18 in [3], we get:

$$
\begin{aligned}
\dot{V}_4 &= \frac{k_e}{2}\sum_{i=1}^{N}\sum_{j=1}^{N}a_{ij}(r_i - r_j)(v_i - v_j)\\
&= \frac{k_e}{2}\sum_{i=1}^{N}\sum_{j=1}^{N}a_{ij}(r_i - r_j)(v_i - \dot{r}_d - (v_j - \dot{r}_d))\\
&= k_e\sum_{i=1}^{N}\sum_{j=1}^{N}a_{ij}(r_i - r_j)(v_i - \dot{r}_d).
\end{aligned} \tag{4.10}
$$

By adding (4.6)–(4.8) and (4.10) together and considering the replacement $e_{ri} = b_i(r_i - r_d) + \sum_{j\in N_i} a_{ij}(r_i - r_j)$ for (4.6), we obtain the expression of the first derivative of V along the trajectories of MVS (4.3) as

$$\dot{V} = -k_e k_\zeta \sum_{i=1}^{N} \zeta_i^2. \tag{4.11}$$

Since $k_e > 0$ and $k_\zeta > 0$, $\dot{V}$ satisfying (4.11) is negative semidefinite and thus $V(t) \leq V(0)$. Hence, $V(t)$ is globally bounded. This implies that the following are all globally bounded:

- $v_i - \dot{r}_d$ and ζ_i for all $i \in \mathcal{I}$
- $\tilde{r}_i$ for all i satisfying $b_i > 0$
- $r_i - r_j$ for all $(i,j) \in \mathcal{E}$

With these results, v_i $(i \in \mathcal{I})$ are globally bounded due to the boundedness of $\dot{r}_d$. Thus, e_{ri} and e_{vi} $(i \in \mathcal{I})$ are globally bounded by definitions (3.47) and (3.48), and $\dot{\zeta}_i$ $(i \in \mathcal{I})$ are bounded due to the first equation in (4.2). This ends the proof of the first statement. □

Second statement proof: With (4.11), the largest invariant set for vehicle i is

$$S_i = \{(v_i - \dot{r}_d, \zeta_i, r_i - r_d, r_i - r_j) \mid \zeta_i = 0\}, \tag{4.12}$$

By definition, V is radially unbounded with respect to its arguments. By applying LaSalle's invariance principle,

$$\lim_{t\to\infty} \zeta_i(t) = 0, i \in \mathcal{I}, \tag{4.13}$$

The boundedness of e_{vi} and $\dot{\zeta}_i$ implies the boundedness of $\ddot{\zeta}_i$ since $\ddot{\zeta}_i = -k_\zeta\dot{\zeta}_i - e_{vi}$ (which follows directly from the first equation included in (4.3)). By applying the well-known Barbalat lemma, we conclude from (4.13) that

$$\lim_{t\to\infty} \dot{\zeta}_i(t) = 0, i \in \mathcal{I}, \tag{4.14}$$

which along with (4.13) further indicates

$$\lim_{t\to\infty} e_{ri}(t) = 0, i \in \mathcal{I}, \tag{4.15}$$

due to the relationship $\dot{\zeta}_i = -k_\zeta \zeta_i - e_{ri}$.

Under Condition 4, matrix $\mathcal{L}_N + \mathcal{B}_N$ has full rank (see Lemma 3.3). Then, (4.15) indicates that $\lim_{t\to\infty} \tilde{r}_i = 0$ $(i \in \mathcal{I})$. Using the results of statement 1 and the boundedness of $\ddot{r}_d(t)$, we readily conclude that $u_i(t)$ and $\ddot{r}_i(t)$ are also bounded. By applying the Barbalat lemma, we further derive that $\lim_{t\to\infty} \tilde{v}_i = 0$ $(i \in \mathcal{I})$, which in turn indicates that $\lim_{x\to\infty} e_{vi}(t) = 0$. This ends the proof of the second statement and the theorem. □

4.4 Bounded Output-feedback Control

Note that the e_{ri} defined by (3.48) are linear functions of $\tilde{r}_i$ and $\tilde{r}_{ij}$. The values of e_{ri} therefore depend linearly on those of $\tilde{r}_i$ and $\tilde{r}_{ij}$. Since the controller (4.2) is a linear function of e_{ri}, it suffers from the limitation that larger $\tilde{r}_i$ or $\tilde{r}_{ij}$ may lead to actuator saturation. An intuitive and useful idea to avoid actuator saturation is to feed back saturated signals. In this section, we apply this idea to construct a bounded output-feedback synchronized tracking controller.

We use the hyperbolic tangent function tanh to modify the Type III LSEs as follows:

$$e_{ri} = b_i \tanh(\tilde{r}_i) + \sum_{j \in N_i} a_{ij} \tanh(\tilde{r}_{ij}), i \in \mathcal{I}. \quad (4.16)$$

Note that, in order to ensure that the e_{ri} defined by (4.16) are bounded, we limit the bounds of components $\tilde{r}_i$ and $\tilde{r}_{ij}$. This is different from the traditional applications of saturation functions, where the signal is not bounded component-wise (see Lemma A.15 in Appendix A.2.5 for the definition and properties of generalized saturation functions). We also note that, as for a higher-dimensional motion, the hyperbolic tangent function $\tanh(\cdot)$ in (4.16) is defined element-wise for vectors $\tilde{r}_i$ and $\tilde{r}_{ij}$.

To make a difference, we refer to the e_{ri} defined by (4.2) and (4.16) as the linear and nonlinear LSEs, respectively. As shown in Lemma 3.3, under Condition 3, the convergence of linear e_{ri} $(i \in \mathcal{I})$ is sufficient and necessary to ensure the convergence of tracking errors $\tilde{r}_i(t)$ $(i \in \mathcal{I})$. To render the design and analysis clear, we expect that this property remains for the nonlinear LSEs. Clearly, if $\tilde{r}_i$ tends to zero for each $i \in \mathcal{I}$, so will the components $\tanh(\tilde{r}_i)$ and $\tanh(\tilde{r}_{ij})$. Then, the remaining question is whether we can derive the convergence of $\tilde{r}_i(t)$ from the convergence of e_{ri} defined by (4.16). The following provides a clear answer to that question.

Lemma 4.2 If $\lim_{t\to\infty} e_{ri}(t) = 0$ for all $i \in \mathcal{I}$ under Condition 4, then $\lim_{t\to\infty} \tilde{r}_i(t) = 0$ for all $i \in \mathcal{I}$, where $e_{ri}(t)$ are defined by (4.16).

Proof: Note that $\tanh(x)$ is an odd, strictly increasing function for all $x \in (-\infty, \infty)$. Thus, $\tanh'(x) =: \frac{\partial \tanh(x)}{\partial x} = sech^2(x) > 0$ for all $x \in (-\infty, \infty)$. For any $\alpha \in (-\infty, \infty)$, by applying the well-known mean value theorem [5], we have $\tanh(\alpha) = sech^2(\alpha^*)\alpha$ and $\tanh(-\alpha) = -\tanh(\alpha) = -sech^2(\alpha^*)\alpha$, where $\alpha^* \in (0, \alpha)$ if $\alpha > 0$, $\alpha^* \in (\alpha, 0)$ if $\alpha < 0$, and $\alpha^* = 0$ if $\alpha = 0$. Accordingly, for any $\tilde{r}_i$ and $\tilde{r}_{ij}$, the nonlinear LSEs defined in (4.16) satisfy

$$e_{ri} = b_i sech^2(\tilde{r}_i^*)\tilde{r}_i + \sum_{j=1}^{N} a_{ij} sech^2(\tilde{r}_{ij}^*)(\tilde{r}_i - \tilde{r}_j), i, j \in \mathcal{I}, \quad (4.17)$$

where the variables $\tilde{r}_i^*$ and $\tilde{r}_{ij}^*$ are defined as α^*.

The N equations in (4.17) can be further written in matrix form as

$$\mathbf{e} = (\mathcal{L}_N^* + \mathcal{B}_N^*)\tilde{\mathbf{r}}, \quad (4.18)$$

where $\mathbf{e} = [e_{r1}, e_{r2}, \cdots, e_{rN}]^T$ with e_{ri} $(i = 1, \cdots, N)$ satisfying (4.17), $\tilde{\mathbf{r}}$ is the tracking error vector defined as in (3.37), and $\mathcal{L}_N^*$ and $\mathcal{B}_N^*$ are defined by

$$\mathcal{L}_N^* = \begin{bmatrix} \sum_{j=1}^N a_{1j} f_{1j}^* & -a_{12} f_{12}^* & \cdots & -a_{1N} f_{1N}^* \\ -a_{21} f_{21}^* & \sum_{j=1}^N a_{2j} f_{2j}^* & \cdots & -a_{2N} f_{2N}^* \\ \vdots & \vdots & \ddots & \vdots \\ -a_{N1} f_{N1}^* & -a_{N2} f_{N2}^* & \cdots & \sum_{j=1}^N a_{Nj} f_{Nj}^* \end{bmatrix} \quad (4.19)$$

$$\mathcal{B}_N^* = \begin{bmatrix} b_1 f_1^* & 0 & \cdots & 0 \\ 0 & b_2 f_2^* & \cdots & 0 \\ \vdots & \vdots & \ddots & \vdots \\ 0 & 0 & \cdots & b_N f_N^* \end{bmatrix}, \quad (4.20)$$

with

$$f_{ij}^* = sech^2(\tilde{r}_{ij}^*) > 0, (i,j) \in \mathcal{E}, \tag{4.21}$$

$$f_i^* = sech^2(\tilde{r}_i^*) > 0, i \in \mathcal{I}. \tag{4.22}$$

Due to the odd property of function tanh, we have $f_{ij}^* = f_{ji}^*$ and $a_{ij}f_{ij}^* = a_{ji}f_{ji}^*$ under Condition 4. Therefore, the graph associated with $\mathcal{L}_N^*$ and $\mathcal{B}_N^*$ has the same structure as the graph associated with $\mathcal{L}_N$ and $\mathcal{B}_N$, and only the (nonzero) weights may be different. Thus, under Condition 4, matrix $\mathcal{L}_N + \mathcal{B}_N$ is nonsingular for any $\tilde{r}_i$ and $\tilde{r}_{ij}$, and so is matrix $\mathcal{L}_N^* + \mathcal{B}_N^*$. This in turn implies that under Condition 4, the convergence of $\mathbf{e}$ is sufficient to ensure the convergence of $\tilde{\mathbf{r}}$. This ends the proof of the lemma. □

In light of the results of Lemma 4.2, the problem of stabilizing $\tilde{r}_i(t)$, $i \in \mathcal{I}$, under Condition 4 can be solved by stabilizing $e_{ri}(t)$, $i \in \mathcal{I}$. Therefore, the remaining task is to design u_i to render $\lim_{t\to\infty} e_{ri}(t) = 0$, $i \in \mathcal{I}$.

Using e_{ri} defined in (4.16), we construct the following control law:

$$\begin{cases} \dot{\zeta}_i = -k_\zeta \tanh(\zeta_i) - e_{ri}, \\ u_i = \ddot{r}_d - k_e e_{ri} + \dot{\zeta}_i, i = 1, \cdots, N, \end{cases} \tag{4.23}$$

where $\zeta_i \in R$, with an arbitrary initial value $\zeta_i(0)$, is the state of the introduced nonlinear auxiliary system, and k_ζ is a strictly positive gain.

From (4.16) and (4.23), we readily derive that

$$\|u_i\|_\infty \le \|\ddot{r}_d\|_\infty + k_\zeta + (k_e+1)\left(b_i + \sum_{j\in N_i} a_{ij}\right), i \in \mathcal{I}. \tag{4.24}$$

This indicates that u_i determined by (4.23) are indeed prior bounded.

The closed-loop dynamics consisting of (4.1) and (4.23) are described by

$$\begin{cases} \dot{\zeta}_i = -k_\zeta \tanh(\zeta_i) - e_{ri}, \\ \ddot{r}_i = -(k_e+1)e_{ri} - k_\zeta \tanh(\zeta_i), i = 1, \cdots, N. \end{cases} \tag{4.25}$$

for which we have the following result.

Theorem 4.3 Consider MVS (4.25) under Condition 4. If $\ddot{r}_d$ are bounded and the parameters k_ζ, b_i and a_{ij}, $i,j \in \mathcal{I}$, satisfy

$$\|\ddot{r}_d\|_\infty + k_\zeta + (k_e+1)\left(b_i + \sum_{j\in N_i} a_{ij}\right) \le u_{\max}, i \in \mathcal{I}, \tag{4.26}$$

then

- $\|u_i\|_\infty \le u_{\max}$, $i \in \mathcal{I}$,
- e_{ri}, e_{vi}, $\tilde{r}_i$, $\tilde{v}_i$, ζ_i and $\dot{\zeta}_i$ are globally bounded,
- $\lim_{t\to\infty}\zeta_i(t) = 0$, $\lim_{t\to\infty}\dot{\zeta}_i(t) = 0$, $\lim_{t\to\infty}e_{ri}(t) = 0$, $\lim_{t\to\infty}e_{vi}(t) = 0$, $\lim_{t\to\infty}\tilde{r}_i(t) = 0$, and $\lim_{t\to\infty}\tilde{v}_i(t) = 0$, $i \in \mathcal{I}$.

Proof: A Lyapunov function based approach and LaSalle's invariance principle have been used in a similar proof, which can be found in the literature [6]. The procedure is also the same as that used in proving Theorem 4.1. □

As for MVS (4.25), it is important to note the following two facts:

- System (4.25) is autonomous, implying that the LaSalle invariance theorem in typical form [7] is applicable.
- The symmetry property of an undirected graph– that is, $a_{ij} = a_{ji}$, $i, j \in \mathcal{I}$ – is used in the proof of Theorem 4.3. This indicates that the result of the theorem cannot be directly extended to the case with a directed subgraph $\mathcal{G}$, despite the fact that the full rank condition on $\mathcal{L}_N + \mathcal{B}_N$ or $\mathcal{L}_N^* + \mathcal{B}_N^*$ does not necessarily require subgraph $\mathcal{G}$ to be undirected.

4.5 Distributed Linear Control

To implement the control law (4.2), the desired acceleration signal $\ddot{r}_d$ is required for each vehicle in the MVS. This is rather restrictive. To remove this limitation, we consider the following control law:

$$\begin{cases} \dot{\zeta}_i = -k_\zeta \zeta_i - e_{ri}, \\ u_i = k_i \left(b_i \ddot{r}_d + \sum_{j \in N_i} a_{ij} u_j - (k_e + 1) e_{ri} - k_\zeta \zeta_i \right), i \in \mathcal{I}, \end{cases} \tag{4.27}$$

where e_{ri}, $i \in \mathcal{I}$, are linear Type III LSEs, and k_e and k_ζ are positive control gains defined as in (4.2).

The differences between the control laws (4.2) and (4.27) are two-fold.

- The term $\ddot{r}_d$ in (4.2) is replaced by $b_i \ddot{r}_d + \sum_{j \in N_i} a_{ij} u_j$ in (4.27).
- All four components of u_i in (4.27) share a common coefficient k_i, whose value depends on the information-exchange weights associated with node i (see Equation (3.63)). However, this is not the case for u_i in (4.2).

The closed-loop dynamics consisting of (4.1) and (4.27) are

$$\begin{cases} \dot{\zeta}_i = -k_\zeta \zeta_i - e_{ri}, \\ \dot{e}_{vi} = -(k_e + 1) e_{ri} - k_\zeta \zeta_i, i \in \mathcal{I}, \end{cases} \tag{4.28}$$

which can be written in a vector form as

$$\begin{bmatrix} \dot{\zeta}_i \\ e_{vi} \\ \dot{e}_{vi} \end{bmatrix} = \underbrace{\begin{bmatrix} -k_\zeta & -1 & 0 \\ 0 & 0 & 1 \\ -k_\zeta & -k_e - 1 & 0 \end{bmatrix}}_{A} \begin{bmatrix} \zeta_i \\ e_{ri} \\ e_{vi} \end{bmatrix}, i \in \mathcal{I}. \tag{4.29}$$

A simple calculation shows that the characteristic polynomial of system matrix A is

$$\lambda(A, s) = s^3 + k_\zeta s^2 + (k_e + 1)s + k_\zeta k_e. \tag{4.30}$$

It is easy to check that the polynomial is Hurwitz provided $k_e > 0$ and $k_\zeta > 0$. Thus we have the following result.

Theorem 4.4 Consider MVS (4.29) under Condition 3. For each $i = 1, \cdots, N$, we have the following:

- There exists a unique solution for u_i satisfying (4.27).

- e_{ri}, e_{vi}, $\tilde{r}_i$, $\tilde{v}_i$, ζ_i and $\dot{\zeta}_i$ are globally bounded.
- $\lim_{t\to\infty}\zeta_i(t) = 0$, $\lim_{t\to\infty}\dot{\zeta}_i(t) = 0$, $\lim_{t\to\infty}e_{ri}(t) = 0$, $\lim_{t\to\infty}e_{vi}(t) = 0$, $\lim_{t\to\infty}\tilde{r}_i(t) = 0$, and $\lim_{t\to\infty}\tilde{v}_i(t) = 0$.

Note that the N individual systems of the form (4.29) are linear. A simple approach to show the convergence of $\tilde{r}_i$ is to analyse the characteristic polynomial of A. If the motion is one dimensional, as discussed here, this approach can be easily implemented because the order of the (node-level) system matrix A is 3rd (relatively lower). For the case of N-dimensional motion with $N \geq 2$, the order of system matrix is $3N$th (not lower than 6th), and it becomes unrealistic to analyse the characteristic polynomial. However, by decoupling the motion and applying the above control law to each dimension, the resulting system matrix for each dimension may have the same form as A and then the approach is still applicable. We note that a Lyapunov function based approach, serving as an alternative way to prove the theorem, is provided in the literature [8]. In the next section, the Lyapunov function based approach is used to prove the stability of a nonlinear variation of system (4.29).

4.6 Distributed Bounded Control

The controller (4.27) is linear and not prior bounded. Suppose σ_1 and σ_2 are two generalized saturation functions (GSFs), as defined in Lemma A.15 in Appendix A.2. To obtain a bounded controller, we modify (4.27) using GSFs as

$$\begin{cases} \dot{\zeta}_i = -k_\zeta \sigma_2(\zeta_i) - \sigma_1(e_{ri}), \\ u_i = k_i \left[b_i \ddot{r}_d + \sum\limits_{j\in N_i} a_{ij}u_j - (k_e+1)\sigma_1(e_{ri}) - k_\zeta \sigma_2(\zeta_i) \right], i \in \mathcal{I}, \end{cases} \tag{4.31}$$

where σ_1 and σ_2 are saturation functions defined element-wise for a vector argument, and σ_1 satisfies the additional condition

$$\int_0^z \sigma_1(\tau)d\tau \to \infty \; as \; |z| \to \infty, \tag{4.32}$$

which is trivially satisfied if $\sigma_1(\tau) = \tanh(\tau)$. This condition ensures that the Lyapunov function candidate considered in the proof of Theorem 4.5 (i.e., Equation 4.35) is radially unbounded.

The closed-loop system consisting of (4.1) and (4.31) is described by

$$\begin{cases} \dot{\zeta}_i = -k_\zeta \sigma_2(\zeta_i) - \sigma_1(e_{ri}), \\ \dot{e}_{vi} = -(k_e+1)\sigma_1(e_{ri}) - k_\zeta \sigma_2(\zeta_i), i \in \mathcal{I}, \end{cases} \tag{4.33}$$

where the gains k_e and k_ζ are defined as in (4.27).

We then have the following result.

Theorem 4.5 Consider MVS (4.33) under Condition 3. We then have the following:

- There exists a unique solution for u_i satisfying (4.31), $i \in \mathcal{I}$, and this solution is prior bounded provided $\ddot{r}_d$ is bounded
- e_{ri}, e_{vi}, $\tilde{r}_i$, $\tilde{v}_i$, ζ_i and $\dot{\zeta}_i$ are globally bounded.
- $\lim_{t\to\infty}\zeta_i(t) = 0$, $\lim_{t\to\infty}\dot{\zeta}_i(t) = 0$, $\lim_{t\to\infty}e_{ri}(t) = 0$, $\lim_{t\to\infty}e_{vi}(t) = 0$, $\lim_{t\to\infty}\tilde{r}_i(t) = 0$, and $\lim_{t\to\infty}\tilde{v}_i(t) = 0$, $i \in \mathcal{I}$.

Proof: From the expressions of u_i given by (4.31), we obtain

$$u = -(\mathcal{L}_N + \mathcal{B}_N)^{-1} W + 1_N \otimes \ddot{r}_d, \tag{4.34}$$

where $u = [u_1, \cdots, u_N]^T \in R^N$ and $W = [w_1, \cdots, w_N]^T \in R^N$ with $w_i = (k_e + 1)\sigma_1(e_{ri}) + k_\zeta \sigma_2(\zeta_i)$. The uniqueness of the solution for u_i is obvious since $(\mathcal{L}_N + \mathcal{B}_N)$ has full rank under Condition 3. It is also clear that if $\ddot{r}_d$ is bounded, u satisfying (4.34) is prior bounded.

Consider the following Lyapunov function candidate for each $i \in \mathcal{I}$:

$$V_i = (e_{vi} - \zeta_i)^2 + k_e \zeta_i^2 + 2k_e \int_0^{e_{ri}} \sigma_1(x)dx. \tag{4.35}$$

A direct calculation shows that the time derivative of V_i along the trajectories of system (4.33) is

$$\dot{V}_i = -2k_e k_\zeta \zeta_i \sigma_2(\zeta_i), \tag{4.36}$$

which implies that $\dot{V}_i$ is negative semidefinite, and that $\dot{V}_i = 0$ if and only if $\zeta_i = 0$. With this result, we can readily show that the last two statements of Theorem 4.5 are true, by carrying out a similar analysis as in the proof of Theorem 4.1. This ends the proof of Theorem 4.5. □

4.7 Simulations

In this section, we verify the theoretical results from the previous sections by simulation. In particular, we present the simulation results of two cases which correspond to the linear control considered in Theorem 4.1 and the prior bounded control considered in Theorem 4.5, respectively. In each case, four followers and a virtual leader are involved, and the motion is planar two-dimensional (i.e., $r_i \in R^2$, $v_i \in R^2$ and $u_i \in R^2$ in (4.1)).

4.7.1 Case 1: Verification of Theorem 4.1

Consider the interaction graph shown in Figure 3.2d (or Figure 4.1), which trivially satisfies Condition 4 with an undirected and connected $\mathcal{G}$ and node 1 having access to the leader. For control law (4.2), we here suppose

$$\mathcal{A}_4 = \begin{bmatrix} 0 & 0.5 & 0 & 0.5 \\ 0.5 & 0 & 0.5 & 0 \\ 0 & 0.5 & 0 & 0.5 \\ 0.5 & 0 & 0.5 & 0 \end{bmatrix}, \quad \mathcal{L}_4 + \mathcal{B}_4 = \begin{bmatrix} 1.5 & -0.5 & 0 & -0.5 \\ -0.5 & 1 & -0.5 & 0 \\ 0 & -0.5 & 1 & -0.5 \\ -0.5 & 0 & -0.5 & 1 \end{bmatrix}, \tag{4.37}$$

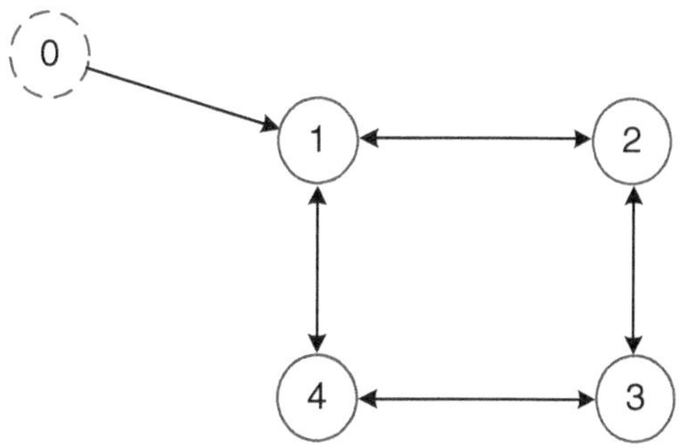

Figure 4.1 The interaction graph for Case 1.

which implies $b_1 = 0.5$ and $b_2 = b_3 = b_4 = 0$. The gains for each dimension of each vehicle are the same, with $k_e = k_\zeta = 0.5$.

The desired trajectories are:

- $r_d = (r_d^x, r_d^y)^T = (\cos(t), \sin(t))^T$
- $\dot{r}_d = (\dot{r}_d^x, \dot{r}_d^y)^T = (-\sin(t), \cos(t))^T$
- $\ddot{r}_d = (\ddot{r}_d^x, \ddot{r}_d^y)^T = (-\cos(t), -\sin(t))^T$.

The initial states of the four vehicles are:

$$
\begin{aligned}
(r_1^x(0), r_1^y(0), v_1^x(0), v_1^y(0)) &= (1, 1, 0, 1)\\
(r_2^x(0), r_2^y(0), v_2^x(0), v_2^y(0)) &= (-1, 1, 0, 1)\\
(r_3^x(0), r_3^y(0), v_3^x(0), v_3^y(0)) &= (-1, -1, 0, 1)\\
(r_4^x(0), r_4^y(0), v_4^x(0), v_4^y(0)) &= (1, -1, 0, 1).
\end{aligned} \tag{4.38}
$$

The initial states of the auxiliary system involved in (4.2) are: $\zeta_i = (0.1, -0.1)^T$, $i = 1, 2, 3, 4$. The Matlab R2014a/simulation solver ode3 (Bogacki-Shampine) with a fixed-step size of 0.01 s is used in the simulation.

The simulation results are presented in Figures 4.2–4.4. As shown in Figures 4.2 and 4.3, the position trajectories of each vehicle converge asymptotically to the desired trajectories, and the settling time is about 28 s to ensure the bounds of tracking errors and synchronization errors are smaller than 0.05. Figure 4.4 shows that the actual trajectory of each vehicle tracks asymptotically the desired 2-D circular trajectory (with radius 1).

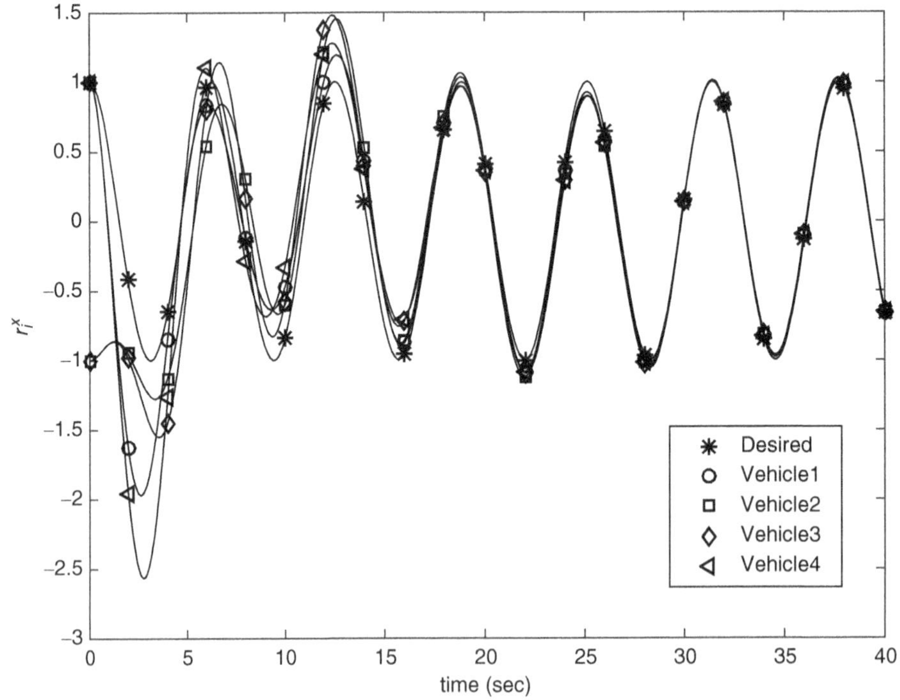

Figure 4.2 The *x*-axis position trajectories for Case 1.

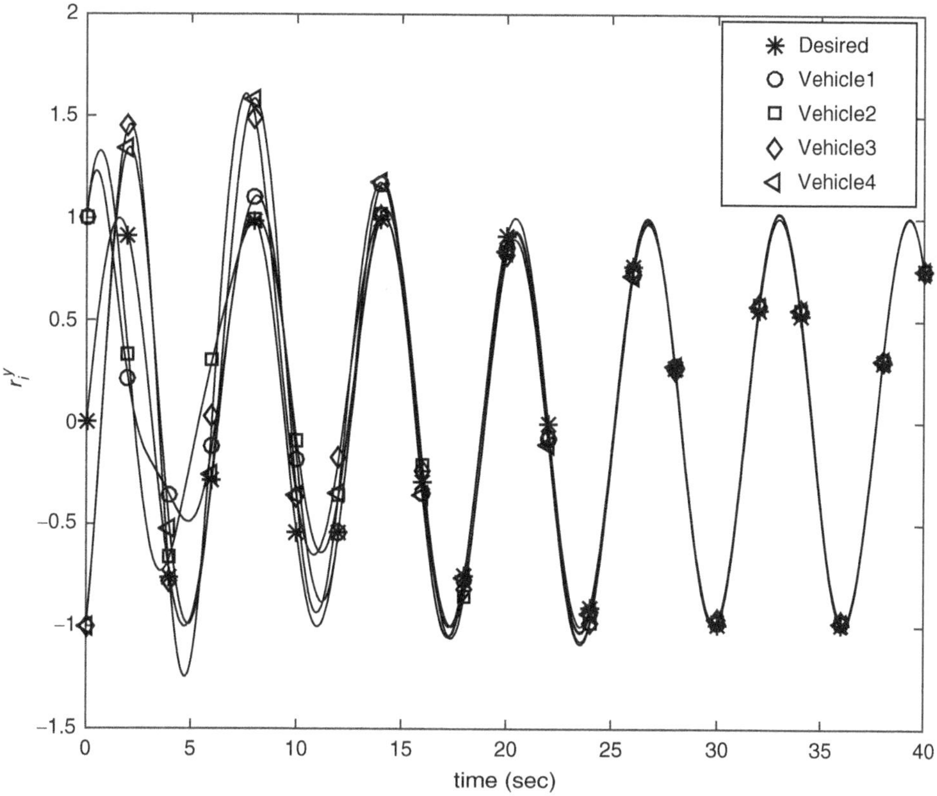

Figure 4.3 The *y*-axis position trajectories for Case 1.

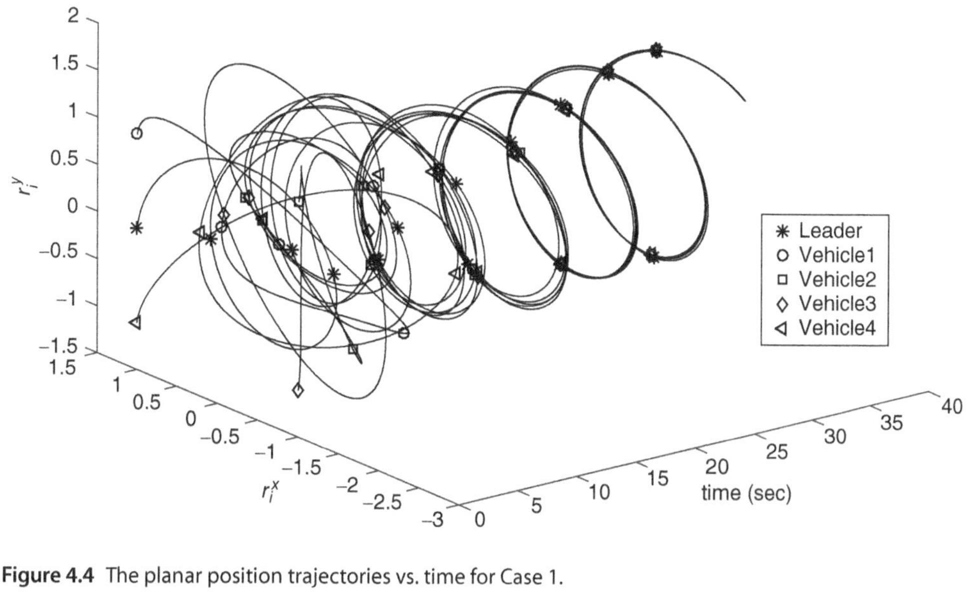

Figure 4.4 The planar position trajectories vs. time for Case 1.

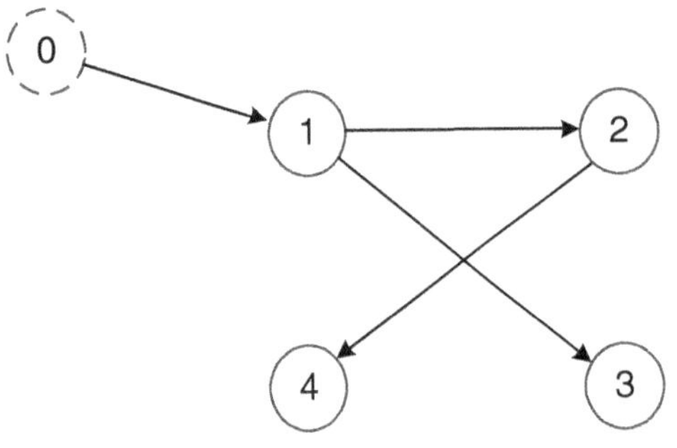

Figure 4.5 The interaction graph for Case 2.

4.7.2 Case 2: Verification of Theorem 4.5

The interaction graph is shown in Figure 4.5. It trivially satisfies Condition 3 with node 0 (the leader) as the root node and a directed subgraph $\mathcal{G}$ describing the interaction among the four vehicles. To be specific, for the bonded controller (4.31) we set

$$\mathcal{A}_4 = \begin{bmatrix} 0 & 0 & 0 & 0 \\ 0.5 & 0 & 0 & 0 \\ 0.5 & 0 & 0 & 0 \\ 0 & 0.5 & 0 & 0 \end{bmatrix}, \quad \mathcal{L}_4 + \mathcal{B}_4 = \begin{bmatrix} 0.5 & 0 & 0 & 0 \\ -0.5 & 0.5 & 0 & 0 \\ -0.5 & 0 & 0.5 & 0 \\ 0 & -0.5 & 0 & 0.5 \end{bmatrix}, \tag{4.39}$$

which implies $b_1 = 0.5$ and $b_2 = b_3 = b_4 = 0$. The saturation functions $\sigma_1 = \sigma_2 = \tanh$ are used, and the gains for each dimension of each vehicle are the same, with $k_e = k_\zeta = 1$. The desired trajectories are:

- $r_d = (r_d^x, r_d^y)^T = (\cos(t), \sin(t))^T$
- $\dot{r}_d = (\dot{r}_d^x, \dot{r}_d^y)^T = (-\sin(t), \cos(t))^T$
- $\ddot{r}_d = (\ddot{r}_d^x, \ddot{r}_d^y)^T = (-\cos(t), -\sin(t))^T$.

The initial states of the four vehicles are:

$$\begin{aligned} (r_1^x(0), r_1^y(0), v_1^x(0), v_1^y(0)) &= (1.5, -0.4, 0.1, -0.02) \\ (r_2^x(0), r_2^y(0), v_2^x(0), v_2^y(0)) &= (0.8, 0.3, 0.2, 0.1) \\ (r_3^x(0), r_3^y(0), v_3^x(0), v_3^y(0)) &= (1.9, -0.6, -0.05, -0.15) \\ (r_4^x(0), r_4^y(0), v_4^x(0), v_4^y(0)) &= (0.2, 0.7, -0.3, 0.2). \end{aligned} \tag{4.40}$$

The initial values of ζ_i $(i = 1, 2, 3, 4)$ are zero in R^2. Matlab R2014a/simulation solver ode3 (Bogacki-Shampine) with a fixed-step size of 0.01 s is used in the simulation.

The simulation results are presented in Figures 4.6–4.8. As shown in Figures 4.6 and 4.7, the position trajectories of each vehicle converge asymptotically to the desired trajectories and the settling time is about 13 s for bounds of tracking errors and synchronization errors smaller than 0.05. Figure 4.8 further shows that the actual planar motion trajectory of each vehicle tracks asymptotically the desired 2-D circular trajectory (with radius 1).

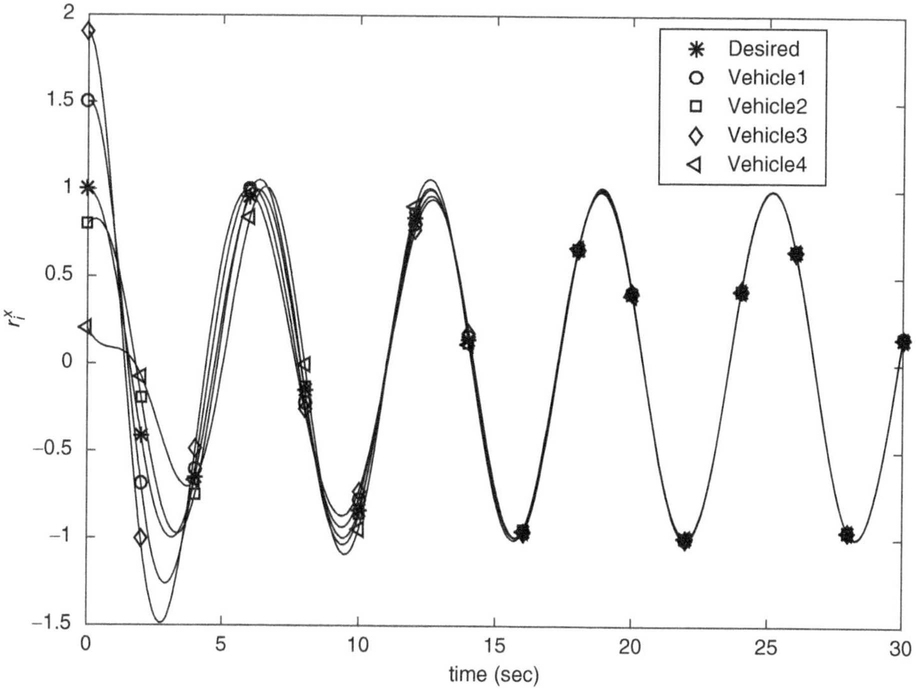

Figure 4.6 The x-axis position trajectories for Case 2.

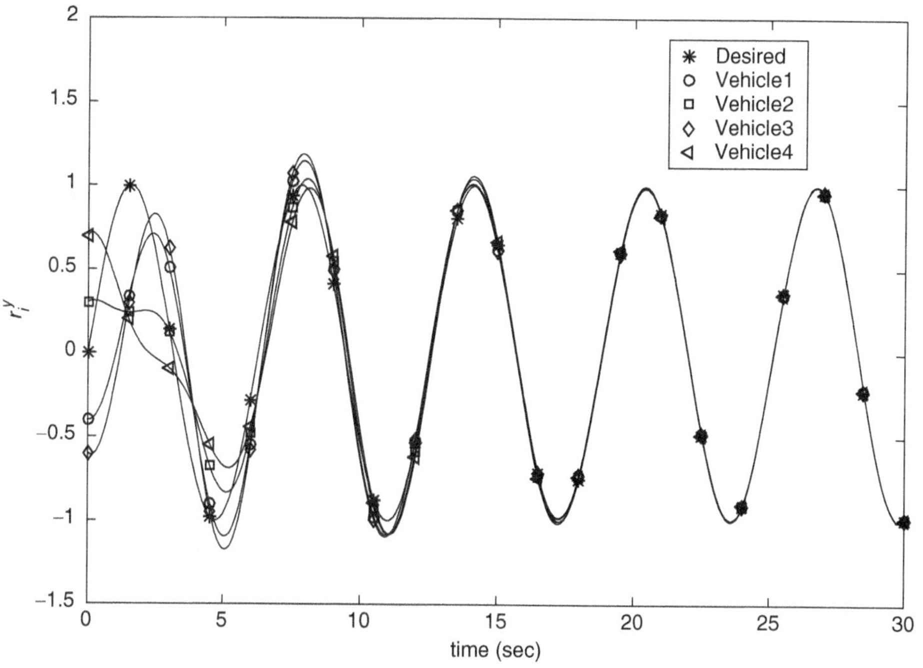

Figure 4.7 The y-axis position trajectories for Case 2.

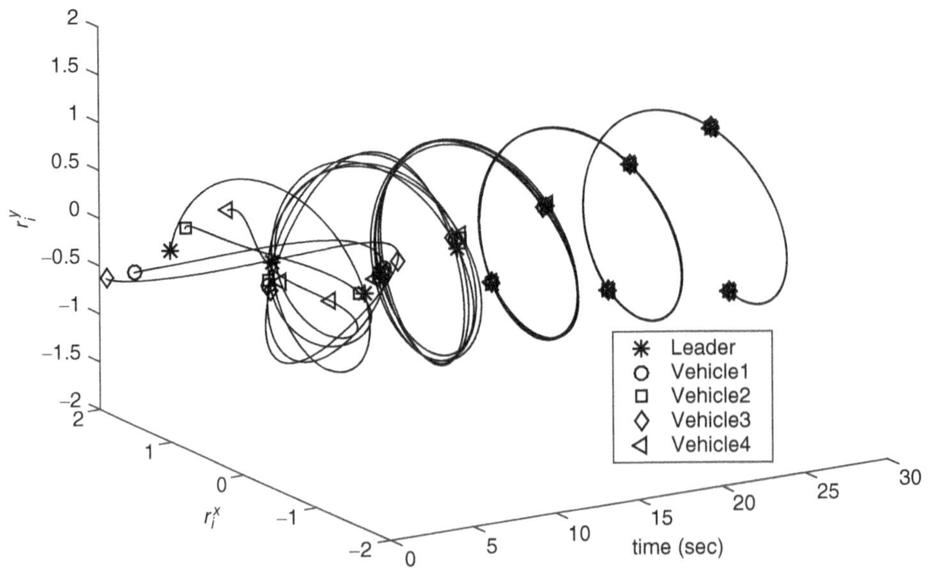

Figure 4.8 The planar position trajectories vs. time for Case 2.

4.8 Summary

Note that all the controllers discussed in Chapter 3 use full state feedback: both position and velocity signals are used in the controller design. In contrast, this chapter discusses the controller design without velocity measurements. In particular, four different output-feedback control laws for synchronized position trajectory tracking are developed using passivity techniques. Once a formation control problem is converted to a synchronized tracking problem (as shown in Chapter 3), all the control laws can be easily applied to achieve formation motion.

These control laws are given in (4.2), (4.23), (4.27) and (4.31), respectively. We also recall that the two control laws given in (4.23) and (4.31), respectively, are non-linear, each of which accounts for the actuator saturation constraint. The non-linear control laws are important in practice since actuator saturation may lead to actuator chattering, or even loss of system stability. Compared with the closed-loop MVS dynamics resulting from the two linear control laws, the closed-loop dynamics resulting from the two non-linear control laws are more complicated. Despite this, the two non-linear control laws ensure that the synchronized tracking objective is asymptotically achieved by prior bounded control.

It is important to note that all the four control laws share the following two common features:

- The signals v_i, $v_i - v_j$ and $v_i - \dot{r}_d$ for $i, j \in \mathcal{I}$, are not used for control design.

- The dimension of the controlled motion described by (4.1) is not limited to a certain number. Therefore, these control laws may be applicable to synchronized tracking in either 1-dimensional, 2-dimensional, or 3-dimensional space.

In the next chapter, we will systematically investigate control design for robustness improvement with respect to input disturbances, which is another issue of practical importance, and is associated with synchronized tracking as well as formation control.

5

Robust and Adaptive Formation Control

A practical issue encountered in the design of formation control algorithms is that real MVSs usually suffer from unpredictable exogenous disturbances as well as different levels of model uncertainties. The implementation performance of formation control algorithms depends on their robustness with respect to exogenous disturbances and model uncertainties. This chapter introduces three different approaches to improve the robustness of synchronized tracking control or formation control. They are based on:

- the active uncertainty and disturbance estimator (UDE), a concept originally proposed by Zhong and Rees [9], and which has recently been applied to MVSs [10]
- sliding-mode control
- adaptive control.

It is important to note that some of the robust control algorithms presented in this chapter are obtained by improving the algorithms presented in last two chapters, and that both state feedback and output feedback approaches are discussed.

5.1 Problem Statement

Consider a group of N identical vehicles. The equation of motion for each vehicle is the well-known second-order nonlinear model:

$$D(\boldsymbol{q}_i)\ddot{\boldsymbol{q}}_i + C(\boldsymbol{q}_i, \dot{\boldsymbol{q}}_i)\dot{\boldsymbol{q}}_i + g(\boldsymbol{q}_i) + F_i^d = Q_i, i \in \mathcal{I} \tag{5.1}$$

where $\boldsymbol{q}_i \in \mathbb{R}^n$ is the (linear or angular) position vector, $\dot{\boldsymbol{q}}_i \in \mathbb{R}^n$ is the (linear or angular) velocity vector, $D(\boldsymbol{q}_i) \in \mathbb{R}^{n\times n}$ is the generalized inertia matrix, $C(\boldsymbol{q}_i, \dot{\boldsymbol{q}}_i) \in \mathbb{R}^{n\times n}$ is the Coriolis matrix, $g(\boldsymbol{q}_i) \in \mathbb{R}^n$ is the gravity vector, $F_i^d \in \mathbb{R}^n$ is the unknown exogenous disturbance vector, and $Q_i \in \mathbb{R}^n$ is the generalized force vector.

Note that Equations (5.1) can be used to describe the dynamics of a wide variety of vehicles, including spacecraft [11, 12], UAVs [13], robot hands (see, for example, Chapter 11 in the book by Ren and Beard [3]), and 3DOF experimental helicopters [14, 15]. The application of robust control algorithms to these MVSs will be presented in Chapters 6–8.

We also remind the reader here that the double integrator model considered in Chapters 3 and 4 is a simplified variation of model (5.1). In fact, adapting model (5.1) by making $D(\boldsymbol{q}_i)$ a constant and setting $C(\boldsymbol{q}_i, \dot{\boldsymbol{q}}_i) = g(\boldsymbol{q}_i) = 0$, gives a standard double integrator model with input disturbance $F_i^d \in \mathbb{R}^n$. Alternatively, by applying the

Formation Control of Multiple Autonomous Vehicle Systems, First Edition. Hugh H.T. Liu and Bo Zhu.

well-known feedback linearization technique [16] to the non-linear model (5.1), we may obtain such a double integrator model, as shown in the following.

Suppose $D(\boldsymbol{q}_i)$ is non-singular in the domain of interest. Let

$$\boldsymbol{d}_i = D^{-1}(\boldsymbol{q}_i)F_i^d, i \in \mathcal{I}, \tag{5.2}$$

and

$$Q_i = C(\boldsymbol{q}_i, \dot{\boldsymbol{q}}_i)\dot{\boldsymbol{q}}_i + g(\boldsymbol{q}_i) + D(\boldsymbol{q}_i)\boldsymbol{u}_i, i \in \mathcal{I}. \tag{5.3}$$

Then, the normalized linearised vehicle model is given by

$$\ddot{\boldsymbol{q}}_i = \boldsymbol{u}_i + \boldsymbol{d}_i, i \in \mathcal{I}, \tag{5.4}$$

where $\boldsymbol{u}_i$ is the virtual control to be designed, and $\boldsymbol{d}_i$ denotes the normalized disturbance for vehicle i. Once $\boldsymbol{u}_i$ is determined, the actual control Q_i is computed by (5.3).

We use $\boldsymbol{q}_d, \dot{\boldsymbol{q}}_d \in \mathbb{R}^n$ to denote the desired position and velocity trajectories, respectively, and $(\boldsymbol{q}_{i0}, \dot{\boldsymbol{q}}_{i0})$ to denote the initial state of vehicle i; that is, $(\boldsymbol{q}_{i0}, \dot{\boldsymbol{q}}_{i0}) = (\boldsymbol{q}_i(0), \dot{\boldsymbol{q}}_i(0))$. We define the position and velocity errors as

$$\tilde{\boldsymbol{q}}_i(t) \triangleq \boldsymbol{q}_i(t) - \boldsymbol{q}_d(t) \in \mathbb{R}^n, i \in \mathcal{I}, \tag{5.5}$$

$$\tilde{\boldsymbol{q}}(t) \triangleq [\tilde{\boldsymbol{q}}_1^T(t), \cdots, \tilde{\boldsymbol{q}}_n^T(t)]^T \in \mathbb{R}^{nN}, \tag{5.6}$$

$$\dot{\tilde{\boldsymbol{q}}}_i(t) \triangleq \dot{\boldsymbol{q}}_i(t) - \dot{\boldsymbol{q}}_d(t) \in \mathbb{R}^n, i \in \mathcal{I}, \tag{5.7}$$

$$\dot{\tilde{\boldsymbol{q}}}(t) \triangleq [\dot{\tilde{\boldsymbol{q}}}_1^T(t), \cdots, \dot{\tilde{\boldsymbol{q}}}_n^T(t)]^T \in \mathbb{R}^{nN}. \tag{5.8}$$

Position and velocity synchronization means that

$$\begin{aligned} &\tilde{\boldsymbol{q}}_1(t) = \tilde{\boldsymbol{q}}_2(t) = \cdots = \tilde{\boldsymbol{q}}_N(t), \\ &\dot{\tilde{\boldsymbol{q}}}_1(t) = \dot{\tilde{\boldsymbol{q}}}_2(t) = \cdots = \dot{\tilde{\boldsymbol{q}}}_N(t), \end{aligned} \tag{5.9}$$

whereas zero-error tracking in both position and velocity requires

$$\begin{aligned} &\tilde{\boldsymbol{q}}_i(t) = 0, i \in \mathcal{I}, \\ &\dot{\tilde{\boldsymbol{q}}}_i(t) = 0, i \in \mathcal{I}. \end{aligned} \tag{5.10}$$

As for the controlled position variables $\boldsymbol{q}_i$, the Type III LSEs are given by

$$\boldsymbol{e}_{qi} \triangleq b_i(\boldsymbol{q}_i - \boldsymbol{q}_d) + \sum_{j \in N_i} a_{ij}(\boldsymbol{q}_i - \boldsymbol{q}_j) \in \mathbb{R}^n, i \in \mathcal{I} \tag{5.11}$$

$$\boldsymbol{e}_q \triangleq \left[\boldsymbol{e}_{q1}^T, \cdots, \boldsymbol{e}_{qN}^T\right]^T \in \mathbb{R}^{nN}. \tag{5.12}$$

With (5.5) and (5.7), the Type III LSEs, $\boldsymbol{e}_{qi}$, satisfy

$$\boldsymbol{e}_{qi} = b_i\tilde{\boldsymbol{q}}_i + \sum_{j \in N_i} a_{ij}(\tilde{\boldsymbol{q}}_i - \tilde{\boldsymbol{q}}_j), i \in \mathcal{I} \tag{5.13}$$

Correspondingly, as for the controlled velocity variables $\dot{\boldsymbol{q}}_i$, the Type III LSEs are given by

$$\dot{\boldsymbol{e}}_{qi} = b_i\dot{\tilde{\boldsymbol{q}}}_i + \sum_{j \in N_i} a_{ij}(\dot{\tilde{\boldsymbol{q}}}_i - \dot{\tilde{\boldsymbol{q}}}_j), i \in \mathcal{I} \tag{5.14}$$

Equations (5.11) and (5.14) can be written in compact form as

$$\boldsymbol{e}_q = ((\mathcal{L}_N + \mathcal{B}_N) \otimes I_n)\tilde{\boldsymbol{q}}, \tag{5.15}$$

$$\dot{\boldsymbol{e}}_q = ((\mathcal{L}_N + \mathcal{B}_N) \otimes I_n)\dot{\tilde{\boldsymbol{q}}}. \tag{5.16}$$

As shown in the last two chapters, the synchronized trajectory tracking problem of MVSs can be converted to the stabilization problem of $\boldsymbol{e}_{qi}$. Since the issue of adaptation to disturbances is also a concern, the objective of this chapter is to develop robust control laws for MVSs (5.4) that can attenuate the effect of disturbances F_i^d and drive $\boldsymbol{e}_{qi}$, $i = 1, \cdots, N$, to the origin – or close to the origin – asymptotically.

5.2 Continuous Control via State Feedback

To design a controller that is capable of stabilizing $\boldsymbol{e}_{qi}$ (as well as $\dot{\boldsymbol{e}}_{qi}$), we first need to derive the control equations of $\boldsymbol{e}_{qi}$, which are expected to show the relationship between control variables $\boldsymbol{e}_{qi}$ and control inputs $\boldsymbol{u}_i$. These equations can be easily derived from (5.4) and (5.11) as

$$\ddot{\boldsymbol{e}}_{qi} = \frac{1}{k_i}\boldsymbol{u}_i - \sum_{j\in N_i} a_{ij}\boldsymbol{u}_j - b_i\ddot{\boldsymbol{q}}_d + \Delta_i, i \in \mathcal{I} \tag{5.17}$$

where k_i are defined in (3.63) and

$$\Delta_i = \frac{1}{k_i}d_i - \sum_{j\in N_i} a_{ij}d_j, i = 1, \cdots, N. \tag{5.18}$$

As shown in (5.17), there are a total of four terms driving the dynamics of $\boldsymbol{e}_{qi}$:

- $\frac{1}{k_i}\boldsymbol{u}_i$
- $-\sum_{j\in N_i} a_{ij}\boldsymbol{u}_j$
- $-b_i\ddot{\boldsymbol{q}}_d$
- Δ_i.

To achieve high-accuracy stabilization of $\boldsymbol{e}_{qi}$, the control input $\boldsymbol{u}_i$ is often required to:

- cancel the (known) term $\sum_{j\in N_i} a_{ij}\boldsymbol{u}_j + b_i\ddot{\boldsymbol{q}}_d$ included in (5.17);
- reject the unknown exogenous disturbance Δ_i;
- feedback $\boldsymbol{e}_{qi}$ and $\dot{\boldsymbol{e}}_{qi}$ to generate a stabilizing control.

In what follows, we construct a simple linear stabilizing control that meets the above three requirements. For simplicity of presentation, we assume that the equation of motion, (5.1), is one dimensional; that is, $n = 1$. However, the control approach to be presented can be easily extended to higher-dimensional motion provided the equation of motion of each dimension can be mutually decoupled and written in the form shown in (5.1).

5.2.1 Controller Development

For the MVS in (5.17), we construct the controller

$$\boldsymbol{u}_i = \boldsymbol{u}_i^0 - \hat{\boldsymbol{d}}_i, i \in \mathcal{I} \tag{5.19}$$

where $\hat{\boldsymbol{d}}_i$ denotes an estimate of $\boldsymbol{d}_i$, and the nominal control $\boldsymbol{u}_i^0$ satisfies

$$\boldsymbol{u}_i^0 = k_i\left(\sum_{j\in N_i} a_{ij}\boldsymbol{u}_j^0 + b_i\ddot{\boldsymbol{q}}_d - k_i^1\boldsymbol{e}_{qi} - k_i^2\dot{\boldsymbol{e}}_{qi}\right), i \in \mathcal{I} \tag{5.20}$$

where k_i^1 and k_i^2 are positive control gains. We denote

$$\hat{\Delta}_i = \frac{1}{k_i}\hat{\boldsymbol{d}}_i - \sum_{j\in N_i} a_{ij}\hat{\boldsymbol{d}}_j, i = 1, \cdots, N. \tag{5.21}$$

Substituting (5.19) into (5.17), gives

$$\begin{aligned}\ddot{\boldsymbol{e}}_{qi} &= \frac{1}{k_i}(\boldsymbol{u}_i^0 - \hat{\boldsymbol{d}}_i) - \sum_{j\in N_i} a_{ij}(\boldsymbol{u}_j^0 - \hat{\boldsymbol{d}}_j) - b_i\ddot{\boldsymbol{q}}_d + \Delta_i,\\ &= \frac{1}{k_i}\boldsymbol{u}_i^0 - \sum_{j\in N_i} a_{ij}\boldsymbol{u}_j^0 - b_i\ddot{\boldsymbol{q}}_d - \hat{\Delta}_i + \Delta_i, i \in \mathcal{I}\end{aligned} \tag{5.22}$$

From (5.20) and (5.22) we see the following three facts:

- the term $\sum_{j\in N_i} a_{ij}\boldsymbol{u}_j^0 - b_i\ddot{\boldsymbol{q}}_d$ in equation (5.22) can be cancelled by the same term as introduced in controller (5.20);
- the effect of the disturbance signal Δ_i can be asymptotically compensated, provided $\hat{\Delta}_i$ is an asymptotic estimate of Δ_i;
- the feedback control $-k_i^1\boldsymbol{e}_{qi} - k_i^2\dot{\boldsymbol{e}}_{qi}$ admits a standard PD control structure, and serves as a linear feedback stabilizer for (5.22).

Note that equation (5.20) does not provide an explicit expression of nominal control $\boldsymbol{u}_i^0$, since nominal controls $\boldsymbol{u}_j^0$ ($j \in N_i$) are used to compute $\boldsymbol{u}_i^0$. To obtain explicit expressions for $\boldsymbol{u}_i^0$ ($i = 1, \cdots, N$), we can write the N equations in (5.20) into the vector form as

$$(\mathcal{L}_N + \mathcal{B}_N)\begin{bmatrix}\boldsymbol{u}_1^0\\ \vdots\\ \boldsymbol{u}_N^0\end{bmatrix} = \begin{bmatrix}b_1\ddot{\boldsymbol{q}}_d - k_1^1\boldsymbol{e}_{q1} - k_1^2\dot{\boldsymbol{e}}_{q1}\\ \vdots\\ b_N\ddot{\boldsymbol{q}}_d - k_N^1\boldsymbol{e}_{qN} - k_N^2\dot{\boldsymbol{e}}_{qN}\end{bmatrix}. \tag{5.23}$$

It is clear from (5.23) that under Condition 3, matrix $(\mathcal{L}_N + \mathcal{B}_N)$ is nonsingular, and there is a unique solution for the $\boldsymbol{u}_i^0$ satisfying (5.20) or (5.23), which can be equivalently computed by

$$\begin{bmatrix}\boldsymbol{u}_1^0\\ \vdots\\ \boldsymbol{u}_N^0\end{bmatrix} = (\mathcal{L}_N + \mathcal{B}_N)^{-1}\begin{bmatrix}b_1\ddot{\boldsymbol{q}}_d - k_1^1\boldsymbol{e}_{q1} - k_1^2\dot{\boldsymbol{e}}_{q1}\\ \vdots\\ b_N\ddot{\boldsymbol{q}}_d - k_N^1\boldsymbol{e}_{qN} - k_N^2\dot{\boldsymbol{e}}_{qN}\end{bmatrix}. \tag{5.24}$$

5.2.2 Analysis of Tracker $\boldsymbol{u}_i^0$

In this section, we consider an important property of the nominal control $\boldsymbol{u}_i^0$ satisfying (5.20): it can achieve zero-error synchronized tracking in the absence of disturbance (i.e., with $\boldsymbol{d}_i = 0$ for each $i \in \mathcal{I}$). To show this point, we first derive the closed-loop error equation in the absence of any exogenous disturbance. This is done by combining (5.19) with (5.20) and setting $\hat{\boldsymbol{d}}_i = 0$, and taking (5.4) with $\boldsymbol{d}_i = 0$. This results in the error equation:

$$\ddot{\boldsymbol{e}}_{qi} = -k_i^1 \boldsymbol{e}_{qi} - k_i^2 \dot{\boldsymbol{e}}_{qi}, i = 1, \cdots, N. \tag{5.25}$$

Note that the N equations (5.25) are linear and second-order. We then have the following lemma.

Lemma 5.1 Consider the control law (5.20) for the nominal MVS $\ddot{\boldsymbol{q}}_i = \boldsymbol{u}_i^0$ $(i = 1, \cdots, N)$. Under Condition 3, all of the error trajectories $\boldsymbol{e}_{qi}(t)$, $\dot{\boldsymbol{e}}_{qi}(t)$, $\tilde{\boldsymbol{q}}(t)$, and $\dot{\tilde{\boldsymbol{q}}}(t)$ are globally bounded and converge exponentially to zero.

Lemma 5.1 indicates that $\boldsymbol{u}_i^0$ $(i = 1, \cdots, N)$ satisfying (5.20) are asymptotic synchronized tracking controllers for the MVS described by nominal double integrators. For this reason, we call $\boldsymbol{u}_i^0$ a nominal control.

According to (5.20), the control signals $\boldsymbol{u}_j^0$, position signals $\boldsymbol{q}_j$ and velocity signals $\dot{\boldsymbol{q}}_j$ $(j \in N_i)$ are required to compute $\boldsymbol{u}_i^0$. This control law is distributed, because only neighbor information is required. The idea of using neighbors' control signals $(\boldsymbol{u}_j^0, j \in N_i)$ to construct a distributed control $(\boldsymbol{u}_i^0, i = 1, \cdots, N)$ can be also found in the literature [3, 17, 18].

For practical applications, the control law (5.20) can be approximately implemented in the following way: use $\boldsymbol{u}_j^0$ obtained during the previous sampling period (i.e., $\boldsymbol{u}_j^0(t - \tau)$, with τ a fixed sampling step), and the current $\ddot{\boldsymbol{q}}_d(t)$, $\boldsymbol{e}_{qi}(t)$, $\dot{\boldsymbol{e}}_{qi}(t)$ to compute the current $\boldsymbol{u}_i^0$. If the sample frequency is sufficiently high, the period τ is sufficiently small and the implementation strategy does not affect the system performance dramatically.

5.2.3 Design of Disturbance Estimators

We shall now introduce an approach to construct $\hat{\boldsymbol{d}}_i$. For simplicity of presentation, the underlying idea of the approach is presented in the frequency domain while the estimation convergence is evaluated in the time domain. We first suppose that the Laplace transform of the exogenous disturbance $\boldsymbol{d}_i$ for each $i = 1, \cdots, N$ exists. This condition is not restrictive: any $\boldsymbol{d}_i(t)$ that is bounded by an exponential function of t satisfies it.

According to the concept of a disturbance estimator (DE) [9, 10], the estimate $\hat{\boldsymbol{d}}_i$ for each $i = 1, \cdots, N$ should satisfy

$$\hat{\boldsymbol{d}}_i(s) = G_f(s)\boldsymbol{d}_i(s) = G_f(s)[\ddot{\boldsymbol{q}}_i(s) - \boldsymbol{u}_i(s)], \tag{5.26}$$

where $G_f(s)$ is the transfer function of a linear filter to be determined. Substituting (5.19) into (5.26), we obtain

$$\hat{\boldsymbol{d}}_i(s) = G_f(s)\left[\ddot{\boldsymbol{q}}_i(s) - \boldsymbol{u}_i^0(s) + \hat{\boldsymbol{d}}_i(s)\right]. \tag{5.27}$$

Solving (5.27) yields

$$\hat{\boldsymbol{d}}_i(s) = \frac{G_f(s)}{1 - G_f(s)}\left[\ddot{\boldsymbol{q}}_i(s) - \boldsymbol{u}_i^0(s)\right]. \tag{5.28}$$

To ensure that Equation (5.28) is implementable in practice, the filter $G_f(s)$ must meet the following two conditions:

- an integral action is included in $\frac{G_f(s)}{1-G_f(s)}$ so that the computation of $\hat{\boldsymbol{d}}_i$ does not need the acceleration signal $\ddot{\boldsymbol{q}}_i$;
- both frequency-domain and time-domain properties of $G_f(s)$ are easily understood.

Clearly, with the first-order filter

$$G_f(s) = \frac{1}{T_q s + 1}, \tag{5.29}$$

we have

$$\frac{G_f(s)}{1 - G_f(s)} = \frac{1}{T_q s}, \tag{5.30}$$

which indeed includes an integral action, and (5.28) becomes

$$\hat{\boldsymbol{d}}_i(s) = \frac{1}{T_q s}\left[\ddot{q}_i(s) - \boldsymbol{u}_i^0(s)\right]. \tag{5.31}$$

Thus, $\hat{\boldsymbol{d}}_i(t)$ (in the time domain) can be calculated as

$$\hat{\boldsymbol{d}}_i(t) = \frac{\dot{\boldsymbol{q}}_i(t) - \dot{\boldsymbol{q}}_i(0) - \int_0^t \boldsymbol{u}_i^0(\tau)d\tau}{T_q}. \tag{5.32}$$

It is seen from (5.32) that velocity signals $\dot{\boldsymbol{q}}_i(t)$ and the integral of nominal control $\int_0^t \boldsymbol{u}_i^0(\tau)d\tau$ are used to compute $\hat{\boldsymbol{d}}_i(t)$. From (5.32), $\hat{\boldsymbol{d}}_i(0) = 0$ and the resulting $\boldsymbol{u}_i(0) = \boldsymbol{u}_i^0(0)$ according to (5.19). This indicates that an initial peak in $\boldsymbol{u}_i$ is avoided when UDE (5.32) is applied.

To evaluate the disturbance estimation performance of (5.32), we define the following estimation errors:

$$\tilde{\boldsymbol{d}}_i \triangleq \boldsymbol{d}_i - \hat{\boldsymbol{d}}_i, i \in \mathcal{I} \tag{5.33}$$

$$\tilde{\mathbf{d}} \triangleq \left[\tilde{\boldsymbol{d}}_1, \cdots, \tilde{\boldsymbol{d}}_N\right]^T \in R^N. \tag{5.34}$$

From (5.26) and (5.33) with $G_f(s) = \frac{1}{T_q s+1}$, we obtain the result that

$$\dot{\tilde{\boldsymbol{d}}}_i = -\frac{\tilde{\boldsymbol{d}}_i}{T_q} + \dot{\boldsymbol{d}}_i. \tag{5.35}$$

We use $\tilde{\boldsymbol{d}}_i(0)$ and $\boldsymbol{d}_i(0)$ to denote the initial values of $\tilde{\boldsymbol{d}}_i(t)$ and $\boldsymbol{d}_i(t)$, respectively. Then, $\hat{\boldsymbol{d}}_i(0) = 0$ implies $\tilde{\boldsymbol{d}}_i(0) = \boldsymbol{d}_i(0)$. For further analysis, we impose the following condition on the first-order derivative of $\boldsymbol{d}_i$.

Assumption 5.2 For each $i = 1, \cdots, N$, $\boldsymbol{d}_i(t)$ is continuously differentiable and its first derivative is bounded and satisfies $\|\dot{\boldsymbol{d}}_i(t)\|_\infty \le \overline{\boldsymbol{d}}_d$.

The following lemma the summarizes the properties of UDE (5.32).

Lemma 5.3 Consider the trajectories of system (5.35) under Assumption 5.2. Then,

- for any positive constant $b_d > T_q\bar{\boldsymbol{d}}_d$, there exists a time $t_d > 0$ such that $\tilde{\boldsymbol{d}}_i(t)$ with any initial value satisfies

$$|\tilde{\boldsymbol{d}}_i(t)| < b_d, \forall t \geq t_d, \tag{5.36}$$

- $\tilde{\boldsymbol{d}}_i(t)$ with any initial value is uniformly bounded by

$$|\tilde{\boldsymbol{d}}_i(t)| \leq \max(|\tilde{\boldsymbol{d}}_i(0)|, T_q\bar{\boldsymbol{d}}_d), \forall t \geq 0, \tag{5.37}$$

- if $\lim_{t\to\infty} \dot{\boldsymbol{d}}_i(t) = 0$, then

$$\lim_{t\to\infty} \tilde{\boldsymbol{d}}_i(t) = 0. \tag{5.38}$$

Proof: The results of this lemma can be easily obtained by applying Lemma A.12 in the appendix to system (5.35) and regarding $\dot{\boldsymbol{d}}_i$ as the system input. □

Inequality (5.36) shows that for any $\boldsymbol{d}_i(t)$ with a bounded derivative $\dot{\boldsymbol{d}}_i(t)$, the resulting disturbance estimation error $\tilde{\boldsymbol{d}}_i(t)$ is (uniformly) ultimately bounded by a bound $T_q\bar{\boldsymbol{d}}_d$, which depends linearly on design parameter T_q. This implies that a smaller T_q results in a smaller ultimate bound of $\tilde{\boldsymbol{d}}_i(t)$, and in particular, if $T_q \to 0$, the ultimate bound of $\tilde{\boldsymbol{d}}_i(t)$ will approach zero. The last statement of the lemma shows an additional condition on $\boldsymbol{d}_i(t)$ to ensure an asymptotic estimation. Clearly, if $\boldsymbol{d}_i(t)$ is a constant, $\lim_{t\to\infty} \dot{\boldsymbol{d}}_i(t) = 0$ is trivially satisfied and $\tilde{\boldsymbol{d}}_i(t)$ converges exactly to zero.

5.2.4 Closed-loop Performance Analysis

Applying the controller (5.19) with (5.20) and (5.32) to (5.17) yields the control configuration for vehicle i, as shown in Figure 5.1. Note that there are a total of three control modules included in the configuration: input transformation (5.3), tracker (5.20) and UDE (5.32).

The closed-loop error dynamics at the node level are

$$\begin{cases} \dot{\tilde{\boldsymbol{d}}}_i = -\dfrac{\tilde{\boldsymbol{d}}_i}{T_q} + \dot{\boldsymbol{d}}_i, \\ \ddot{\boldsymbol{e}}_{qi} = -k_i^1 \boldsymbol{e}_{qi} - k_i^2 \dot{\boldsymbol{e}}_{qi} + \tilde{\Delta}_i, i \in \mathcal{I}, \end{cases} \tag{5.39}$$

where

$$\tilde{\Delta}_i = \Delta_i - \hat{\Delta}_i = \frac{1}{k_i}\tilde{\boldsymbol{d}}_i - \sum_{j\in N_i} a_{ij}\tilde{\boldsymbol{d}}_j. \tag{5.40}$$

We define

$$A_i \triangleq \begin{bmatrix} 0 & 1 \\ -k_i^1 & -k_i^2 \end{bmatrix}, i \in \mathcal{I}, \tag{5.41}$$

$$\tilde{\boldsymbol{\Delta}} \triangleq [\tilde{\Delta}_1, \cdots, \tilde{\Delta}_N]^T \in R^N. \tag{5.42}$$

Clearly, matrix A_i with positive entries k_i^1 and k_i^2 is Hurwitz. Moreover, with (5.40), the relationship between $\tilde{\mathbf{d}}$ and $\tilde{\boldsymbol{\Delta}}$ is

$$\tilde{\boldsymbol{\Delta}}(t) = (\mathcal{L}_N + \mathcal{B}_N)\tilde{\mathbf{d}}(t), \forall t \geq 0. \tag{5.43}$$

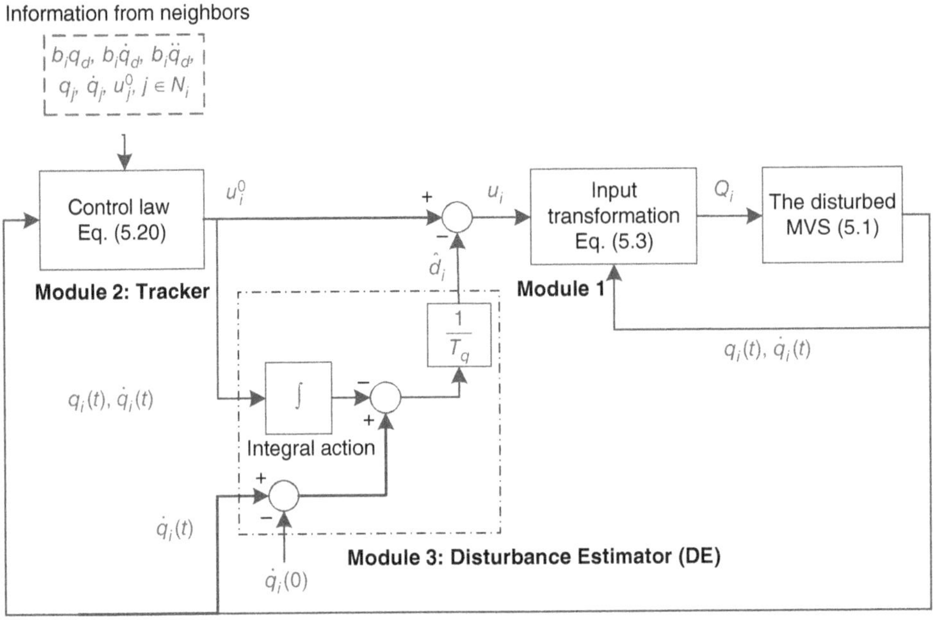

Figure 5.1 The proposed robust control configuration for vehicle i.

It then follows that

$$\|\tilde{\mathbf{\Delta}}(t)\|_\infty \leq \|\mathcal{L}_N + \mathcal{B}_N\|_\infty \|\tilde{\mathbf{d}}(t)\|_\infty, \forall t \geq 0, \tag{5.44}$$

which will be used to derive the result of Theorem 5.6.

Under Condition 3, $\mathcal{L}_N + \mathcal{B}_N$ has a full rank and the following statements are true:

- Because the first-order equation in (5.39), with $\dot{\boldsymbol{d}}_i$ as the input, is globally input-to-state stable (GISS; see Definition A.5 in the appendix for a detailed explanation of this concept). Thus, if each $\dot{\boldsymbol{d}}_i$ is globally bounded, so will be each $\tilde{\boldsymbol{d}}_i$ and each $\tilde{\Delta}_i$.
- Since matrix A_i is Hurwitz, the second-order equation in (5.39), with $\tilde{\Delta}_i$ as the input, is also GISS. Thus, if each $\dot{\boldsymbol{d}}_i$ is globally bounded, so will be each error vector $(\boldsymbol{e}_{qi}, \dot{\boldsymbol{e}}_{qi})^T$.

Further, we have the following results.

Lemma 5.4 Under Condition 3, $\tilde{\mathbf{\Delta}}(t) \to \mathbf{0}_N$ as $t \to \infty$ if and only if each $\tilde{\mathbf{d}}(t) \to \mathbf{0}_N$ as $t \to \infty$.

Theorem 5.5 Consider the closed-loop system (5.39) under Condition 3 and Assumption 5.2. If $\lim_{t\to\infty} \dot{\boldsymbol{d}}_i(t) = 0$ for each $i \in \mathcal{I}$, then:

- $\tilde{\mathbf{\Delta}}$ is continuously differentiable and globally bounded, and $\tilde{\mathbf{\Delta}}(t) \to \mathbf{0}_N$ as $t \to \infty$;
- the LNSE trajectories $\mathbf{e}_q(t)$, $\dot{\mathbf{e}}_q(t)$ are globally bounded, and $\mathbf{e}_q(t) \to \mathbf{0}_N$, $\dot{\mathbf{e}}_q(t) \to \mathbf{0}_N$ as $t \to \infty$;
- the tracking error trajectories $\tilde{\boldsymbol{q}}(t)$, $\dot{\tilde{\boldsymbol{q}}}(t)$ are globally bounded, and $\tilde{\boldsymbol{q}}(t) \to \mathbf{0}_N$, $\dot{\tilde{\boldsymbol{q}}}(t) \to \mathbf{0}_N$ as $t \to \infty$;

Theorem 5.5 applies to the situation with constant disturbances, and shows that under this situation, zero-error estimation of $\boldsymbol{d}_i(t)$ and zero-error tracking of $\boldsymbol{q}_d(t)$ are both achieved for each vehicle. For the general situation with non-vanishing $\dot{\boldsymbol{d}}_i(t)$, we have the following result.

Theorem 5.6 Consider the closed-loop system (5.39) under Condition 3 and Assumption 5.2. Then:

- $\tilde{\Delta}_i$ are bounded, and for any positive constant $b_{\Delta q} > \|\mathcal{L}_N + \mathcal{B}_N\|_\infty T_q \bar{d}_d$, there exists a time $t_d > 0$ such that
$$|\tilde{\Delta}_i(t)| < b_{\Delta q}, \forall t \geq t_d, i \in \mathcal{I}; \tag{5.45}$$
- the error trajectories $\mathbf{e}_q(t)$, $\dot{\mathbf{e}}_q(t)$, $\tilde{\boldsymbol{q}}(t)$, and $\dot{\tilde{\boldsymbol{q}}}(t)$ are globally bounded;
- there exists a time $t_q > 0$ such that under any initial condition,
$$\|\mathbf{e}_q(t)\|_2 \leq b\left(T_q \bar{d}_d\right), \forall t \geq t_q, \tag{5.46}$$
$$\|\tilde{\boldsymbol{q}}(t)\|_2 \leq \|(\mathcal{L}_N + \mathcal{B}_N)^{-1}\|_2 b\left(T_q \bar{d}_d\right), \forall t \geq t_q, \tag{5.47}$$

where
$$b(T\bar{d}_d) = \frac{2\sqrt{N}\|\mathcal{L}_N + \mathcal{B}_N\|_\infty T_q \bar{d}_d}{\theta} \max_{i \in \mathcal{I}} \left(\lambda_{\max}(P_i) \sqrt{\frac{\lambda_{\max}(P_i)}{\lambda_{\min}(P_i)}} \right), \tag{5.48}$$

θ can be any number in the set $\{\theta \mid 0 < \theta < 1\}$, and P_i are the solutions to the Lyapunov equations
$$P_i A_i + A_i^T P_i = -I_2, i \in \mathcal{I}. \tag{5.49}$$

Proof: The proof is given in Section A.3.1 of the appendix. □

Theorem 5.6 shows that for each vehicle i, both the signal $\tilde{\Delta}_i$ (which disturbs the dynamics of $\boldsymbol{e}_{qi}$) and the error trajectory $\tilde{\boldsymbol{q}}_i$ are globally bounded, with an ultimate bound depending linearly on the UDE parameter T_q. More specifically, for each vehicle i, the ultimate bounds of both $\tilde{\Delta}_i$ and $\tilde{q}_i$ can be made arbitrarily small by selecting a sufficiently small T_q, and in particular, the ultimate bounds of both $\tilde{\Delta}_i$ and $\tilde{q}_i$ approach zero if $T_q \to 0$. Thus, the steady-state tracking accuracy can be easily controlled by tuning the UDE parameter.

In contrast, if UDE is not used, we need turn to choosing suitable gains k_i^1 and k_i^2 to control the steady-state tracking accuracy. However, the choice of k_i^1 and k_i^2 is not easy because the relationship between them and the ultimate bound of the tracking error, as given by (5.41) and (5.46)–(5.49), is non-linear and not clear. To show this point more clearly, we give the following explanation.

Applying the nominal control (5.20) to the disturbed vehicle dynamics (5.17) results in the error equations:
$$\ddot{\boldsymbol{e}}_{qi} = -k_i^1 \boldsymbol{e}_{qi} - k_i^2 \dot{\boldsymbol{e}}_{qi} + \Delta_i, i \in \mathcal{I}. \tag{5.50}$$
where Δ_i, $i \in \mathcal{I}$, are defined in (5.18). Note that one can also obtain (5.50) by setting $\hat{\boldsymbol{d}}_i \equiv 0$, $i \in \mathcal{I}$, in (5.39).

System (5.50) with Δ_i as the disturbance input is also GISS. Moreover, by applying Lemma A.16 to system (5.50), we readily conclude that the ultimate bound of the trajectories of system (5.50) depends linearly on the bound of Δ_i and the value of $\lambda_{\max}(P_i)\sqrt{\frac{\lambda_{\max}(P_i)}{\lambda_{\min}(P_i)}}$ (which is determined by the values of k_i^1 and k_i^2). Since $\boldsymbol{d}_i$ is an unknown uncontrolled exogenous signal for each vehicle, the ultimate bound of each Δ_i is also uncontrolled. Therefore, the only possible way to obtain a smaller ultimate bound of the state trajectories of (5.50) is to choose values for the gains k_i^1 and k_i^2 such that $\lambda_{\max}(P_i)\sqrt{\frac{\lambda_{\max}(P_i)}{\lambda_{\min}(P_i)}}$ is small enough. Note that the relationship between k_i^1 and k_i^2 and $\lambda_{\max}(P_i)\sqrt{\frac{\lambda_{\max}(P_i)}{\lambda_{\min}(P_i)}}$, given by (5.41) and (5.49), is typically obscure. It is not easy to implement this idea to meet a specified requirement for the ultimate bound of the tracking error.

5.3 Bounded State Feedback Control

5.3.1 Design of Bounded State Feedback

This section will introduce the controller design under the input saturation constraint:

$$\boldsymbol{u}_i \leq \overline{u}, i \in \mathcal{I}, \tag{5.51}$$

where $\overline{u} \in R^+$ denotes a given upper bound for each $\boldsymbol{u}_i$.

To ensure the problem is solvable, we make the following assumptions about the exogenous disturbances $\boldsymbol{d}_i$, $i \in \mathcal{I}$, and the desired trajectory $q_d(t)$:

Assumption 5.7 For each $i \in \mathcal{I}$, $\boldsymbol{d}_i(t)$ is twice continuously differentiable almost everywhere, and both $\boldsymbol{d}_i(t)$ and $\dot{\boldsymbol{d}}_i(t)$ are bounded by known positive constants, i.e., $|\boldsymbol{d}_i(t)| \leq \overline{d}$ and $|\dot{\boldsymbol{d}}_i(t)| \leq \overline{d}_d$ for all $t \geq 0$, where $\overline{d} \in R^+, \overline{d}_d \in R^+$.

Assumption 5.8 $\boldsymbol{q}_d(t)$ is twice continuously differentiable, and $\ddot{\boldsymbol{q}}_d(t)$ is bounded by a known positive constant $\overline{q}_{dd}$ smaller than $\overline{u}$, i.e., $|\ddot{\boldsymbol{q}}_d(t)| \leq \overline{q}_{dd} < \overline{u}$ for all $t \geq 0$.

The objective of this section is to design $\boldsymbol{u}_i$ for MVS (5.4) under the constraint $|\boldsymbol{u}_i(t)| \leq \overline{u}$ and Assumptions 5.7 and 5.8, such that robust synchronized position tracking is achieved; that is,

$$\lim_{t\to\infty} |\tilde{\boldsymbol{q}}_i(t)| \leq \epsilon_q, i \in \mathcal{I}, \tag{5.52}$$

where ϵ_q is a positive constant that can be arbitrarily small.

Remark 5.9 The boundedness condition on $\boldsymbol{d}_i$ and $\ddot{\boldsymbol{q}}_d$, as imposed by Assumptions 5.7 and 5.8, seems a little restrictive. However, this condition is necessary to ensure that the constraint (5.51) may be satisfied. If this constraint is not taken into account, the boundedness condition on $\boldsymbol{d}_i$ and $\ddot{\boldsymbol{q}}_d$ can be removed from these assumptions.

In what follows, the controller (5.19) with (5.20) and (5.32) will be modified to satisfy (5.51). Consider the two-component structure as in (5.19) but with a minor revision on the nominal control $\boldsymbol{u}_i^0$:

$$\boldsymbol{u}_i^0 = k_i \left[\sum_{j=1}^{n} a_{ij}\boldsymbol{u}_j^0 + b_i \ddot{\boldsymbol{q}}_d - k_i^1 \sigma_1(\boldsymbol{e}_{qi}) - k_i^2 \sigma_2(\dot{\boldsymbol{e}}_{qi}) \right], i \in \mathcal{I}, \tag{5.53}$$

where k_i are defined in (3.63), k_i^1 and k_i^2 are the same as in (5.20), and $\sigma_1(\cdot)$ and $\sigma_2(\cdot)$ are two generalized saturation functions, defined in Lemma A.15 in Appendix A.2.

In (5.53), $\sigma_1(\cdot)$ is required to be strictly increasing and continuously differentiable, to satisfy the condition

$$\int_0^z \sigma_1(\tau)d\tau \to \infty as|z| \to \infty, \tag{5.54}$$

(a detailed explanation can be found in the literature [19]). Under this condition, the Lyapunov function candidates to be considered in Lemmas 5.10 and 5.11 and Theorem 5.12 below are radially unbounded.

The control law (5.53) can be equivalently given by

$$(\mathcal{L}_N + \mathcal{B}_N)\begin{bmatrix} \boldsymbol{u}_1^0 \\ \vdots \\ \boldsymbol{u}_N^0 \end{bmatrix} = \begin{bmatrix} b_1\ddot{\boldsymbol{q}}_d - k_1^1\sigma_1(\boldsymbol{e}_{q1}) - k_1^2\sigma_2(\dot{\boldsymbol{e}}_{q1}) \\ \vdots \\ b_N\ddot{\boldsymbol{q}}_d - k_N^1\sigma_1(\boldsymbol{e}_{qN}) - k_N^2\sigma_2(\dot{\boldsymbol{e}}_{qN}) \end{bmatrix}. \tag{5.55}$$

Use $\overline{\sigma}_1$ and $\overline{\sigma}_2$ to denote the upper bounds of σ_1 and σ_2, respectively. From (5.55) we readily see that:

- when $(\mathcal{L}_N + \mathcal{B}_N)$ is non-singular, there is a unique solution for $\boldsymbol{u}_i^0$ satisfying (5.53), determined explicitly by

$$\begin{bmatrix} \boldsymbol{u}_1^0 \\ \vdots \\ \boldsymbol{u}_N^0 \end{bmatrix} = (\mathcal{L}_N + \mathcal{B}_N)^{-1}\begin{bmatrix} b_1\ddot{\boldsymbol{q}}_d - k_1^1\sigma_1(\boldsymbol{e}_{q1}) - k_1^2\sigma_2(\dot{\boldsymbol{e}}_{q1}) \\ \vdots \\ b_N\ddot{\boldsymbol{q}}_d - k_N^1\sigma_1(\boldsymbol{e}_{qN}) - k_N^2\sigma_2(\dot{\boldsymbol{e}}_{qN}) \end{bmatrix}, \tag{5.56}$$

- $\boldsymbol{u}_i^0$, $i \in \mathcal{I}$, satisfying (5.53) are prior bounded if $\ddot{\boldsymbol{q}}_d$ is, and the bounds depend on $\overline{q}_{dd}$, $\overline{\sigma}_1$, $\overline{\sigma}_2$, k_i^1 and k_i^2, as well as information-exchange matrices $\mathcal{L}_N$ and $\mathcal{B}_N$.

Note that the linear nominal control developed in Section 5.2.1 is capable of (globally) asymptotically stabilizing error $\boldsymbol{e}_{qi}$ under the condition $\boldsymbol{d}_i = 0$. A natural question then arises: Does the prior bounded nominal control given by (5.53) have such an ability? The answer is that it does. To show this point, we first derive the resulting error equation. Applying the nominal control (5.53) to the nominal MVS (i.e., system (5.4) with $\boldsymbol{d}_i = 0$), yields

$$\ddot{\boldsymbol{e}}_{qi} = -k_i^1\sigma_1(\boldsymbol{e}_{qi}) - k_i^2\sigma_2(\dot{\boldsymbol{e}}_{qi}), i \in \mathcal{I}, \tag{5.57}$$

which are non-linear second-order equations with prior bounded right-hand sides (different from the linear equations (5.25)).

Using the Lyapunov function approach and LaSalle's invariance principle to analyse the stability of (5.57), we readily derive Lemma 5.10, where the design parameters a_{ij}, k_i^1 and k_i^2 ensure that Condition (5.51) is met.

Lemma 5.10 Consider MVS (5.4) with (5.53), and suppose $\boldsymbol{d}_i = 0$ for each $i \in \mathcal{I}$. For each vehicle $i \in \mathcal{I}$,

- the error equation (5.57) has a globally uniformly asymptotically stable equilibrium point at the origin $(\boldsymbol{e}_{qi}, \dot{\boldsymbol{e}}_{qi}) = (0, 0)$;
- the error trajectories $\tilde{\boldsymbol{q}}_i$ and $\dot{\tilde{\boldsymbol{q}}}_i$ are bounded for all $t \geq 0$, and converge asymptotically to the origin; that is, $\lim_{t\to\infty} \tilde{\boldsymbol{q}}_i(t) = \lim_{t\to\infty} \dot{\tilde{\boldsymbol{q}}}(t) = 0$;

- the nominal control $\boldsymbol{u}_i^0$, $i \in \mathcal{I}$, satisfy

$$|\boldsymbol{u}_i^0(t)| \leq \|(\mathcal{L}_N + \mathcal{B}_N)^{-1}\|_\infty \max_{i \in \mathcal{I}}(b_i\overline{q}_{dd} + k_i^1\overline{\sigma}_1 + k_i^2\overline{\sigma}_2), \forall t \geq 0, \tag{5.58}$$

and thus if parameters b_i, k_i^1 and k_i^2 satisfy

$$\|(\mathcal{L}_N + \mathcal{B}_N)^{-1}\|_\infty \max_{i \in \mathcal{I}}(b_i\overline{q}_{dd} + k_i^1\overline{\sigma}_1 + k_i^2\overline{\sigma}_2) \leq \overline{u}, \tag{5.59}$$

then $|\boldsymbol{u}_i^0(t)| \leq \overline{u}$ for all $t \geq 0$.

Proof: The proof is given in Section A.3 of the appendix. □

The results of Lemma 5.10 indicate that the bounded control (5.53) can stabilize the error dynamics for the ideal case without disturbances. A further question is: if the UDEs (5.32) are also used, is the same level of robustness with respect to input disturbances achieved? The following section will give the answer to this question.

5.3.2 Robustness Analysis

Applying controller (5.19) with (5.53) and (5.32) to MVS (5.4) yields the following node-level closed-loop error dynamics:

$$\begin{cases} \dot{\tilde{\boldsymbol{d}}}_i = -\dfrac{1}{T_q}\tilde{\boldsymbol{d}}_i + \dot{\boldsymbol{d}}_i, \\ \ddot{\boldsymbol{e}}_{qi} = -k_i^1\sigma_1(\boldsymbol{e}_{qi}) - k_i^2\sigma_2(\dot{\boldsymbol{e}}_{qi}) + \tilde{\Delta}_i, i \in \mathcal{I}, \end{cases} \tag{5.60}$$

where the disturbance estimation errors $\tilde{\Delta}_i$ are defined by (5.40).

The node-level MVS (5.60) is a feedback connection of a first-order and a second-order equation, with $\dot{\boldsymbol{d}}_i$ as the exogenous disturbance input. As for the second-order equation with vanishing $\tilde{\Delta}_i$, we have the following result.

Lemma 5.11 Consider the second equation in (5.60). If $\tilde{\Delta}_i(t)$ is bounded for all t and $\lim_{t\to\infty} \tilde{\Delta}_i(t) = 0$, then trajectories $\boldsymbol{e}_{qi}(t)$ and $\dot{\boldsymbol{e}}_{qi}(t)$ are globally bounded for all t and $\lim_{t\to\infty} \boldsymbol{e}_{qi}(t) = \lim_{t\to\infty} \dot{\boldsymbol{e}}_{qi}(t) = 0$.

Proof: See Lemma 1 in [20] for a similar proof, where the special case $\sigma_1(\cdot) = \sigma_2(\cdot) = \tanh(\cdot)$ was discussed. It is important to note that the proof therein also indicates tha: under any initial condition $(\boldsymbol{e}_{qi}(0), \dot{\boldsymbol{e}}_{qi}(0))$ and any bounded $\tilde{\delta}_i$, the resulting state trajectories $(\boldsymbol{e}_{qi}, \dot{\boldsymbol{e}}_{qi})$ do not have a finite escape time, that is, $(\boldsymbol{e}_{qi}, \dot{\boldsymbol{e}}_i)$ remains bounded in finite time. □

With Lemma 5.11, the following result is true concerning the performance of MVS (5.60).

Theorem 5.12 Consider the closed-loop MVS described by (5.60) under Condition 3 and Assumptions 5.7 and 5.8. If $\lim_{t\to\infty} \dot{\boldsymbol{d}}_i(t) = 0$ for each $i \in \mathcal{I}$, then we have the following results for each vehicle $i \in \mathcal{I}$:

- the disturbance estimation errors $\tilde{\Delta}_i(t)$ are bounded for all t, and $\lim_{t\to\infty} \tilde{\Delta}_i(t) = 0$;

- the trajectories $\boldsymbol{e}_{qi}(t)$, $\dot{\boldsymbol{e}}_{qi}(t)$ are globally bounded for all t and $\lim_{t\to\infty} \boldsymbol{e}_{qi}(t) = \lim_{t\to\infty} \dot{\boldsymbol{e}}_{qi}(t) = 0$;
- the trajectories $\tilde{\boldsymbol{q}}_i(t)$, $\dot{\tilde{\boldsymbol{q}}}_i(t)$ are globally bounded for all t, and moreover, $\lim_{t\to\infty} \tilde{\boldsymbol{q}}_i(t) = \lim_{t\to\infty} \dot{\tilde{\boldsymbol{q}}}_i(t) = 0$.

Proof: Note that under the theorem conditions, matrix $\mathcal{L}_N + \mathcal{B}_N$ is invertible and the results of Lemma 5.11 hold. With these results in hand, it is easy to prove the statements of this theorem. □

As for the general situation where $\tilde{\Delta}_i$ does not vanish, the trajectories of (5.60) do not converge to the origin, and, in particular, if $\tilde{\Delta}_i$ is large enough to cover the effect of the bounded feedback term $k_i^1\sigma_1(\boldsymbol{e}_{qi})$ or $-k_i^2\sigma_2(\dot{\boldsymbol{e}}_{qi})$, the trajectories may even tend to infinity as time t increases. So, we tend to expect that MVS (5.60) is input-to-state stable (with or without restrictions on the initial condition and the input disturbance $\tilde{\Delta}_i$). The following lemma summarizes the property of the second equation in (5.60), where the concepts of small-signal L_∞ stability (see Definition 5.2 in the book by Khalil [7]) and local input-to-state stability (LISS; see Definition A.6 in the appendix or Definition A.2 in Khalil [7]) are used.

Lemma 5.13 The second equation in (5.60) with $\tilde{\Delta}_i$ as disturbance inputs and $(\boldsymbol{e}_{qi}, \dot{\boldsymbol{e}}_{qi})^T$ as outputs, is small-signal L_∞ stable and LISS.

Proof: The proof is provided in Section A.3 of the appendix. □

To better understand the results of Lemma 5.13, we give the following explanations:

- The notion of small-signal L_∞ stable means that for every sufficiently small $\tilde{\Delta}_i$, the output $(\boldsymbol{e}_{qi}, \dot{\boldsymbol{e}}_{qi})^T$ is bounded. Small-signal L_∞ stability is weaker than L_∞ stability (which is a familiar notion of bounded-input bounded-output stability), in the sense that the input is required to be sufficiently small to ensure this boundedness of the system trajectories.
- Since the system outputs are exactly the system state, small-signal L_∞ stability of the system further implies that the system state is also bounded for any sufficiently small input $\tilde{\Delta}_i$. Thus the system is also LISS.
- The second-order equation is not GISS or L_∞ stable because, to ensure the boundedness of state $(\boldsymbol{e}_{qi}, \dot{\boldsymbol{e}}_{qi})^T$, the magnitude of $\tilde{\Delta}_i(t)$ can not be arbitrarily large. In fact, if the magnitude of $\tilde{\Delta}_i(t)$ is larger than $k_i^1\overline{\sigma}_r$ (or $k_i^2\overline{\sigma}_v$), the effect of the negative feedback term $-k_i^1\sigma_1(\boldsymbol{e}_{qi})$ (or $-k_i^2\sigma_v(\dot{\boldsymbol{e}}_{qi})$) will be completely covered by that of $\tilde{\Delta}_i(t)$, and the resulting $\boldsymbol{e}_{qi}(t)$ and $\dot{\boldsymbol{e}}_{qi}(t)$ may tend to infinity as t tends to infinity.

Remark 5.14 While the linear system (5.39) with $\dot{d}_i$ as the inputs is GISS and L_∞ stable, the non-linear system (5.60) with $\dot{d}_i$ as the inputs is only LISS and small-signal L_∞ stable. This implies that actuator saturation can be avoided by bounded control but that bounded control will result in degraded closed-loop stability.

5.3.3 The Effect of UDE on Stability

We shall further show the effect of UDE on the boundedness of system trajectories by a comparison study. Since the performance of a system with UDEs has been demonstrated, this section is mainly concerned with the performance of a system without UDEs. Applying nominal control to the disturbed MVS results in the following closed-loop error dynamics:

$$\ddot{\boldsymbol{e}}_{qi} = -k_i^1\sigma_1(\boldsymbol{e}_{qi}) - k_i^2\sigma_2(\dot{\boldsymbol{e}}_{qi}) + \Delta_i, i \in \mathcal{I}, \tag{5.61}$$

with Δ_i given by (5.18).

Equations (5.61) are also small-signal L_∞ stable with Δ_i as inputs, noting that the second-order equations in (5.60) are small-signal L_∞ stable, and that the only difference between them and (5.61) is the inputs. Despite this slight difference, the condition on exogenous disturbances $\boldsymbol{d}_i$ ($i \in \mathcal{I}$) ensuring that (5.61) are small-signal L_∞ stable is much stronger. This point can be seen from the fact that any constant $\boldsymbol{d}_i$ ($i \in \mathcal{I}$) with large magnitudes will render the trajectories $(\boldsymbol{e}_{qi}, \dot{\boldsymbol{e}}_{qi})^T$ of (5.61) unbounded. This is clearly different from the case with UDEs since, as shown in Theorem 5.12, the effect of any constant $\boldsymbol{d}_i$ on the ultimate bound of $(\boldsymbol{e}_{qi}, \dot{\boldsymbol{e}}_{qi})^T$ can be completely asymptotically compensated using the outputs of UDEs.

For the case where $\dot{\boldsymbol{d}}_i(t)$ do not vanish as t increases, the bounds of Δ_i depend on the bounds of $\boldsymbol{d}_i$, which are not controlled because $\boldsymbol{d}_i$ are exogenous disturbance signals. In contrast, the disturbance estimation errors $\tilde{\Delta}_i$, which drive the second equation in (5.60), are closely related to the UDE parameter T_p and thus can be controlled. More specifically, the smaller T_p is, the smaller the ultimate bounds of both $\tilde{\boldsymbol{d}}$ and Δ_i are. Thus, as for the case with UDEs, by tuning T_p, it is easy to render $\tilde{\Delta}_i$ as small as possible, so as to satisfy the small-signal condition on input and to ensure the boundedness of $(\boldsymbol{e}_{qi}, \dot{\boldsymbol{e}}_{qi})^T$.

5.3.4 The Effect of UDE on the Bounds of Control

In the following, we briefly evaluate the effect of UDE on the bounds of $\boldsymbol{u}_i$. Using the result of Lemma 5.3, we readily obtain

$$|\hat{\boldsymbol{d}}_i(t)| = |\boldsymbol{d}_i(t) - \tilde{\boldsymbol{d}}_i(t)| \leq \bar{d} + \max\left(|\tilde{d}_{i0}|, T_q\bar{d}_d\right), i \in \mathcal{I}. \tag{5.62}$$

Combining (5.58) and (5.62), yields

$$\|\boldsymbol{u}_i\|_\infty \leq \|(\mathcal{L}_N + \mathcal{B}_N)^{-1}\|_\infty \max_{i\in\mathcal{I}} \left(b_i\bar{q}_{dd} + k_i^1\bar{\sigma}_1 + k_i^2\bar{\sigma}_2\right) + \bar{d} + \max\left(|\tilde{\boldsymbol{d}}_{i0}|, T_q\bar{d}_d\right). \tag{5.63}$$

To meet the constraint (5.51), parameters b_i, k_i^1, k_i^2 and T_q need to satisfy

$$\|(\mathcal{L}_N + \mathcal{B}_N)^{-1}\|_\infty \max_{i\in\mathcal{I}} \left(b_i\bar{q}_{dd} + k_i^1\bar{\sigma}_1 + k_i^2\bar{\sigma}_2\right) + \bar{d} + \max\left(|\tilde{\boldsymbol{d}}_{i0}|, T_q\bar{d}_d\right) \leq \bar{u}. \tag{5.64}$$

If the initial disturbance estimation error $\tilde{\boldsymbol{d}}_{i0}$ is small enough that $|\tilde{\boldsymbol{d}}_{i0}| \leq T_q\bar{d}_d$ for each i, then the inequality (5.64) is reduced to

$$\|(\mathcal{L}_N + \mathcal{B}_N)^{-1}\|_\infty \max_{i\in\mathcal{I}} \left(b_i\bar{q}_{dd} + k_i^1\bar{\sigma}_1 + k_i^2\bar{\sigma}_2\right) + \bar{d} + T_q\bar{d}_d \leq \bar{u}. \tag{5.65}$$

If the reference acceleration $\ddot{\boldsymbol{q}}_d$ is identically zero or vanishing (i.e., $\lim_{t\to\infty} \ddot{\boldsymbol{q}}_d(t) = 0$), we can drop the term $b_i\ddot{\boldsymbol{q}}_d$ in $\boldsymbol{u}_i^0$, and then the control bound (5.65) is further reduced to

$$\|(\mathcal{L}_N + \mathcal{B}_N)^{-1}\|_\infty \max_{i\in\mathcal{I}} \left(k_i^1\overline{\sigma}_1 + k_i^2\overline{\sigma}_2\right) + \overline{d} + T_q\overline{d}_d \leq \overline{u}. \tag{5.66}$$

It is seen from inequality (5.64) that the bound of $\boldsymbol{u}_i$ depends on $\overline{q}_{dd}$, $\overline{d}$, and $\overline{d}_d$ as well as parameters b_i, k_i^1, k_i^2, and T_q. Moreover, the values of $\overline{q}_{dd}$, $\overline{d}$, and $\overline{d}_d$ are associated with the properties of exogenous disturbances. Consider the following three conditions on the exogenous disturbances:

- $\overline{q}_{dd}$ is too large, such that $\|(\mathcal{L}_N + \mathcal{B}_N)^{-1}\|_\infty\overline{q}_{dd}\max_{i\in\mathcal{I}} b_i \geq \overline{\mu}$,
- $\overline{d}$ is too large, such that $\overline{d} \geq \overline{\mu}$,
- $\overline{d}_d$ is too large, such that $T_q\overline{d}_d \geq \overline{\mu}$,

From inequality (5.65) we readily conclude that if the exogenous disturbances satisfy one of the above three conditions, the input constraint (5.51) will be violated. This in turn indicates that the control design under the constraint (5.51) is solvable if the disturbances do not satisfy any one of these conditions.

5.4 Continuous Control via Output Feedback

This section will introduce how to apply the UDE-based approach to the problem of output-feedback synchronized tracking for robustness improvement. We still use the two-component control structure given by (5.19). However, the design of each component ($\boldsymbol{u}_i^0$ or $\hat{\boldsymbol{d}}_i$) in Sections 5.2 and 5.3 needs velocity signals $\dot{\boldsymbol{e}}_{qi}$, $\dot{\boldsymbol{q}}$ or $\dot{\tilde{\boldsymbol{q}}}$, $i \in \mathcal{I}$. Therefore, under the output-feedback condition, both of the components must be redesigned to avoid using these velocity signals.

5.4.1 Design of $\boldsymbol{u}_i^0$ and $\hat{\boldsymbol{d}}_i$

We here apply the control law (4.27) presented in Section 4.5 to design nominal control $\boldsymbol{u}_i^0$. Specifically, we consider the $\boldsymbol{u}_i^0$ satisfying

$$\begin{cases} \dot{\zeta}_i = -k_\zeta\zeta_i - \boldsymbol{e}_{qi}, \\ \boldsymbol{u}_i^0 = k_i\left(b_i\ddot{r}_d + \sum\limits_{j\in N_i} a_{ij}\boldsymbol{u}_j^0 - (k_e+1)\boldsymbol{e}_{qi} - k_\zeta\zeta_i\right), i \in \mathcal{I}, \end{cases} \tag{5.67}$$

where $\boldsymbol{e}_{qi}$, $i \in \mathcal{I}$, are the linear LNSEs, k_i are as defined in (3.63), and k_ζ and k_e are the positive control gains, as in (4.27). Note that the controller (5.67) is distributed in the sense that only local information interaction is required.

In contrast to the UDE design in the state-feedback case, the UDE design under output feedback should not use velocities $\dot{\boldsymbol{q}}(t)$ and $\dot{\boldsymbol{q}}(0)$. For this reason, we need to redesign the filter $G_f(s)$ for the relationship equation

$$\hat{\boldsymbol{d}}_i(s) = \frac{G_f(s)}{1 - G_f(s)}[\ddot{\boldsymbol{q}}_i(s) - \boldsymbol{u}_i^0(s)]. \tag{5.68}$$

It is clear from (5.68) that if $\frac{G_f(s)}{1-G_f(s)}$ includes a double integral operation, $\frac{1}{s^2}$, and has a relative order not smaller than 2, then the computation of $\hat{\boldsymbol{d}}_i$ by (5.68) does not require $\dot{\boldsymbol{q}}_i$ since the second-order derivative action on $\boldsymbol{q}_i$ (to obtain $\ddot{\boldsymbol{q}}_i$) can be exactly cancelled by the double integral action included in $\frac{G_f(s)}{1-G_f(s)}$. Moreover, it is easy to verify that $G_f(s)$, which results in such a $\frac{G_f(s)}{1-G_f(s)}$, is required to be at least second-order.

As shown in our previous work [8], a simple second-order filter satisfying the above requirements is given by:

$$G_f(s) = \frac{1}{a_1 s^2 + a_0 s + 1}, \tag{5.69}$$

which results in

$$\frac{G_f(s)}{1 - G_f(s)} = \frac{s}{s^2(a_1 s + a_0)}, \tag{5.70}$$

where a_1 and a_0 may be any positive scalars; for practical implementation, they can be chosen to satisfy an additional requirement on filter bandwidth.

Substituting (5.70) into (5.68) gives

$$\begin{aligned}\hat{\boldsymbol{d}}_i(s) &= \frac{s}{s^2(a_1 s + a_0)}[\ddot{\boldsymbol{q}}_i(s) - \boldsymbol{u}_i^0(s)],\\ &= G_{fh}(s)\left[\frac{1}{s^2}\ddot{\boldsymbol{q}}_i(s)\right] - G_{fl}(s)\left[\frac{1}{s}\boldsymbol{u}_i^0(s)\right],\end{aligned} \tag{5.71}$$

where $G_{fh}(s)$ and $G_{fl}(s)$ are respectively a high-pass filter and a low-pass filter, defined by

$$G_{fh}(s) = \frac{s}{a_1 s + a_0},\; G_{fl}(s) = \frac{1}{a_1 s + a_0}. \tag{5.72}$$

According to (5.71), $\hat{\boldsymbol{d}}_i(t)$ can be computed in the time domain by

$$\hat{\boldsymbol{d}}_i(t) = G_{fh}(s)[\boldsymbol{q}_i(t) - \boldsymbol{q}_i(0)] - G_{fl}(s)\int_0^t \boldsymbol{u}_i^0(\tau)d\tau \tag{5.73}$$

From (5.73), $\hat{\boldsymbol{d}}_i(0) = 0$, and thus $\boldsymbol{u}_i(0) = \boldsymbol{u}_i^0(0)$. This implies that the use of UDE (5.73) will not introduce a large initial peak in $\boldsymbol{u}_i$.

5.4.2 Stability Analysis

Owing to the filter (5.69), the actual disturbances and their estimates satisfy

$$\hat{\boldsymbol{d}}_i(s) = \frac{1}{a_1 s^2 + a_0 s + 1}\boldsymbol{d}_i(s), i \in \mathcal{I}, \tag{5.74}$$

and the estimation errors, $\tilde{\boldsymbol{d}}_i$ $(i \in \mathcal{I})$, evolve according to

$$a_1\ddot{\tilde{\boldsymbol{d}}}_i + a_0\dot{\tilde{\boldsymbol{d}}}_i + \tilde{\boldsymbol{d}}_i = a_1\ddot{\boldsymbol{d}}_i + a_0\dot{\boldsymbol{d}}_i, i \in \mathcal{I}. \tag{5.75}$$

Regarding $\dot{\boldsymbol{d}}_i$ (instead of $\boldsymbol{d}_i$) as the inputs of equations (5.75), and $\tilde{\boldsymbol{d}}_i$ as the outputs, the transfer function describing equations (5.75) is

$$\overline{G}(s) = \frac{a_1 s + a_0}{a_1 s^2 + a_0 s + 1}. \tag{5.76}$$

We use $k_{\overline{G}}$ to denote the L_∞ norm of $\overline{G}(s)$; that is $k_{\overline{G}} = \|\overline{G}(s)\|_\infty$ (see, for instance, Chapter 4 of the book by Zhou *et al.* [21] for the definition of the L_∞ norm of linear operators). Then, the estimation performance of UDE (5.73) is summarized in the following lemma.

Lemma 5.15 Under Assumption 5.7, the trajectories of (5.75) satisfy the following:

- $\tilde{\boldsymbol{d}}_i$, $i \in \mathcal{I}$, are bounded by $k_{\overline{G}}\overline{d}_d$ under the zero initial condition;
- the gain $k_{\overline{G}}$ may be arbitrarily small by choosing $\frac{1}{a_1} \gg \frac{a_0}{a_1} \gg 1$;
- if $\dot{\boldsymbol{d}}_i(t) \to 0$ as $t \to \infty$, then $\left(a_1\ddot{\boldsymbol{d}}_i(t) + a_0\dot{\boldsymbol{d}}_i(t)\right) \to 0$ and $\tilde{\boldsymbol{d}}_i(t) \to 0$ as $t \to \infty$.

Proof: The first statement of this lemma is clear from the definition of $k_{\overline{G}}$. The second statement can be proven using the approach presented on Page 613 of [7]. The last statement is straightforward since $\ddot{\boldsymbol{d}}_i$ is bounded under Assumption 5.7, and, by the well known Barbalat lemma, the condition $\dot{\boldsymbol{d}}_i \to 0$ is sufficient to ensure $\ddot{\boldsymbol{d}}_i \to 0$. □

Consider the controller (5.19) for MVS (5.4) with components given by (5.67) and (5.73), respectively. The resulting closed-loop system is

$$\begin{cases} \ddot{\tilde{\boldsymbol{d}}}_i = -\dfrac{a_0}{a_1}\dot{\tilde{\boldsymbol{d}}}_i - \dfrac{1}{a_1}\tilde{\boldsymbol{d}}_i + \ddot{\boldsymbol{d}}_i + \dfrac{a_0}{a_1}\dot{\boldsymbol{d}}_i, \\ \dot{\zeta}_i = -k_\zeta\zeta_i - \boldsymbol{e}_{qi}, \\ \ddot{\boldsymbol{e}}_{qi} = -(k_e+1)\boldsymbol{e}_{qi} - k_\zeta\zeta_i + \tilde{\Delta}_i, i \in \mathcal{I}, \end{cases} \tag{5.77}$$

where the first equation describes the dynamics of disturbance estimation error associated with vehicle i, and the last two equations together describe the dynamics of LNSE associated with vehicle i.

In the light of Lemma 5.15, we readily derive the following theorem.

Theorem 5.16 Consider MVS (5.77) under Condition 3 and Assumption 5.7. Then,

- if $\dot{\boldsymbol{d}}_i(t) \to 0$ for each $i \in \mathcal{I}$, the error trajectories $\boldsymbol{e}_{qi}(t)$, $\dot{\boldsymbol{e}}_{qi}(t)$, $\tilde{\boldsymbol{q}}_i(t)$, $\dot{\tilde{\boldsymbol{q}}}_i(t)$ are globally bounded for all t, and, moreover, $\lim_{t\to\infty} \boldsymbol{e}_{qi}(t) = 0$, $\lim_{t\to\infty} \dot{\boldsymbol{e}}_{qi}(t) = 0$, $\lim_{t\to\infty} \tilde{\boldsymbol{q}}_i(t) = 0$, $\lim_{t\to\infty} \dot{\tilde{\boldsymbol{q}}}_i(t) = 0$; that is, zero-error synchronized tracking is asymptotically achieved.
- for general disturbances satisfying Assumption 5.7 (i.e., with bounded derivatives), the error trajectories $\boldsymbol{e}_{qi}$, $\dot{\boldsymbol{e}}_{qi}$, $\tilde{\boldsymbol{q}}_i$, and $\dot{\tilde{\boldsymbol{q}}}_i$ are globally bounded, and may be arbitrarily small provided the UDE parameters satisfy $\frac{1}{a_1} \gg \frac{a_0}{a_1} \gg 1$.

Proof: This theorem can be proven with the results of Lemma 5.15, so for simplicity, we omit it here. □

5.5 Discontinuous Control via Output Feedback

This section will introduce a sliding-mode control approach that improves robustness in synchronized tracking. Disturbances are compensated using a discontinuous solution, which is very different from the continuous UDE-based solution developed in the foregoing section. We assume that velocity signals are not accessible for control design.

5.5.1 Controller Design

Recall the equations

$$\ddot{\boldsymbol{e}}_{qi} = \frac{1}{k_i}\boldsymbol{u}_i - \sum_{j\in N_i} a_{ij}\boldsymbol{u}_j - b_i\ddot{\boldsymbol{q}}_d + \Delta_i, i \in \mathcal{I}, \tag{5.78}$$

which describe the dynamics of LNSEs (also given in (5.17)). We here suppose that for each i, $\boldsymbol{e}_{qi} \in \mathbb{R}^n$.

For (5.78), consider a decentralized control law of the form:

$$\boldsymbol{u}_i = k_i\left(b_i\ddot{\boldsymbol{q}}_d + \sum_{j\in N_i} a_{ij}\boldsymbol{u}_j - \boldsymbol{e}_{qi} + \boldsymbol{\Gamma}_i - \hat{\Delta}_i\right), i \in \mathcal{I}, \tag{5.79}$$

where k_i are defined as in (3.63). This results in the following closed-loop second-order equations:

$$\ddot{\boldsymbol{e}}_{qi} = -\boldsymbol{e}_{qi} + \boldsymbol{\Gamma}_i - \hat{\Delta}_i + \Delta_i, i \in \mathcal{I}. \tag{5.80}$$

To better understand the control law (5.79), we give the following explanations:

- the control component $b_i\ddot{\boldsymbol{q}}_d + \sum_{j\in N_i} a_{ij}\boldsymbol{u}_j$ is introduced to cancel the same term as included in controlled dynamics (5.78);
- the control component $-\boldsymbol{e}_{qi}$ serves as a negative feedback of LNSE;
- the control component $\boldsymbol{\Gamma}_i$ (to be determined later) is introduced to inject damping;
- the signal $\hat{\boldsymbol{\Delta}}_i$ denotes an estimate of Δ_i to be determined later, and is used to achieve disturbance compensation.

The control task is to determine the $\boldsymbol{\Gamma}_i$ and $\hat{\Delta}_i$ for (5.80) that makes the trajectories $(\boldsymbol{e}_{qi}, \dot{\boldsymbol{e}}_{qi})^T$ converge asymptotically to the origin. The difficulty stems from the control condition that velocity signals $\dot{\boldsymbol{e}}_{qi}$ cannot be used to determine either $\boldsymbol{\Gamma}_i$ or $\hat{\Delta}_i$.

We define the auxiliary variables

$$\xi_i \triangleq \boldsymbol{e}_{qi} + k_{Ii}\int \boldsymbol{e}_{qi}\mathrm{d}t, i \in \mathcal{I}, \tag{5.81}$$

where the constant gain $k_{Ii} > 0$ for each i. Then, using ξ_i $(i \in \mathcal{I})$ as inputs, we construct the following auxiliary systems and auxiliary variables:

$$\begin{cases} \dot{\boldsymbol{z}}_{1i} = -\boldsymbol{z}_{1i} + h_{2i}\boldsymbol{z}_{2i} + w_{1i}\xi_i \\ \dot{\boldsymbol{z}}_{2i} = h_{3i}\boldsymbol{z}_{1i} - h_{4i}\boldsymbol{z}_{2i} - \xi_i \end{cases} \tag{5.82}$$

$$\boldsymbol{r}_{fi} \triangleq \dot{\boldsymbol{z}}_{1i} + k_{0i}\boldsymbol{z}_{1i} \tag{5.83}$$

$$\boldsymbol{r}_i \triangleq \dot{\xi}_i + k_{0i}\xi_i + \boldsymbol{r}_{fi}, \tag{5.84}$$

where

- variables $\boldsymbol{z}_{1i}$, $\boldsymbol{z}_{2i}$ are the states of the second-order auxiliary system with zero initial condition; that is, $\boldsymbol{z}_{1i}(0) = \boldsymbol{z}_{2i}(0) = 0$;
- parameters h_{2i}, h_{3i}, h_{4i}, and w_{1i} will be determined later, but satisfy the condition

$$h_{4i} > 0, h_{2i}h_{3i} < h_{4i}, i \in \mathcal{I} \tag{5.85}$$

  which is sufficient to render the auxiliary systems (5.82) input-to-state stable (ISS) with respect to input ξ_i;

- both variables $\boldsymbol{r}_{fi}$ and $\boldsymbol{r}_i$ are filtered signals, and are defined in (5.83) and (5.84); by the definition, $\boldsymbol{r}_{fi}$ can be used for control design whereas $\boldsymbol{r}_i$ cannot, because, according to (5.81) and (5.84), velocity error signals $\dot{\boldsymbol{e}}_{qi}$ are needed to compute $\dot{\boldsymbol{\xi}}_i$ and $\boldsymbol{r}_i$.
- the parameter k_{0i} involved in (5.84) is a positive constant for each i, ensuring that the convergence of $\boldsymbol{\xi}_i$ can be derived from the convergence of $\boldsymbol{r}_i - \boldsymbol{r}_{fi}$.

Substituting (5.83) into (5.84), gives the equation

$$\boldsymbol{r}_i = (\dot{\boldsymbol{\xi}}_i + \dot{\boldsymbol{z}}_{1i}) + k_{0i}(\boldsymbol{\xi}_i + \boldsymbol{z}_{1i}), \tag{5.86}$$

which is crucial to derive the result of Lemma 5.17 below.

In addition, we can readily derive from (5.82)–(5.84) that:

$$\begin{aligned} \dot{\boldsymbol{r}}_{fi} &= -\boldsymbol{r}_{fi} + w_{1i}\boldsymbol{r}_i - [(k_{0i} - 1)w_{1i} - h_{2i}h_{3i}]\boldsymbol{z}_{1i} \\ &\quad + (k_{0i} - w_{1i} - h_{4i})h_{2i}\boldsymbol{z}_{2i} - (h_{2i} + w_{1i}^2)\boldsymbol{\xi}_i, \end{aligned} \tag{5.87}$$

$$\begin{aligned} \dot{\boldsymbol{r}}_i &= (k_{Ii} + k_{0i} + w_{1i})\boldsymbol{r}_i - (k_{Ii} + k_{0i} + 1)\boldsymbol{r}_{fi} \\ &\quad - [(k_{0i} - 1)w_{1i} - h_{2i}h_{3i}]\boldsymbol{z}_{1i} + [(k_{0i} - w_{1i})h_{2i} - h_{2i}h_{4i}]\boldsymbol{z}_{2i} \\ &\quad - (h_{2i} + w_{1i}^2 + k_{Ii}k_{0i} + k_{0i}^2)\boldsymbol{\xi}_i - k_{Ii}^2\boldsymbol{e}_{qi} + \ddot{\boldsymbol{e}}_{qi}. \end{aligned} \tag{5.88}$$

We choose parameters k_{0i}, w_{1i} and h_{4i}, such that

$$k_{0i} - w_{1i} - h_{4i} = 0, \tag{5.89}$$

and k_{Ii}, k_{0i}, and w_{1i} such that

$$k_{2i} = -(k_{Ii} + k_{0i} + w_{1i}) \tag{5.90}$$

are positive constants.

Under conditions (5.89) and (5.90), the terms associated with $\boldsymbol{z}_{2i}$ in (5.87)–(5.88) are naturally zero, and (5.87)–(5.88) are further reduced to:

$$\dot{\boldsymbol{r}}_{fi} = -\boldsymbol{r}_{fi} + w_{1i}\boldsymbol{r}_i - [(k_{0i} - 1)w_{1i} - h_{2i}h_{3i}]\boldsymbol{z}_{1i} - (h_{2i} + w_{1i}^2)\boldsymbol{\xi}_i, \tag{5.91}$$

$$\begin{aligned} \dot{\boldsymbol{r}}_i &= -k_{2i}\boldsymbol{r}_i - (k_{Ii} + k_{0i} + 1)\boldsymbol{r}_{fi} - [(k_{0i} - 1)w_{1i} - h_{2i}h_{3i}]\boldsymbol{z}_{1i} \\ &\quad - (h_{2i} + w_{1i}^2 + k_{Ii}k_{0i} + k_{0i}^2)\boldsymbol{\xi}_i - k_{Ii}^2\boldsymbol{e}_{qi} + \ddot{\boldsymbol{e}}_{qi}. \end{aligned} \tag{5.92}$$

Substituting (5.80) into (5.92), yields

$$\begin{aligned} \dot{\boldsymbol{r}}_i &= -k_{2i}\boldsymbol{r}_i - (k_{Ii} + k_{0i} + 1)\boldsymbol{r}_{fi} - [(k_{0i} - 1)w_{1i} - h_{2i}h_{3i}]\boldsymbol{z}_{1i} \\ &\quad - (h_{2i} + w_{1i}^2 + k_{Ii}k_{0i} + k_{0i}^2)\boldsymbol{\xi}_i - (k_{Ii}^2 + 1)\boldsymbol{e}_{qi} + \boldsymbol{\Gamma}_i - \hat{\Delta}_i + \Delta_i. \end{aligned} \tag{5.93}$$

Note that both the control components $\boldsymbol{\Gamma}_i$ and $-\hat{\Delta}_i$ appear in (5.93). Since $\boldsymbol{r}_i$ and $\Delta_i - \hat{\Delta}_i$ are expected to be asymptotically convergent, a simple and natural idea to determine $\boldsymbol{\Gamma}_i$ is to choose

$$\begin{aligned} \boldsymbol{\Gamma}_i &= (k_{Ii}^2 + 1)\boldsymbol{e}_{qi} + (k_{Ii} + k_{0i} + 1)\boldsymbol{r}_{fi} + [(k_{0i} - 1)w_{1i} - h_{2i}h_{3i}]\boldsymbol{z}_{1i} \\ &\quad + (h_{2i} + w_{1i}^2 + k_{Ii}k_{0i} + k_{0i}^2)\boldsymbol{\xi}_i, \end{aligned} \tag{5.94}$$

such that all the other terms included in the right-hand side of (5.93) are exactly cancelled. In fact, with (5.94), (5.93) is reduced to

$$\dot{\boldsymbol{r}}_i = -k_{2i}\boldsymbol{r}_i - \hat{\Delta}_i + \Delta_i, \tag{5.95}$$

which is GISS with $k_{2i} > 0$ and $\Delta_i - \hat{\Delta}_i$ as the input. For $\boldsymbol{r}_i$ satisfying (5.95), we readily conclude that if $\Delta_i - \hat{\Delta}_i$ converges to zero, $\boldsymbol{r}_i$ will be too.

Using the (accessible) signals $\boldsymbol{\xi}_i$, $\boldsymbol{z}_{1i}$, $\boldsymbol{r}_{fi}$, $\boldsymbol{\xi}_i$, and $\boldsymbol{e}_{qi}$, we construct the following control law for $\hat{\Delta}_i$:

$$\hat{\Delta}_i = k_{1i}\,\text{sign}\,(\boldsymbol{\xi}_i + \boldsymbol{z}_{1i}) + w_{1i}\boldsymbol{r}_{fi} + \boldsymbol{\xi}_i + \boldsymbol{e}_{qi}, \tag{5.96}$$

where k_{1i} are positive constants for all $i \in \mathcal{I}$, and sign$(\cdot)$ is the standard signum function. With (5.96), (5.95) becomes

$$\dot{\boldsymbol{r}}_i = -k_{2i}\boldsymbol{r}_i - k_{1i}\,\text{sign}\,(\boldsymbol{\xi}_i + \boldsymbol{z}_{1i}) - w_{1i}\boldsymbol{r}_{fi} - \boldsymbol{\xi}_i - \boldsymbol{e}_{qi} + \Delta_i, \tag{5.97}$$

5.5.2 Stability Analysis

Before presenting the stability result, we introduce the following lemma, which will be invoked later on.

Lemma 5.17 Define auxiliary functions $L_i(t) \in \mathbb{R}$, $i \in \mathcal{I}$, as

$$L_i(t) \triangleq \boldsymbol{r}_i^T(\Delta_i - k_{1i}\,\text{sgn}\,(\boldsymbol{\xi}_i + \boldsymbol{z}_{1i})). \tag{5.98}$$

If the gains k_{1i} $(i \in \mathcal{I})$ introduced in (5.96) satisfy the condition

$$k_{1i} > \|\Delta_i(t)\|_\infty + \frac{1}{k_{0i}}\|\dot{\Delta}_i(t)\|_\infty, \tag{5.99}$$

then the integration of $L_i(t)$ along time t is upper bounded by a positive constant

$$\zeta_{bi} \triangleq \sum_{k=1}^{n} k_{1i}\xi_{ik}(0) - \boldsymbol{\xi}_i^T(0)\Delta_i(0), \tag{5.100}$$

that is,

$$\int_0^t L_i(\tau)d\tau \le \zeta_{bi}, \tag{5.101}$$

where $\|\cdot\|_\infty$ denotes the infinity norm, and ξ_{ik} denotes the kth entry of vector $\boldsymbol{\xi}_i$.

Proof: This lemma can be proven by substituting equation (5.86) into the expression of L_i and then following the arguments used in the proof of Lemma 1 in the paper by Xian *et al.* [22]. □

We define a function $P_i(t) \in \mathbb{R}$ for each i as

$$P_i(t) \triangleq \zeta_{bi} - \int_0^t L_i(\tau)d\tau, \tag{5.102}$$

which is positive under the parameter condition (5.99), and will be used to construct a Lyapunov candidate in the following theorem.

Theorem 5.18 Consider controller (5.79) with (5.94) and (5.96) for MVS (5.4). All error signals are bounded and the objective of synchronized tracking defined by (5.9) is globally asymptotically achieved, provided the control parameters satisfy (5.89), (5.90), (5.99) and

$$h_{2i} + w_{1i}^2 = -1 \tag{5.103}$$

$$(k_{0i} - 1)w_{1i} - h_{2i}h_{3i} = 1 \tag{5.104}$$

where k_{Ii}, k_{0i} are chosen such that $k_{Ii} \geq \frac{k_{0i}+1}{2}$.

Proof: Define a domain $D \in \mathbb{R}^{5\times n+1}$ which contains the origin $\boldsymbol{y}_i(t) = 0$, and where $\boldsymbol{y}_i(t) \triangleq [\boldsymbol{y}_{i1}^T(t) \quad y_{i2}(t)]^T \in \mathbb{R}^{5\times 2+1}$, with $\boldsymbol{y}_{i1} \triangleq [\boldsymbol{e}_{qi}^T \quad \boldsymbol{\xi}_i^T \quad \boldsymbol{z}_{1i}^T \quad \boldsymbol{r}_{fi}^T \quad \boldsymbol{r}_i^T]^T \in \mathbb{R}^{5\times n}$ and $y_{i2} = \sqrt{P_i(t)} \in \mathbb{R}$.

Consider a continuously differentiable positive-definite function $V_i(t, \boldsymbol{y}_i) : \mathbb{R}_+ \times D \rightarrow \mathbb{R}_+$ defined by

$$V_i(t, \boldsymbol{y}_i) = \frac{1}{2}\boldsymbol{e}_{qi}^T\boldsymbol{e}_{qi} + \frac{1}{2}\boldsymbol{\xi}_i^T\boldsymbol{\xi}_i + \frac{1}{2}\boldsymbol{z}_{1i}^T\boldsymbol{z}_{1i} + \frac{1}{2}\boldsymbol{r}_{fi}^T\boldsymbol{r}_{fi} + \frac{1}{2}\boldsymbol{r}_i^T\boldsymbol{r}_i + P_i(t), \tag{5.105}$$

which is bounded by

$$\frac{1}{2}\|\boldsymbol{y}_i\|^2 \leq V_i \leq \|\boldsymbol{y}_i\|^2, \tag{5.106}$$

provided the condition (5.99) is satisfied.

The first-order derivative of (5.105) along the closed-loop trajectories is

$$\dot{V}_i = \boldsymbol{e}_{qi}^T\dot{\boldsymbol{e}}_{qi} + \boldsymbol{\xi}_i^T\dot{\boldsymbol{\xi}}_i + \boldsymbol{z}_{1i}^T\dot{\boldsymbol{z}}_{1i} + \boldsymbol{r}_{fi}^T\dot{\boldsymbol{r}}_{fi} + \boldsymbol{r}_i^T\dot{\boldsymbol{r}}_i + \dot{P}_i(t). \tag{5.107}$$

Substitute (5.102), (5.95), and (5.91) into the above equation, and use the relationships (5.83), (5.84), (5.103), (5.104), and $\dot{\boldsymbol{e}}_{qi} = \boldsymbol{\xi}_i - k_{Ii}\boldsymbol{e}_{qi}$. Finally, we derive

$$\dot{V}_i = -k_{Ii}\boldsymbol{e}_{qi}^T\boldsymbol{e}_{qi} - k_{0i}\boldsymbol{\xi}_i^T\boldsymbol{\xi}_i - k_{0i}\boldsymbol{z}_{1i}^T\boldsymbol{z}_{1i} - \boldsymbol{r}_{fi}^T\boldsymbol{r}_{fi} - k_{2i}\boldsymbol{r}_i^T\boldsymbol{r}_i - k_{0i}\boldsymbol{\xi}_i^T\boldsymbol{e}_{qi} - \boldsymbol{r}_{fi}^T\boldsymbol{e}_{qi}. \tag{5.108}$$

Since $-\boldsymbol{\xi}_i^T\boldsymbol{e}_{qi}$ and $-\boldsymbol{r}_{fi}^T\boldsymbol{e}_{qi}$ are upper bounded by

$$-\boldsymbol{\xi}_i^T\boldsymbol{e}_{qi} \leq \frac{1}{2}\boldsymbol{\xi}_i^T\boldsymbol{\xi}_i + \frac{1}{2}\boldsymbol{e}_{qi}^T\boldsymbol{e}_{qi}, \tag{5.109}$$

$$-\boldsymbol{r}_{fi}^T\boldsymbol{e}_{qi} \leq \frac{1}{2}\boldsymbol{r}_{fi}^T\boldsymbol{r}_{fi} + \frac{1}{2}\boldsymbol{e}_{qi}^T\boldsymbol{e}_{qi}, \tag{5.110}$$

we have

$$\dot{V}_i \leq -\left(k_{Ii} - \frac{k_{0i}+1}{2}\right)\boldsymbol{e}_{qi}^T\boldsymbol{e}_{qi} - \frac{1}{2}k_{0i}\boldsymbol{\xi}_i^T\boldsymbol{\xi}_i - k_{0i}\boldsymbol{z}_{1i}^T\boldsymbol{z}_{1i} - \frac{1}{2}\boldsymbol{r}_{fi}^T\boldsymbol{r}_{fi} - k_{2i}\boldsymbol{r}_i^T\boldsymbol{r}_i. \tag{5.111}$$

We define

$$\lambda \triangleq \min\left\{k_{Ii} - \frac{k_{0i}+1}{2}, \frac{1}{2}k_{0i}, \frac{1}{2}, k_{2i}\right\} \in \mathbb{R}. \tag{5.112}$$

Under the theorem conditions, we have $\lambda > 0$, and thus (5.111) indicates that:

$$\dot{V}_i \leq -\lambda\|\boldsymbol{y}_{i1}\|^2 \leq 0. \tag{5.113}$$

From (5.106) and (5.113), the lower and upper bounds for (5.105) can be written as

$$W_1(\boldsymbol{y}_i) = \frac{1}{2}\|\boldsymbol{y}_i\|^2, \quad W_2(\boldsymbol{y}_i) = \|\boldsymbol{y}_i\|^2, \tag{5.114}$$

and an upper bound of $\dot{V}_i$ can be chosen as

$$W(y_i) = \lambda \|y_{i1}\|^2. \tag{5.115}$$

Moreover, these bounds are independent of the initial condition of the system and $W_1(y_i)$ is radially unbounded.

From (5.106) and (5.113), we have $V_i(t, y_i) \in \mathcal{L}_\infty$ and $e_{qi}, \xi_i, z_{1i}, r_{fi}, r_i \in \mathcal{L}_\infty$. Thus, $\dot{z}_{1i}$, $\dot{\xi}_i$, $\dot{r}_{fi} \in \mathcal{L}_\infty$, owing to the equation relationships (5.83), (5.84), and (5.91). Since $\dot{e}_{qi} = \xi_i - k_{li}e_{qi}$, $\dot{e}_{qi} \in \mathcal{L}_\infty$. Further, under Assumption 5.8, with (5.80), (5.92), (5.94) and (5.96), we conclude $\ddot{e}_{qi} \in \mathcal{L}_\infty$ and $\dot{r}_i \in \mathcal{L}_\infty$.

The above boundedness statements also indicate that $\dot{W}(y_i) \in \mathcal{L}_\infty$. Thus, $W(y_i)$ is uniformly continuous. Now we can invoke Theorem 8.4 from Khalil [7] to state that $\lambda\|y_{i1}\|^2 \to 0$ as $t \to \infty$, for all $y_i(0) \in \mathbb{R}^{5\times2+1}$. From the definition of y_{i1}, we conclude that $e_{qi}(t), \xi_i(t), z_{1i}(t), r_{fi}(t), r_i(t) \to 0$ as $t \to \infty$, for all $y_i(0)$. Finally, we can use (5.84) and the equation $\dot{e}_{qi} = \xi_i - k_{li}e_{qi}$ again to conclude $\dot{e}_{qi}(t) \to 0$ as $t \to \infty$, for all $y_i(0)$. Thus, both position synchronization and velocity synchronization are asymptotically achieved. This ends the proof of the theorem. □

Remark 5.19 According to (5.89), (5.90), (5.103), and (5.104), the parameters of auxiliary systems for each vehicle can be directly calculated from gains k_{li}, k_{0i} and k_{2i} by the following expressions:

$$w_{1i} = -(k_{li} + k_{0i} + k_{2i}), \tag{5.116}$$

$$h_{2i} = -1 - (k_{li} + k_{0i} + k_{2i})^2, \tag{5.117}$$

$$h_{3i} = \frac{1 + (k_{0i} - 1)(k_{li} + k_{0i} + k_{2i})}{1 + (k_{li} + k_{0i} + k_{2i}^2)}, \tag{5.118}$$

$$h_{4i} = k_{li} + 2k_{0i} + k_{2i}. \tag{5.119}$$

Therefore, once k_{li}, k_{0i} and k_{2i} are chosen, the other parameters w_{1i}, h_{2i}, h_{3i} and h_{4i} are determined accordingly.

Remark 5.20 From (5.99), it can be seen that parameters k_{1i} need to be sufficiently large to ensure closed-loop stability, and that the minimum value of k_{1i} depends on the bounds of $\Delta_i(t)$ and $\dot{\Delta}_i(t)$. Since $\Delta_i(t)$ and $\dot{\Delta}_i(t)$ are prior unknown, k_{1i} are tuned by trial and error. However, we cannot simply choose very large values for k_{1i} because this may cause actuator saturation or serious chattering. In practical applications, we choose relatively small positive numbers as the initial values of k_{1i}, and then enlarge them depending on the obtained results.

5.6 GSE-based Synchronization Control

Synchronization errors are used to evaluate the performance of a synchronization controller: how the trajectory of each vehicle converges with respect to the others. There are various ways to choose the synchronization error. For example, we may include the error information of all vehicles in the synchronization error of each vehicle. However, when the number of vehicles involved is large, choosing this kind of synchronization error for control design will lead to intensive on-line computational work. For this reason, we introduce here how to use Type II LSEs and coupled errors to design synchronization controllers. As shown in Section 3.3.2, the Type II LSE is a linear combination of partial standard synchronization errors. The coupled position error links position tracking error and the integration of Type II position LSE into one variable (see Equation (5.132) below). Similarly, the coupled velocity error vector links velocity tracking error and the integration of Type II velocity LSE into one variable (see Equation (5.133)). Using these error vectors, we can obtain synchronization tracking controllers that are applicable to MVSs with any number of vehicles.

This section will introduce the cross-coupling concept [23–26] to the synchronized tracking problem of MVS, and present a controller using GSE vector. The controller guarantees that relative position errors among vehicles converge to the origin at the same rate.

5.6.1 Coupled Errors

The control design presented in Sections 5.2–5.5 is at node-level; that is, the design yields $\boldsymbol{u}_i$ for each i. In contrast, the control design to be introduced here is at network-level and gives the expression of vector

$$\boldsymbol{u} = \left[\boldsymbol{u}_1^T \cdots \boldsymbol{u}_N^T\right]^T. \tag{5.120}$$

For simplicity, we first consider the situation without input disturbances. In this situation, system (5.4) can be written in vector form as:

$$\ddot{\boldsymbol{q}} = \boldsymbol{u}. \tag{5.121}$$

We denote

$$\tilde{\boldsymbol{q}}_{ij} = \boldsymbol{q}_i - \boldsymbol{q}_j, \tag{5.122}$$

$$\dot{\tilde{\boldsymbol{q}}}_{ij} = \dot{\boldsymbol{q}}_i - \dot{\boldsymbol{q}}_j, \tag{5.123}$$

where $i, j = 1, \ldots, N$, $i \neq j$. From (5.5), (5.8), (5.122) and (5.123),

$$\tilde{\boldsymbol{q}}_{ij} = \tilde{\boldsymbol{q}}_i - \tilde{\boldsymbol{q}}_j, \tag{5.124}$$

$$\dot{\tilde{\boldsymbol{q}}}_{ij} = \dot{\tilde{\boldsymbol{q}}}_i - \dot{\tilde{\boldsymbol{q}}}_j. \tag{5.125}$$

Using the position and velocity variables $\boldsymbol{q}_j$ and $\dot{\boldsymbol{q}}_j$, we define the following Type II LSEs:

$$\varepsilon_i(t) = \sum_{j\in N_i} a_{ij}\tilde{\boldsymbol{q}}_{ij} = \sum_{j\in N_i} a_{ij}(\tilde{\boldsymbol{q}}_i - \tilde{\boldsymbol{q}}_j), i \in \mathcal{I}, \tag{5.126}$$

$$\dot{\varepsilon}_i(t) = \sum_{j\in N_i} a_{ij}\dot{\tilde{\boldsymbol{q}}}_{ij} = \sum_{j\in N_i} a_{ij}(\dot{\tilde{\boldsymbol{q}}}_i - \dot{\tilde{\boldsymbol{q}}}_j), i \in \mathcal{I}. \tag{5.127}$$

where a_{ij} are the weights associated with the adjacency matrix $\mathcal{A}$. Then, let

$$\Xi(t) \triangleq [\varepsilon_1(t) \ \cdots \ \varepsilon_i(t) \ \cdots \ \varepsilon_N(t)]^T, \tag{5.128}$$

$$\dot{\Xi}(t) \triangleq [\dot{\varepsilon}_1(t) \ \cdots \ \dot{\varepsilon}_i(t) \ \cdots \ \dot{\varepsilon}_N(t)]^T. \tag{5.129}$$

Then, according to (5.126) and (5.127),

$$\Xi(t) = \mathcal{L}_N \tilde{\boldsymbol{q}}(t) \tag{5.130}$$

$$\dot{\Xi}(t) = \mathcal{L}_N \dot{\tilde{\boldsymbol{q}}}(t). \tag{5.131}$$

We construct the coupled position error vector $\boldsymbol{E}^*(t)$ as

$$\boldsymbol{E}^*(t) = \tilde{\boldsymbol{q}}(t) + \boldsymbol{B}\mathcal{L}_N^T \int_0^t \Xi(\tau)d\tau \tag{5.132}$$

where $\boldsymbol{E}^* \triangleq [E_1^* \quad E_2^* \quad \cdots \quad E_N^*]^T$, and $\boldsymbol{B} \triangleq \overline{b}\mathbf{I}_N$ is a scalar matrix (a diagonal matrix with all the main diagonal entries equal) with $\overline{b} > 0$.

From (5.132), the tracking error vector $\tilde{\boldsymbol{q}}(t)$ and the integration of Type II position LSE, $\Xi(t)$, are linked into one variable $\boldsymbol{E}^*(t)$. Correspondingly, the coupled velocity error vector satisfies

$$\dot{\boldsymbol{E}}^*(t) = \dot{\tilde{\boldsymbol{q}}}(t) + \boldsymbol{B}\mathcal{L}_N^T \Xi(t), \tag{5.133}$$

which indicates that if both $\dot{\boldsymbol{E}}^*(t)$ and $\Xi(t)$ converge asymptotically to zero, so will velocity error $\dot{\tilde{\boldsymbol{q}}}(t)$. This fact will be used in the proof of Theorem 5.21 below to give equation (5.160).

From (5.132), the coupled position errors with $\mathcal{L}_N$ given by (3.44) are

$$\begin{aligned}
E_1^*(t) &= \tilde{q}_1(t) + \overline{b}\int_0^t (\varepsilon_1(\tau) - \varepsilon_N(\tau))d\tau \\
E_2^*(t) &= \tilde{q}_2(t) + \overline{b}\int_0^t (\varepsilon_2(\tau) - \varepsilon_1(\tau))d\tau \\
&\vdots \\
E_N^*(t) &= q_N(t) + \overline{b}\int_0^t (\varepsilon_N(\tau) - \varepsilon_{N-1}(\tau))d\tau.
\end{aligned} \tag{5.134}$$

The coupled position errors with $\mathcal{L}_N$ given by (3.45) are

$$\begin{aligned}
E_1^*(t) &= \tilde{q}_1(t) + \overline{b}\int_0^t (2\varepsilon_1(\tau) - \varepsilon_2(\tau) - \varepsilon_N(\tau))d\tau \\
E_2^*(t) &= \tilde{q}_2(t) + \overline{b}\int_0^t (2\varepsilon_2(\tau) - \varepsilon_3(\tau) - \varepsilon_1(\tau))d\tau \\
&\vdots \\
E_N^*(t) &= \tilde{q}_N(t) + \overline{b}\int_0^t (2\varepsilon_N(\tau) - \varepsilon_{N-1}(\tau) - \varepsilon_1(\tau))d\tau.
\end{aligned} \tag{5.135}$$

The coupled position errors with $\mathcal{L}_N$ given by (3.46) are

$$\begin{aligned}
E_1^*(t) &= \tilde{q}_1(t) + \overline{b}\int_0^t (3\varepsilon_1(\tau) - \varepsilon_2(\tau) - \varepsilon_{N-1}(\tau) - \varepsilon_N(\tau))d\tau \\
E_2^*(t) &= \tilde{q}_2(t) + \overline{b}\int_0^t (3\varepsilon_2(\tau) - \varepsilon_3(\tau) - \varepsilon_N(\tau) - \varepsilon_1(\tau))d\tau \\
&\vdots \\
E_N^*(t) &= \tilde{q}_N(t) + \overline{b}\int_0^t (3\varepsilon_N(\tau) - \varepsilon_1(\tau) - \varepsilon_{N-2}(\tau) - \varepsilon_{N-1}(\tau))d\tau.
\end{aligned} \tag{5.136}$$

We see from (5.134)–(5.136) that synchronization error ε_i appears in E_i^* and E_{i+1}^* with opposite sign. Therefore, the coupled errors are driven in opposite directions by ε_i. This property contributes to the elimination of ε_i.

We construct an auxiliary vector $\boldsymbol{\xi}(t) \triangleq [\xi_1(t) \;\; \xi_2(t) \;\; \cdots \;\; \xi_N(t)]^T$ as

$$\boldsymbol{\xi}(t) = \dot{\boldsymbol{E}}^*(t) + \boldsymbol{\Lambda}\boldsymbol{E}^*(t). \tag{5.137}$$

Equation (5.137), along with (5.132) and (5.133), indicates that

$$\dot{\boldsymbol{E}}^*(t) = -\boldsymbol{\Lambda}\boldsymbol{E}^*(t) + \boldsymbol{\xi}(t), \tag{5.138}$$

$$\boldsymbol{\xi}(t) = \boldsymbol{\Lambda}\tilde{\boldsymbol{q}}(t) + \boldsymbol{\Lambda}\boldsymbol{B}\int_0^t \mathcal{L}_N^T \boldsymbol{\Xi}(\tau)d\tau + \dot{\tilde{\boldsymbol{q}}}(t) + \boldsymbol{B}\mathcal{L}_N^T\boldsymbol{\Xi}(t). \tag{5.139}$$

From (5.137), we see that $\boldsymbol{E}^*(t)$ and $\dot{\boldsymbol{E}}^*(t)$ are used to construct $\boldsymbol{\xi}(t)$. Moreover, since $\boldsymbol{\Lambda}$ is a positive-definite diagonal matrix,

$$\boldsymbol{\xi}(t) \to \boldsymbol{0} \Rightarrow \boldsymbol{E}^*(t) \to \boldsymbol{0}. \tag{5.140}$$

Differentiating (5.137) with respect to t and considering (5.133), we get

$$\dot{\boldsymbol{\xi}} = \ddot{\boldsymbol{E}}^* + \boldsymbol{\Lambda}\dot{\boldsymbol{E}}^* = \ddot{\tilde{\boldsymbol{q}}} + \boldsymbol{B}\mathcal{L}_N^T\dot{\boldsymbol{\Xi}} + \boldsymbol{\Lambda}\dot{\boldsymbol{E}}^*. \tag{5.141}$$

In what follows, the vectors of the second type of LSEs, $\boldsymbol{\Xi}$ and $\dot{\boldsymbol{\Xi}}$, as well as the coupled position and velocity vectors, $\boldsymbol{E}^*$ and $\dot{\boldsymbol{E}}^*$, will be used to construct a synchronized tracking controller. In addition, the system stability will be shown using the Lyapunov function approach.

5.6.2 Controller Design and Convergence Analysis

Consider the following control law for (5.121):

$$\boldsymbol{u} = \ddot{\boldsymbol{q}}_d - \mathbf{K}_P\boldsymbol{E}^* - (\boldsymbol{K}_D + \boldsymbol{\Lambda})\dot{\boldsymbol{E}}^* - \boldsymbol{K}_S\mathcal{L}_N^T\boldsymbol{\Xi} - \boldsymbol{B}\mathcal{L}_N^T\dot{\boldsymbol{\Xi}}, \tag{5.142}$$

where $\mathbf{K}_P = \mathrm{diag}(K_{P1}, \cdots, K_{PN}) \in \mathbb{R}^{N\times N}$, $\mathbf{K}_D = \mathrm{diag}\,(K_{D1}, \cdots, K_{DN}) \in \mathbb{R}^{N\times N}$, and $\boldsymbol{K}_S = \mathrm{diag}\,(K_{S1}, \cdots, K_{SN}) \in \mathbb{R}^{N\times N}$ are diagonal positive-definite constant matrices, and

$$\boldsymbol{\Lambda} = [K_{P1}/K_{D1} \;\; \cdots \;\; K_{Pi}/K_{Di} \;\; \cdots \;\; K_{PN}/K_{DN}] \in \mathbb{R}^{N\times N}. \tag{5.143}$$

Thus,

$$\mathbf{K}_P = \mathbf{K}_D\boldsymbol{\Lambda}. \tag{5.144}$$

Using (5.137) and (5.144), we rewrite (5.142) as

$$\begin{aligned}\boldsymbol{u} &= \ddot{\boldsymbol{q}}_d - \boldsymbol{\Lambda}\dot{\boldsymbol{E}}^* - (\boldsymbol{K}_D\dot{\boldsymbol{E}}^* + \mathbf{K}_P\boldsymbol{E}^*) - \boldsymbol{K}_S\mathcal{L}_N^T\boldsymbol{\Xi} - \boldsymbol{B}\mathcal{L}_N^T\dot{\boldsymbol{\Xi}} \\ &= \ddot{\boldsymbol{q}}_d - \boldsymbol{\Lambda}\dot{\boldsymbol{E}}^* - \mathbf{K}_D\boldsymbol{\xi} - \boldsymbol{K}_S\mathcal{L}_N^T\boldsymbol{\Xi} - \boldsymbol{B}\mathcal{L}_N^T\dot{\boldsymbol{\Xi}},\end{aligned} \tag{5.145}$$

Then, the trajectories of system (5.121) with controller (5.145) satisfy

$$\ddot{\tilde{\boldsymbol{q}}} = -\boldsymbol{\Lambda}\dot{\boldsymbol{E}}^* - \mathbf{K}_D\boldsymbol{\xi} - \boldsymbol{K}_S\mathcal{L}_N^T\boldsymbol{\Xi} - \mathbf{B}\mathbf{L}^T\dot{\boldsymbol{\Xi}} \tag{5.146}$$

Substituting (5.146) into (5.141) gives

$$\dot{\boldsymbol{\xi}} = -\mathbf{K}_D\boldsymbol{\xi} - \boldsymbol{K}_S\mathcal{L}_N^T\boldsymbol{\Xi}. \tag{5.147}$$

We are now ready to present the main result.

Theorem 5.21 Consider MVS (5.121) with control law (5.142) and a communication graph satisfying Condition 1. If parameter matrices $\boldsymbol{K}_S$ and $\boldsymbol{\Lambda}$ satisfy the commuting matrix conditions:

$$\mathcal{L}_N^T \boldsymbol{K}_S = \boldsymbol{K}_S \mathcal{L}_N^T, \tag{5.148}$$

$$(\boldsymbol{\Lambda K}_S)\mathcal{L}_N^T = \mathcal{L}_N^T(\boldsymbol{\Lambda K}_S), \tag{5.149}$$

then controller (5.142) ensures that both tracking errors and synchronization errors converge to zero globally and asymptotically, i.e.

$$\lim_{t\to\infty} \tilde{\boldsymbol{q}}(t),\ \dot{\tilde{\boldsymbol{q}}}(t),\ \Xi(t),\ \dot{\Xi}(t) = 0 \tag{5.150}$$

Proof: Consider the Lyapunov function candidate

$$V(\xi,\Xi) = \frac{1}{2}\xi^T\xi + \frac{1}{2}\Xi^T \boldsymbol{K}_S \Xi + \frac{1}{2}\left(\int \mathcal{L}_N^T \Xi(\tau)d\tau\right)^T \boldsymbol{B\Lambda K}_S \left(\int \mathcal{L}_N^T \Xi(\tau)d\tau\right). \tag{5.151}$$

Since $\boldsymbol{K}_S$, $\boldsymbol{B}$, $\boldsymbol{\Lambda}$ are all positive-definite matrices, $V(\xi,\Xi)$ is a positive-definite function. Differentiating (5.151) with respect to t, yields

$$\dot{V} = \xi^T\dot{\xi} + \Xi^T \boldsymbol{K}_S \dot{\Xi} + \left(\int \mathcal{L}_N^T \Xi(\tau)d\tau\right)^T \boldsymbol{B\Lambda} K_S \mathcal{L}_N^T \Xi. \tag{5.152}$$

By substituting (5.147) into (5.152), we obtain

$$\dot{V} = -\xi^T \mathrm{K}_D \xi - \xi^T \boldsymbol{K}_S \mathcal{L}_N^T \Xi + \Xi^T \boldsymbol{K}_S \dot{\Xi} + \left(\int \mathcal{L}_N^T \Xi(\tau)d\tau\right)^T \boldsymbol{B\Lambda} K_S \mathcal{L}_N^T \Xi. \tag{5.153}$$

Substituting (5.139) into $\xi^T \boldsymbol{K}_S \mathcal{L}_N^T \Xi$ and considering the relationships $\boldsymbol{\Lambda}^T = \boldsymbol{\Lambda}$ and $\boldsymbol{B}^T = \boldsymbol{B}$, we then obtain

$$\begin{aligned}\dot{V} &= -\xi^T \mathrm{K}_D \xi - \left(\boldsymbol{\Lambda}\tilde{\boldsymbol{q}} + \boldsymbol{\Lambda B}\int_0^t \mathcal{L}_N^T \Xi(\tau)d\tau + \dot{\tilde{\boldsymbol{q}}} + \boldsymbol{B}\mathcal{L}_N^T \Xi\right)^T \boldsymbol{K}_S \mathcal{L}_N^T \Xi \\ &\quad + \Xi^T \boldsymbol{K}_S \dot{\Xi} + \left(\int \mathcal{L}_N^T \Xi(\tau)d\tau\right)^T \boldsymbol{B\Lambda} K_S \mathcal{L}_N^T \Xi \\ &= -\xi^T \mathrm{K}_D \xi - (\dot{\tilde{\boldsymbol{q}}} + \boldsymbol{B}\mathcal{L}_N^T \Xi + \boldsymbol{\Lambda}\tilde{\boldsymbol{q}})^T \boldsymbol{K}_S \mathcal{L}_N^T \Xi + \Xi^T \boldsymbol{K}_S \dot{\Xi} \\ &= -\xi^T \mathrm{K}_D \xi - \dot{\tilde{\boldsymbol{q}}}^T \boldsymbol{K}_S \mathcal{L}_N^T \Xi - (\mathcal{L}_N^T \Xi)^T \boldsymbol{B K}_S (\mathcal{L}_N^T \Xi) \\ &\quad - \tilde{\boldsymbol{q}}^T \boldsymbol{\Lambda K}_S \mathcal{L}_N^T \Xi + \Xi^T \boldsymbol{K}_S \dot{\Xi}.\end{aligned} \tag{5.154}$$

Note that $\dot{\Xi}(t) = \mathcal{L}_N \dot{\tilde{\boldsymbol{q}}}(t)$ and $\boldsymbol{K}_S = \boldsymbol{K}_S^T$. Thus, under (5.148),

$$\Xi^T \boldsymbol{K}_S \dot{\Xi} = \Xi^T \boldsymbol{K}_S (\mathcal{L}_N \dot{\tilde{\boldsymbol{q}}}) = \dot{\tilde{\boldsymbol{q}}}^T \mathcal{L}_N^T \boldsymbol{K}_S^T \Xi = \dot{\tilde{\boldsymbol{q}}}^T \boldsymbol{K}_S \mathcal{L}_N^T \Xi, \tag{5.155}$$

and the term $\Xi^T \boldsymbol{K}_S \dot{\Xi}$ in (5.154) is cancelled by $-\dot{\tilde{\boldsymbol{q}}}^T \boldsymbol{K}_S \mathcal{L}_N^T \Xi$. Using equations (5.149), $(\boldsymbol{\Lambda K}_S)^T = \boldsymbol{\Lambda K}_S$ and $\Xi(t) = \mathcal{L}_N \tilde{\boldsymbol{q}}(t)$, we obtain

$$\tilde{\boldsymbol{q}}^T \boldsymbol{\Lambda} K_S \mathcal{L}_N^T \Xi = \tilde{\boldsymbol{q}}^T \mathcal{L}_N^T \boldsymbol{\Lambda K}_S \Xi = (\mathcal{L}_N \tilde{\boldsymbol{q}})^T (\boldsymbol{\Lambda K}_S)\Xi = \Xi^T \boldsymbol{\Lambda K}_S \Xi. \tag{5.156}$$

With relationships (5.155) and (5.156), (5.154) is further reduced to

$$\dot{V} = -\xi^T \mathrm{K}_D \xi - (\mathcal{L}_N^T \Xi)^T \boldsymbol{B} \boldsymbol{K}_S^T (\mathcal{L}_N^T \Xi) - \Xi^T \boldsymbol{\Lambda} \boldsymbol{K}_S \Xi \\ \leq 0 \tag{5.157}$$

Since $V(\xi, \Xi)$ is non-negative and $\dot{V}(\xi, \Xi) \leq 0$, we readily conclude that $V(\xi, \Xi) \in \mathcal{L}_\infty$, $\xi(t) \in \mathcal{L}_\infty$, $\Xi(t) \in \mathcal{L}_\infty$, and $\int \mathcal{L}_N^T \Xi(\tau) d\tau \in \mathcal{L}_\infty$. Using $\xi(t) \in \mathcal{L}_\infty$, we conclude from (5.137) that $\boldsymbol{E}^*(t) \in \mathcal{L}_\infty$, $\dot{\boldsymbol{E}}^*(t) \in \mathcal{L}_\infty$. Thus, $\tilde{\boldsymbol{q}}(t) \in \mathcal{L}_\infty$ due to (5.132), $\dot{\tilde{\boldsymbol{q}}}(t) \in \mathcal{L}_\infty$ due to (5.133), and $\dot{\Xi}(t) \in \mathcal{L}_\infty$ due to $\dot{\Xi} = \mathcal{L}_N \dot{\tilde{\boldsymbol{q}}}$. Since $\boldsymbol{q}_d(t)$ and $\dot{\boldsymbol{q}}_d(t)$ are assumed bounded, the boundedness of $\boldsymbol{q}(t)$ and $\dot{\boldsymbol{q}}(t)$ follows from that of $\tilde{\boldsymbol{q}}(t)$ and $\dot{\tilde{\boldsymbol{q}}}(t)$. The preceding information and equation (5.141) can also be used to conclude that $\ddot{\boldsymbol{q}}(t) \in \mathcal{L}_\infty$ and $\dot{\xi}(t) \in \mathcal{L}_\infty$.

Further, Barbalat's lemma can be invoked to conclude from (5.157) that

$$\lim_{t\to\infty} \xi(t) = 0, \lim_{t\to\infty} \mathcal{L}_N^T \Xi(t) = 0, \lim_{t\to\infty} \Xi(t) = 0. \tag{5.158}$$

This result along with relationship (5.140) indicates that

$$\lim_{t\to\infty} \boldsymbol{E}^*(t) = 0, \lim_{t\to\infty} \dot{\boldsymbol{E}}^*(t) = 0. \tag{5.159}$$

Due to (5.133), the convergence of $\dot{\boldsymbol{E}}^*(t)$ and $\Xi(t)$ ensures

$$\lim_{t\to\infty} \dot{\tilde{\boldsymbol{q}}}(t) = 0. \tag{5.160}$$

Moreover, under Condition 1, the convergence of $\Xi(t)$ implies

$$\tilde{\boldsymbol{q}}_1(\infty) = \cdots = \tilde{\boldsymbol{q}}_N(\infty), \tag{5.161}$$

It follows from (5.132) that

$$\mathbf{1}_N^T \boldsymbol{E}^*(t) = \mathbf{1}_N^T \tilde{\boldsymbol{q}}(t) + \mathbf{1}_N^T \boldsymbol{B} \mathcal{L}_N^T \int_0^t \Xi(\tau) d\tau, \tag{5.162}$$

Since $\mathbf{1}_N^T \boldsymbol{B} \mathcal{L}_N^T = \mathbf{0}_N^T$,

$$\mathbf{1}_N^T \boldsymbol{E}^*(t) = \mathbf{1}_N^T \tilde{\boldsymbol{q}}(t), \tag{5.163}$$

From (5.161) and the convergence of $\boldsymbol{E}^*(t)$, we then conclude that

$$\tilde{\boldsymbol{q}}_1(\infty) = \cdots = \tilde{\boldsymbol{q}}_N(\infty) = 0. \tag{5.164}$$

Thus, synchronized tracking is achieved. This ends the proof of the theorem. □

Remark 5.22 As for the theorem conditions, it is worthwhile noting the following facts:

- Condition 1 ensures that $\tilde{\boldsymbol{q}}_1(\infty) = \cdots = \tilde{\boldsymbol{q}}_N(\infty)$ can be derived from the result $\Xi(\infty) = \mathbf{0}_N$ (see Lemma 3.2).
- The conditions (5.148) and (5.149) ensure equations (5.155) and (5.156) hold, and so we can derive (5.157) from (5.154).
- The conditions (5.148) and (5.149) are trivially satisfied if both $\boldsymbol{K}_S$ and $\boldsymbol{\Lambda}$ are scalar matrices (with all the main diagonal entries equal).

5.7 GSE-based Adaptive Formation Control

To achieve robust synchronized tracking, an integral control approach and a sliding-mode control approach are presented in Sections 5.4 and 5.5, respectively. This section will introduce an alternative: adaptive control. This is a typical approach to address uncertainties. The adaptive control to be shown is linked with Type II LSEs, and serves as a leader–follower formation control solution for a class of MVSs with unknown inertial parameters and subject to exogenous disturbance. According to the analysis in Section 3.6, this formation-control solution can be easily reduced to a robust synchronized trajectory-tracking solution.

In the formation-control problem, the controlled motion for each vehicle is the relative position motion between the vehicle and leader, and the trajectory to be tracked for each vehicle is the desired position deviation of this vehicle from the leader. For this reason, we here consider relative motion equations rather than absolute motion equations in the control design. This is different from the approach in the previous sections, but renders the presentation of the adaptive control concept clear and easily understood.

5.7.1 Problem Statement

The relative motion for vehicle i is described by

$$m\ddot{q}_i + mC_0\dot{q}_i + mg(q_i) + F_i^d = F_i, i \in \mathcal{I}, \tag{5.165}$$

which is a special form of the second-order non-linear equation (5.4). Here, we suppose $n = 1$ for simplicity (the extension to three-dimensional motion is presented in the next chapter). Thus, $q_i \in \mathbb{R}$ denotes the 1-D relative position and $\dot{q}_i \in \mathbb{R}$ the 1-D relative velocity, the vehicle mass $m > 0$ is an unknown constant, mC_0 is the Coriolis-like coefficient, $mg(q_i)$ is the gravity, $F_i^d \in \mathbb{R}$ is the *unknown* constant disturbance, and $F_i \in \mathbb{R}$ is the control force. Note that there are two unknown parameters in the relative motion: m and F_i^d.

We rewrite the N equations in (5.165) into the vector parameterised form as follows:

$$W(\ddot{\boldsymbol{q}}, \dot{\boldsymbol{q}}, \boldsymbol{q})\boldsymbol{\Theta} = \boldsymbol{u} \tag{5.166}$$

where the regression matrix $W(\ddot{\boldsymbol{q}}, \dot{\boldsymbol{q}}, \boldsymbol{q}) \in \mathbb{R}^{N\times(N+1)}$, the position trajectory vector $\boldsymbol{q} \in \mathbb{R}^{N\times 1}$, the unknown parameter vector $\boldsymbol{\Theta} \in \mathbb{R}^{(N+1)\times 1}$, and the control input vector $\boldsymbol{u} \in \mathbb{R}^{N\times 1}$ are given by

$$W(\cdot) = \begin{bmatrix} \ddot{q}_1 + C_0\dot{q}_1 + g(q_1) & 1 & 0 & \cdots & 0 \\ \ddot{q}_2 + C_0\dot{q}_2 + g(q_2) & 0 & 1 & \ddots & 0 \\ \vdots & \vdots & \ddots & \ddots & \ddots \\ \ddot{q}_N + C_0\dot{q}_N + g(q_N) & 0 & 0 & \cdots & 1 \end{bmatrix}, \tag{5.167}$$

$$\boldsymbol{q} = \begin{bmatrix} q_1 & q_2 & \cdots & q_N \end{bmatrix}^T, \tag{5.168}$$

$$\boldsymbol{\Theta} = \begin{bmatrix} m & F_1^d & F_2^d & \cdots & F_N^d \end{bmatrix}^T, \tag{5.169}$$

$$\boldsymbol{u} = \begin{bmatrix} F_1 & F_2 & \cdots & F_N \end{bmatrix}^T. \tag{5.170}$$

Consider a desired relative position trajectory $\boldsymbol{q}_d(t) \in R^N$ for MVS (5.166). Suppose the first two derivatives of $\mathbf{q}_d(t)$ are bounded functions of time. The position tracking error vector, $\tilde{\boldsymbol{q}}(t) = (\tilde{q}_1, \tilde{q}_2, \cdots, \tilde{q}_N)^T \in R^N$, is defined as

$$\tilde{\boldsymbol{q}}(t) = \boldsymbol{q}(t) - \boldsymbol{q}_d(t), \tag{5.171}$$

where $\boldsymbol{q}(t)$ satisfies the equation (5.166).

Note that the desired trajectories for the N vehicles involved in the formation network are generally not identical – that is, $q_{1d}(t)$, $q_{2d}(t)$, $\cdots$, $q_{Nd}(t)$ – as the entries of vector $\boldsymbol{q}_d(t)$ are generally different from each other. This kind of desired trajectory for an MVS physically represents the expected position deviations of the vehicles from the leader's position trajectory, and is specified according to the requirement of a formation pattern. This is different to the synchronized tracking problem (where all involved vehicles share the same desired trajectory).

5.7.2 Controller Development

This section will propose a formation control strategy for MVS (5.166), where the cross-coupling concept introduced in Section 5.6 will be incorporated into the design of the adaptive synchronization controller. For control design and stability analysis, the following auxiliary variables are introduced (as in Section 5.6).

$$\begin{cases} \Xi(t) = \mathcal{L}_N \tilde{\boldsymbol{q}}(t), \\ \dot{\Xi}(t) = \mathcal{L}_N \dot{\tilde{\boldsymbol{q}}}(t), \\ \boldsymbol{E}^*(t) = \tilde{\boldsymbol{q}}(t) + \boldsymbol{B}\mathcal{L}_N^T \int_0^t \Xi(\tau) d\tau, \\ \dot{\boldsymbol{E}}^*(t) = \dot{\tilde{\boldsymbol{q}}}(t) + \boldsymbol{B}\mathcal{L}_N^T \Xi(t), \\ \xi(t) = \dot{\boldsymbol{E}}^*(t) + \boldsymbol{\Lambda}\boldsymbol{E}^*(t), \\ \dot{\xi}(t) = \ddot{\tilde{\boldsymbol{q}}}(t) + \boldsymbol{B}\mathcal{L}_N^T \dot{\Xi}(t) + \boldsymbol{\Lambda}\dot{\boldsymbol{E}}^*(t). \end{cases} \tag{5.172}$$

Using these variables, we construct the following adaptive control law:

$$\begin{cases} \dot{\hat{\boldsymbol{\Theta}}} = -\Gamma W^T(\boldsymbol{\zeta}, \dot{\boldsymbol{q}}, \boldsymbol{q})\xi, \\ \boldsymbol{u} = W(\boldsymbol{\zeta}, \dot{\boldsymbol{q}}, \boldsymbol{q})\hat{\boldsymbol{\Theta}}_i - \boldsymbol{K}_D \xi - \boldsymbol{K}_S {\mathcal{L}_N}^T \Xi, \end{cases} \tag{5.173}$$

where $\hat{\boldsymbol{\Theta}}$ denotes the estimate of $\boldsymbol{\Theta}$, $\Gamma \in R^{(N+1)\times(N+1)}$ is a constant, diagonal, positive-definite, adaptation gain matrix, $\boldsymbol{K}_D \in R^{N\times N}$, $\mathcal{L}_N \in R^{N\times N}$, $\boldsymbol{K}_S \in R^{N\times N}$, $\boldsymbol{B} \in R^{N\times N}$, and $\boldsymbol{\Lambda} \in R^{N\times N}$ are defined as in Section 5.6, and $W(\boldsymbol{\zeta}, \dot{\boldsymbol{q}}, \boldsymbol{q})$ is defined as in (5.167) with the dummy variable $\boldsymbol{\zeta}$ given by

$$\boldsymbol{\zeta}(t) = \ddot{\boldsymbol{q}}_d(t) - \boldsymbol{\Lambda}\dot{\boldsymbol{E}}^*(t) - \boldsymbol{B}{\mathcal{L}_N}^T \dot{\Xi}(t). \tag{5.174}$$

Note that the first equation in (5.173) gives an adaptation on-line estimation law for unknown parameter Θ_i, and that the second equation gives the control law with feedback control terms. Moreover, one readily observes from (5.173) and (5.174) that there are a total of five gains to be determined for the adaptive controller: Γ, $\boldsymbol{K}_D$, $\boldsymbol{K}_S$, $\boldsymbol{\Lambda}$, and $\boldsymbol{B}$.

Denote the parameter estimation error by variable $\tilde{\Theta}(t)$ as

$$\tilde{\boldsymbol{\Theta}}(t) = \boldsymbol{\Theta}(t) - \hat{\boldsymbol{\Theta}}(t). \tag{5.175}$$

From the first equation in (5.173),

$$\dot{\hat{\Theta}} = \Gamma W^T(\zeta, \dot{q}, q)\xi. \tag{5.176}$$

since Θ is a constant vector. By combining (5.166) and (5.173) we obtain the closed-loop equation:

$$W(\ddot{q}, \dot{q}, q)\Theta = W(\zeta, \dot{q}, q)\hat{\Theta} - K_D\xi - K_S\mathcal{L}_N^T\Xi. \tag{5.177}$$

From the definition of $W(\ddot{q}, \dot{q}, q)$, we readily get

$$\begin{aligned} & W(\zeta, \dot{q}, q)\hat{\Theta} - W(\ddot{q}, \dot{q}, q)\Theta \\ & = W(\zeta, \dot{q}, q)\Theta - W(\ddot{q}, \dot{q}, q)\Theta - W(\zeta, \dot{q}, q)\tilde{\Theta} \\ & = m(-\ddot{\tilde{q}} - \Lambda\dot{E}^* - B\mathcal{L}_N^T\dot{\Xi}) - W(\zeta, \dot{q}, q)\tilde{\Theta} \\ & = -m\dot{\xi} - W(\zeta, \dot{q}, q)\tilde{\Theta}, \end{aligned}$$

where the last equation in (5.172) is used. Then, the resulting closed-loop equation is

$$\begin{cases} \dot{\tilde{\Theta}} = \Gamma W^T(\zeta, \dot{q}, q)\xi \\ m\dot{\xi} = -W(\zeta, \dot{q}, q)\tilde{\Theta} - K_D\xi - K_S{\mathcal{L}_N}^T\Xi. \end{cases} \tag{5.178}$$

By analysing system (5.178) using a Lyapunov function based approach, we obtain the following result.

Theorem 5.23 Consider MVS (5.166) with the dynamic control law (5.173) and a communication graph satisfying Condition 1. If the parameters K_S and Λ satisfy the commuting matrix conditions (5.148) and (5.149), then both position tracking error $\tilde{q}(t)$ and position synchronization error $\Xi(t)$ converge globally to zero:

$$\lim_{t\to\infty} \tilde{q}(t) = 0_N, \lim_{t\to\infty} \Xi(t) = 0_N. \tag{5.179}$$

Proof: Clearly, this theorem can be proven if we can find a Lyapunov function candidate with a first derivative satisfying (5.153). To this end, we consider the Lyapunov function candidate:

$$\begin{aligned} V(\xi, \tilde{\Theta}, \Xi) = & \frac{1}{2}\xi m\xi + \frac{1}{2}\tilde{\Theta}\Gamma^{-1}\tilde{\Theta} + \frac{1}{2}\Xi^T K_S\Xi + \\ & \frac{1}{2}\left(\int {\mathcal{L}_N}^T\Xi(\tau)d\tau\right)^T B\Lambda K_S \int {\mathcal{L}_N}^T\Xi(\tau)d\tau. \end{aligned} \tag{5.180}$$

Since Γ^{-1}, K_S, B, and Λ are all positive-definite matrices, $V(\xi, \tilde{\Theta}, \Xi)$ is also a positive-definite function. Differentiating (5.180) with respect to time t yields

$$\dot{V} = \xi^T m\dot{\xi} + \tilde{\Theta}^T\Gamma^{-1}\dot{\tilde{\Theta}} + \Xi^T K_S\dot{\Xi} + \left(\int {\mathcal{L}_N}^T\Xi d\tau\right)^T B\Lambda K_S{\mathcal{L}_N}^T\Xi \tag{5.181}$$

By substituting the two equations in (5.178) into (5.181), we obtain

$$\begin{aligned} \dot{V} = & \xi^T(-W(\zeta, \dot{q}, q)\tilde{\Theta} - K_D\xi - K_S L^T\Xi) + \tilde{\Theta}^T W^T(\zeta, \dot{q}, q)\xi \\ & + \Xi^T K_S\dot{\Xi} + \left(\int {\mathcal{L}_N}^T\Xi(\tau)d\tau\right)^T B\Lambda K_S{\mathcal{L}_N}^T\Xi \end{aligned}$$

$$= -\xi^T K_D \xi - \xi^T K_S \mathcal{L}_N{}^T \Xi + \Xi^T K_S \dot{\Xi}$$
$$+ \left(\int \mathcal{L}_N{}^T \Xi(\tau) d\tau \right)^T B \Lambda K_S \mathcal{L}_N{}^T \Xi. \tag{5.182}$$

Note that $\dot{V}$ in (5.182) has the same expression as $\dot{V}$ in (5.153), and the parameter conditions of Theorems 5.21 and 5.23 are also the same. Therefore, using similar arguments as in the proof of Theorem 5.21, we readily derive the result (5.157) for (5.180). With this result, all the signals $\xi(t)$, $\tilde{\Theta}(t)$, $\Xi(t)$ are bounded. Moreover, the signals $\dot{\xi}(t)$, $\dot{\Xi}(t)$, $\tilde{q}(t)$ and $\dot{\tilde{q}}(t)$ are bounded. In addition, Barbalat's lemma can be invoked to conclude that

$$\lim_{t\to\infty} \xi(t) = \mathbf{0}_N, \lim_{t\to\infty} \mathcal{L}_N^T \Xi(t) = \mathbf{0}_N, \lim_{t\to\infty} \Xi(t) = \mathbf{0}_N, \tag{5.183}$$

which can be further used to show that

$$\lim_{t\to\infty} \mathbf{E}^*(t) = \mathbf{0}_N, \lim_{t\to\infty} \dot{\mathbf{E}}^*(t) = \mathbf{0}_N, \lim_{t\to\infty} \dot{\xi}(t) = \mathbf{0}_N,$$
$$\lim_{t\to\infty} \int_0^t \mathcal{L}_N^T \Xi(\tau) d\tau = \mathbf{0}_N, \lim_{t\to\infty} \tilde{q}(t) = \mathbf{0}_N, \lim_{t\to\infty} \dot{\tilde{q}}(t) = \mathbf{0}_N. \tag{5.184}$$

This ends the proof of the theorem. □

The result of Theorem 5.23 implies that the proposed control strategy can achieve synchronized position trajectory tracking (with different desired trajectories for the involved vehicles) or leader–follower formation motion in the presence of unknown constant parameters m and F_i^d, $i = 1, \ldots, N$. We also note that this approach can be further applied to the position synchronization of multiple motion axes, and to the synchronous attitude rotation of multiple spacecraft about single/multiple given axes (see Chapter 6 for a spacecraft application).

5.8 Summary

In this chapter, we have introduced three approaches to the design of robust position synchronized tracking or leader–follower formation control:

- uncertainty and disturbance estimators (UDE) under state feedback or output feedback;
- sliding-mode control under output feedback;
- adaptive control.

It is important to note that various UDEs are adopted to improve the controllers developed in Chapters 3 and 4. Moreover, UDE-based robust synchronized tracking approaches can be easily applied to solve leader–follower formation control problems. For simplicity of presentation, we do not present these detailed applications here.

In the following chapters, we will introduce how to apply these approaches to four typical vehicle systems: spacecraft, fixed-wing aircraft, unmanned ground vehicles, and 3-DOF experimental helicopters.

Bibliography

1 D. Luenberger, "Observers for multivariable systems," *IEEE Transactions on Automatic Control*, vol. 11, no. 2, pp. 190–197, 1966.
2 H. Zhang, F. L. Lewis, and A. Das, "Optimal design for synchronization of cooperative systems: state feedback, observer and output feedback," *IEEE Transactions on Automatic Control*, vol. 56, no. 8, pp. 1948–1952, 2011.
3 W. Ren and R. W. Beard, *Distributed Consensus in Multi-vehicle Cooperative Control*. Springer, 2008.
4 J. R. Lawton, R. W. Beard, and B. J. Young, "A decentralized approach to formation maneuvers," *IEEE Transactions on Robotics and Automation*, vol. 19, no. 6, pp. 933–941, 2003.
5 G. Strang, *Calculus*. Wellesley-Cambridge, 2010.
6 B. Zhu, W. Sun, and C. Meng, "Position tracking of multi double-integrator dynamics by bounded distributed control without velocity measurements," in *American Control Conference*, (Washington, DC), pp. 4033–4038, June 17–19 2013.
7 H. K. Khalil, *Nonlinear Systems*, vol. 3. Prentice Hall, 2002.
8 B. Zhu, Z. Li, H. H. T. Liu, and H. Gao, "Robust second-order tracking of multi-vehicle systems without velocity measurements on directed communication topologies," in *American Control Conference*, (Portland, Oregon), pp. 5414–5419, June 4–6 2014.
9 Q.-C. Zhong and D. Rees, "Control of uncertain LTI systems based on an uncertainty and disturbance estimator," *Journal of Dynamic Systems, Measurement, and Control*, vol. 126, no. 4, pp. 34–44, 2004.
10 B. Zhu, H. H.-T. Liu, and Z. Li, "Robust distributed attitude synchronization of multiple three-DOF experimental helicopters," *Control Engineering Practice*, vol. 36, pp. 87–99, 2015.
11 S. R. Ploen, F. Y. Hadaegh, and D. P. Scharf, "Rigid body equations of motion for modeling and control of spacecraft formations. Part 1: Absolute equations of motion," in *American Control Conference, 2004. Proceedings of the 2004*, vol. 4, pp. 3646–3653, IEEE, 2004.
12 S.-J. Chung, U. Ahsun, and J.-J. E. Slotine, "Application of synchronization to formation flying spacecraft: Lagrangian approach," *Journal of Guidance, Control, and Dynamics*, vol. 32, no. 2, pp. 512–526, 2009.
13 H. H. Liu and J. Shan, "Adaptive synchronized attitude angular velocity tracking control of multi-UAVs," in *Proceedings of the American Control Conference*, vol. 1, p. 128, 2005.

Formation Control of Multiple Autonomous Vehicle Systems, First Edition. Hugh H.T. Liu and Bo Zhu.

14 T. Kiefer, A. Kugi, and W. Kemmetmüller, "Modeling and flatness-based control of a 3DOF helicopter laboratory experiment," in *6th IFAC Symposium on Nonlinear Control Systems*, vol. 9, 2004.
15 Z. Li, H. H. T. Liu, B. Zhu, and H. Gao, "Robust second-order consensus tracking of multiple 3-DOF laboratory helicopters via output-feedback," *IEEE Transactions on Mechatronics*, vol. 20, no. 5, pp. 2538–2549, 2015.
16 A. Isidori, *Nonlinear Control Systems*. Elsevier, 1995.
17 S. Liu, L. Xie, and F. L. Lewis, "Synchronization of multi-agent systems with delayed control input information from neighbors," *Automatica*, vol. 47, no. 10, pp. 2152–2164, 2011.
18 G. Hu, "Robust consensus tracking of a class of second-order multi-agent dynamic systems," *Systems & Control Letters*, vol. 61, no. 1, pp. 134–142, 2012.
19 E. Aguiñaga-Ruiz, A. Zavala-Río, V. Santibáñez, and F. Reyes, "Global trajectory tracking through static feedback for robot manipulators with bounded inputs," *IEEE Transactions on Control Systems Technology*, vol. 17, no. 4, pp. 934–944, 2009.
20 A. Abdessameud and A. Tayebi, "On consensus algorithms design for double integrator dynamics," *Automatica*, vol. 49, no. 1, pp. 253–260, 2013.
21 K. Zhou, J. C. Doyle, K. Glover, *et al.*, *Robust and Optimal Control*, vol. 40. Prentice-Hall, 1996.
22 B. Xian, M. S. De Queiroz, D. M. Dawson, and M. L. McIntyre, "Output feedback variable structure-like control of nonlinear mechanical systems," in *Proceedings of the 42nd IEEE Conference on Decision and Control (CDC)*, pp. 368–373, 2003.
23 D. Sun, *Synchronization and Control of Multiagent Systems*, vol. 41. CRC Press, 2010.
24 J. Shan, H. H. T. Liu, and S. Nowotny, "Synchronized trajectory tracking control of multiple 3-DOF experimental helicopters," *IEE Proceedings of Control Theory and Applications*, vol. 152, no. 6, pp. 683–692, 2005.
25 Y. Koren, "Cross-coupled biaxial computer control for manufacturing systems," *ASME – Journal of Dynamic Systems Measurement and Control*, vol. 102, no. 4, pp. 265–272, 1980.
26 D. Sun, "Position synchronization of multiple motion axes with adaptive coupling control," *Automatica*, vol. 39, no. 6, pp. 997–1005, 2003.

Part III

Formation Control: Case Studies

6

Formation Control of Space Systems

This chapter will introduce how to apply the adaptive formation control strategy proposed in Section 5.7 to formations of spacecraft, and show the simulation results of two typical applications: motion synchronization among multiple axes of one spacecraft, and/or any given axis of multiple spacecraft while realizing the convergence of position tracking errors.

The proposed synchronization framework uses a Lagrangian formulation because of its simplicity when dealing with complex systems involving multiple dynamics. We first briefly recall the derivation of equations of the Lagrangian form (6.1), and then show that the rotational dynamics of a rigid spacecraft can be written in this form, which implies that the approaches introduced in the foregoing chapter can be applied to the rotational synchronized tracking problem of multiple spacecraft. However, we do not dwell on such applications since we are mainly concerned with the formation control problem in this chapter. To address the position synchronization of formations of spacecraft, we first set out the relative translational dynamics of multiple spacecraft in the Lagrangian form, and then apply the control strategy proposed in Section 5.7. Simulation results are presented to show the effectiveness of the strategy.

6.1 Lagrangian Formulation of Spacecraft Formation

6.1.1 Lagrangian Formulation

Consider the Euler–Lagrange equations

$$\frac{d}{dt}\frac{\partial L_i}{\partial \dot{\boldsymbol{q}}_i} - \frac{\partial L_i}{\partial \boldsymbol{q}_i} = \tau_i, i \in \mathcal{I}, \tag{6.1}$$

where the Lagrangian function

$$L_i(\boldsymbol{q}_i, \dot{\boldsymbol{q}}_i) = \frac{1}{2}\dot{\boldsymbol{q}}_i^T M_i(\boldsymbol{q}_i)\dot{\boldsymbol{q}}_i - V_i(\boldsymbol{q}_i) \tag{6.2}$$

is a real-valued function with continuous first partial derivatives, i ($i \in \mathcal{I}$) denotes the index of a spacecraft in a formation-flight network, $\boldsymbol{q}_i \in R^m$ are generalized coordinates, $\dot{\boldsymbol{q}}_i \in R^m$ are the generalized velocities, and $\boldsymbol{\tau}_i$ is a generalized force or torque acting on the ith spacecraft.

Formation Control of Multiple Autonomous Vehicle Systems, First Edition. Hugh H.T. Liu and Bo Zhu.

From (6.2) with a symmetric matrix $M_i(\boldsymbol{q}_i)$, we obtain

$$\frac{d}{dt}\frac{\partial L_i}{\partial \dot{\boldsymbol{q}}_i} = \frac{d}{dt}(M_i(\boldsymbol{q}_i)\dot{\boldsymbol{q}}_i) = \dot{M}_i(\boldsymbol{q}_i)\dot{\boldsymbol{q}}_i + M_i(\boldsymbol{q}_i)\ddot{\boldsymbol{q}}_i,$$

$$\frac{\partial L_i}{\partial \boldsymbol{q}_i} = \frac{1}{2}\dot{\boldsymbol{q}}_i^T\frac{\partial M_i(\boldsymbol{q}_i)}{\partial \boldsymbol{q}_i}\dot{\boldsymbol{q}}_i - \frac{\partial V_i(\boldsymbol{q}_i)}{\partial \boldsymbol{q}_i}, i \in \mathcal{I}. \tag{6.3}$$

Substituting (6.3) into (6.1), yields

$$M_i(\boldsymbol{q}_i)\ddot{\boldsymbol{q}}_i + C_i(\boldsymbol{q}_i, \dot{\boldsymbol{q}}_i)\dot{\boldsymbol{q}}_i + G_i(\boldsymbol{q}_i) = \tau_i, i \in \mathcal{I}, \tag{6.4}$$

with

$$G_i(\boldsymbol{q}_i) = \frac{\partial V_i(\boldsymbol{q}_i)}{\partial \boldsymbol{q}_i}, i \in \mathcal{I}. \tag{6.5}$$

We assume that the spacecraft system (6.4) is fully actuated; that is, the number of control inputs is equal to the dimension of the configuration manifold m.

6.1.2 Attitude Dynamics of Rigid Spacecraft

In this section, we shall show that the rotational dynamics of a rigid spacecraft can be described in the Lagrangian form (6.4) with $m = 3$. More detailed derivations can be found in the literature [1].

According to the Euler rotational equations of motion, the angular motion equation of spacecraft i can be described in body-fixed coordinates by

$$J_i\dot{\omega}_i - (J_i\omega_i) \times \omega_i = \boldsymbol{u}_i + \boldsymbol{d}_{ri}, i = 1, \cdots, N, \tag{6.6}$$

where $J_i \in R^{3\times 3}$ is the inertial matrix expressed in its body frame, $\omega_i \in R^3$ is the angular velocity vector, $\mathbf{u}_i$ is the internal control torque, and $\mathbf{d}_{ri}$ denotes the external disturbance torque. We assume that J_i is a symmetric positive definite matrix for each i.

In general, $\boldsymbol{d}_{ri}$ arises due to the aerodynamic drag torque and the gravity gradient torque, and $\boldsymbol{u}_i$ is generated either by a variable speed control moment gyroscope or a control moment gyroscope [1].

To avoid the singularity issue associated with the Euler angular representation, we may use a quaternion to represent an angular orientation between two different coordinate frames:

$$\beta_{i1} = e_{i1}\sin\frac{\theta_i}{2}, \beta_{i2} = e_{i2}\sin\frac{\theta_i}{2}, \beta_{i3} = e_{i3}\sin\frac{\theta_i}{2}, \beta_{i4} = \cos\frac{\theta_i}{2}. \tag{6.7}$$

where $\boldsymbol{e}_i = (e_{i1}, e_{i2}, e_{i3})^T$ is the Euler axis of rotation expressed in the body frame and θ_i is the rotation angle about $\boldsymbol{e}_i$.

The modified Rodrigues parameters can be written as

$$\boldsymbol{q}_i = (q_{i1}, q_{i2}, q_{i3})^T = \boldsymbol{e}_i \tan\frac{\theta_i}{4}. \tag{6.8}$$

The angular position trajectories of the ith spacecraft evolve according to

$$\dot{\boldsymbol{q}}_i = \mathbf{Z}(\boldsymbol{q}_i)\omega_i, \tag{6.9}$$

where

$$\mathbf{Z}(\boldsymbol{q}_i) = \frac{1}{2}\left[\mathbf{I}\left(\frac{1 - \boldsymbol{q}_i^T\boldsymbol{q}_i}{2}\right) + \boldsymbol{q}_i\boldsymbol{q}_i^T + \mathbf{S}(\boldsymbol{q}_i)\right]$$

$$= \frac{1}{4}\begin{bmatrix} (1+q_{i1}^2-q_{i2}^2-q_{i3}^2) & 2(q_{i1}q_{i2}-q_{i3}) & 2(q_{i1}q_{i3}+q_{i2}) \\ 2(q_{i1}q_{i2}+q_{i3}) & (1-q_{i1}^2+q_{i2}^2-q_{i3}^2) & 2(q_{i2}q_{i3}-q_{i1}) \\ 2(q_{i1}q_{i3}-q_{i2}) & 2(q_{i2}q_{i3}+q_{i1}) & (1-q_{i1}^2-q_{i2}^2+q_{i3}^2). \end{bmatrix}$$

with the skew-symmetric matrix function $\mathbf{S}(\boldsymbol{q}_i)$ for an arbitrary $\boldsymbol{q}_i \in R^3$ defined as

$$\mathbf{S}(\boldsymbol{q}_i) = \begin{bmatrix} \mathbf{0} & -q_{i3} & q_{i2} \\ q_{i3} & 0 & -q_{i1} \\ -q_{i2} & q_{i1} & 0 \end{bmatrix}. \tag{6.10}$$

The quaternion defined in (6.7) can be obtained from the modified Rodrigues parameters by the following transformation:

$$\beta_{ij} = \frac{2q_{ij}}{1+\boldsymbol{q}_i^T\boldsymbol{q}_i}, j = 1,2,3,$$
$$\beta_{i4} = \frac{1-\boldsymbol{q}_i^T\boldsymbol{q}_i}{1+\boldsymbol{q}_i^T\boldsymbol{q}_i}. \tag{6.11}$$

Its inverse transformation is

$$q_{ij} = \frac{\beta_{ij}}{1+\beta_{i4}}, j = 1,2,3. \tag{6.12}$$

From (6.9),

$$\omega_i = \mathbf{Z}^{-1}(\boldsymbol{q}_i)\dot{\boldsymbol{q}}_i, \tag{6.13}$$

which further indicates that

$$\dot{\omega}_i = \mathbf{Z}^{-1}(\boldsymbol{q}_i)\ddot{\boldsymbol{q}}_i + \frac{d}{dt}\mathbf{Z}^{-1}(\boldsymbol{q}_i)\dot{\boldsymbol{q}}_i. \tag{6.14}$$

Substituting the relationship equation

$$\frac{d}{dt}\mathbf{Z}^{-1}(\boldsymbol{q}_i) = -\mathbf{Z}^{-1}(\boldsymbol{q}_i)\left(\frac{d}{dt}\mathbf{Z}(\boldsymbol{q}_i)\right)\mathbf{Z}^{-1}(\boldsymbol{q}_i) \tag{6.15}$$

into (6.14), yields

$$\dot{\omega}_i = \mathbf{Z}^{-1}(\boldsymbol{q}_i)\ddot{\boldsymbol{q}}_i - \mathbf{Z}^{-1}(\boldsymbol{q}_i)\left(\frac{d}{dt}\mathbf{Z}(\boldsymbol{q}_i)\right)\mathbf{Z}^{-1}(\boldsymbol{q}_i)\dot{\boldsymbol{q}}_i. \tag{6.16}$$

For simplicity, denote $\mathbf{Z} = \mathbf{Z}(\boldsymbol{q}_i)$, $\mathbf{Z}^{-1} = \mathbf{Z}^{-1}(\boldsymbol{q}_i)$, and $\dot{\mathbf{Z}} = \frac{d}{dt}\mathbf{Z}(\boldsymbol{q}_i)$. Then, with (6.13) and (6.16), Equation (6.6) becomes

$$J_i\mathbf{Z}^{-1}\ddot{\boldsymbol{q}}_i - J_i\mathbf{Z}^{-1}\dot{\mathbf{Z}}\mathbf{Z}^{-1}\dot{\boldsymbol{q}}_i - (J_i\mathbf{Z}^{-1}\dot{\boldsymbol{q}}_i)\times(\mathbf{Z}^{-1}\dot{\boldsymbol{q}}_i) = \boldsymbol{u}_i + \boldsymbol{d}_{ri}, \tag{6.17}$$

which is clearly equivalent to

$$\mathbf{Z}^{-T}J_i\mathbf{Z}^{-1}\ddot{\boldsymbol{q}}_i - \mathbf{Z}^{-T}\left[J_i\mathbf{Z}^{-1}\dot{\mathbf{Z}}\mathbf{Z}^{-1} + S(J_i\omega_i)\mathbf{Z}^{-1}\right]\dot{\boldsymbol{q}}_i = \mathbf{Z}^{-T}(\boldsymbol{u}_i + \boldsymbol{d}_{ri}). \tag{6.18}$$

where the relation equations

$$S(J_i\omega_i) = S(J_i\mathbf{Z}^{-1}\dot{\boldsymbol{q}}_i), \tag{6.19}$$
$$(J_i\mathbf{Z}^{-1}\dot{\boldsymbol{q}}_i)\times(\mathbf{Z}^{-1}\dot{\boldsymbol{q}}_i) = S(J_i\mathbf{Z}^{-1}\dot{\boldsymbol{q}}_i)\mathbf{Z}^{-1}\dot{\boldsymbol{q}}_i, \tag{6.20}$$

are used.

Let

$$M_i(\boldsymbol{q}_i) = \mathbf{Z}^{-T} J_i \mathbf{Z}^{-1},$$
$$\boldsymbol{\tau}_i = \mathbf{Z}^{-T} \boldsymbol{u}_i,$$
$$\boldsymbol{f}_i = \mathbf{Z}^{-T} \boldsymbol{d}_{ri},$$
$$C_i(\boldsymbol{q}_i, \dot{\boldsymbol{q}}_i) = -\mathbf{Z}^{-T} J_i \mathbf{Z}^{-1} \dot{\mathbf{Z}} \mathbf{Z}^{-1} - \mathbf{Z}^{-T} S(J_i \omega_i) \mathbf{Z}^{-1}.$$

Then, Equation (6.18) can be written in the form

$$M_i(\boldsymbol{q}_i)\ddot{\boldsymbol{q}}_i + C_i(\boldsymbol{q}_i, \dot{\boldsymbol{q}}_i)\dot{\boldsymbol{q}}_i = \boldsymbol{\tau}_i + \boldsymbol{f}_i. \tag{6.21}$$

which is a Lagrangian formulation for the attitude dynamics of a rigid spacecraft.

Note that all the terms in (6.18) are left multiplied by $\mathbf{Z}^{-T}(\boldsymbol{q}_i)$. This ensures that the resulting $M_i(\boldsymbol{q}_i)$ is symmetric. Moreover, since the spacecraft attitude dynamics (6.21) have the form (5.1), we can apply the control approaches developed in Chapter 5 to the problem of synchronized tracking of angular position in a multi-spacecraft system.

In addition, the most important feature of Equation (6.21) is that the matrix $\dot{M}_i - 2C_i$ is skew symmetric due to energy conservation. Since $S(J_i\omega_i)$ is skew symmetric, we can easily verify, using equation (6.15), that matrix

$$\begin{aligned}\dot{M}_i - 2C_i &= \frac{d\mathbf{Z}^{-T}}{dt} J_i \mathbf{Z}^{-1} + \mathbf{Z}^{-T} J_i \frac{d\mathbf{Z}^{-1}}{dt} \\ &\quad + 2\mathbf{Z}^{-T} J_i \mathbf{Z}^{-1} \dot{\mathbf{Z}} \mathbf{Z}^{-1} + 2\mathbf{Z}^{-T} S(J_i\omega_i) \mathbf{Z}^{-1} \\ &= \frac{d\mathbf{Z}^{-T}}{dt} J_i \mathbf{Z}^{-1} - \mathbf{Z}^{-T} J_i \frac{d\mathbf{Z}^{-1}}{dt} + 2\mathbf{Z}^{-T} S(J_i\omega_i) \mathbf{Z}^{-1}\end{aligned} \tag{6.22}$$

is skew symmetric.

6.1.3 Relative Translational Dynamics

If the effect of attitude dynamics on the translational dynamics is weak and ignored, the translational dynamics can be modeled as double integrators, which can be easily augmented with the attitude dynamics (6.21). Alternatively, the coupled translational and rotational motions of formation spacecraft can be written in the Lagrangian form (6.4).

We use $\boldsymbol{x}_i$ to denote a state vector of biased variables constructed from the position vector r_i such that

$$\boldsymbol{r}_i(t) = \boldsymbol{x}_i(t) + \boldsymbol{b}_i(t), i \in \mathcal{I}, \tag{6.23}$$

with the separation vector $\boldsymbol{b}_i(t)$ independent of the dynamics. For arbitrary translational dynamics, state synchronization corresponds to

$$\boldsymbol{x}_1 = \boldsymbol{x}_2 = \cdots = \boldsymbol{x}_N, \tag{6.24}$$

In what follows, we present relative translational dynamics applicable to formation flight on a circular or spiral configuration in low Earth orbit, and then develop a formation controller based on the relative dynamics with respect to the desired formation center of mass. For the simplicity of deriving a control law, we consider a non-inertial orbital coordinate system, FRO. The orbital frame FRO is defined as follows: its origin is attached to the center of mass of the formation with its y axis aligned such that:

- the position vector R_0 represents the position of the formation center of mass in the Earth-centred inertial frame

- the z axis points along the orbital plane normal
- the x axis completes a right-hand system.

For simplicity, we assume a circular reference orbit for the formation center of mass. The angular velocity of the reference orbit is simply

$$\omega_0 = \sqrt{\frac{\mu_e}{R_0^3}}, \tag{6.25}$$

where $\mu_e = 3.98645 \times 10^{14}$ m^3/s^2 is the gravitational constant of Earth and R_0 is the radius of the orbit of the formation center of mass.

Let $R_k = (0, r_k, 0)^T \in R^3$ denote the position vector from the origin of the inertial coordinate frame to the kth spacecraft with a known orbital angular speed $\omega_k(t) \in R$. Let $\boldsymbol{q}_{ik}$ denote the position vector from the kth spacecraft to the ith spacecraft:

$$\boldsymbol{q}_{ik} = (\tilde{x}_{ik}, \tilde{y}_{ik}, \tilde{z}_{ik})^T. \tag{6.26}$$

Then, the relative dynamics of the ith spacecraft with respect to the kth spacecraft in the orbital frame FRO can be written as

$$m_i\left(\ddot{\tilde{x}}_{ik} - 2\omega_k\dot{\tilde{y}}_{ik} - \omega_k^2\tilde{x}_{ik} - \dot{\omega}_k\tilde{y}_{ik} + \frac{\mu_e\tilde{x}_{ik}}{\|R_k + \boldsymbol{q}_{ik}\|^3} + \frac{F_{kx}}{m_k}\right) + F_i^{dx} = F_{ix}$$

$$m_i\left(\ddot{\tilde{y}}_{ik} + 2\omega_k\dot{\tilde{x}}_{ik} - \omega_k^2\tilde{y}_{ik} + \dot{\omega}_k\tilde{x}_{ik} + \frac{\mu_e(\tilde{y}_{ik} + r_k)}{\|R_k + \boldsymbol{q}_{ik}\|^3} - \frac{\mu_e}{r_k^2} + \frac{F_{ky}}{m_k}\right) + F_i^{dy} = F_{iy}$$

$$m_i\left(\ddot{\tilde{z}}_{ik} + \frac{\mu_e\tilde{z}_{ik}}{\|R_k + \boldsymbol{q}_{ik}\|^3} + \frac{F_{kz}}{m_k}\right) + F_i^{dz} = F_{iz}, \tag{6.27}$$

where m_i and m_k are the masses of the ith and kth spacecraft, respectively, $\boldsymbol{F}_i^d = (F_i^{dx}, F_i^{dy}, F_i^{dz})^T \in R^3$ are the constant disturbance vectors, and $\boldsymbol{F}_i = (F_{ix}, F_{iy}, F_{iz})^T \in R^3$ and $\boldsymbol{F}_k = (F_{kx}, F_{ky}, F_{kz})^T \in R^3$ are the control vectors for the ith and kth spacecraft, respectively. Then, it is straightforward to check that the relative equation (6.27) can be written in the compact form

$$m_i\ddot{\boldsymbol{q}}_{ik} + C_i(\omega_k)\dot{\boldsymbol{q}}_{ik} + \boldsymbol{N}_i(\boldsymbol{q}_{ik}, \omega_k, \dot{\omega}_k, R_k, \boldsymbol{F}_k) + \boldsymbol{F}_i^d = \boldsymbol{F}_i, \tag{6.28}$$

where $C_i(\omega_k) \in R^{3\times3}$ is the Coriolis-like matrix given by

$$C_i(\omega_k) = 2m_i\omega_k\begin{bmatrix} 0 & -1 & 0 \\ 1 & 0 & 0 \\ 0 & 0 & 0 \end{bmatrix}, \tag{6.29}$$

and $\boldsymbol{N}_i$ is a nonlinear term consisting of gravitational effects and inertial forces, defined as

$$\mathbf{N}_i(\cdot) = m_i\begin{bmatrix} -\omega_k^2\tilde{x}_{ik} - \dot{\omega}_k\tilde{y}_{ik} + \dfrac{\mu_e\tilde{x}_{ik}}{\|R_k + \boldsymbol{q}_{ik}\|^3} + \dfrac{F_{kx}}{m_k} \\ -\omega_k^2\tilde{y}_{ik} + \dot{\omega}_k\tilde{x}_{ik} + \dfrac{\mu_e(\tilde{y}_{ik} + r_k)}{\|R_k + \boldsymbol{q}_{ik}\|^3} - \dfrac{\mu_e}{r_k^2} + \dfrac{F_{ky}}{m_k} \\ \dfrac{\mu_e\tilde{z}_{ik}}{\|R_k + \boldsymbol{q}_{ik}\|^3} + \dfrac{F_{kz}}{m_k} \end{bmatrix}. \tag{6.30}$$

The left-hand side of equation (6.28) can be linearly parametrized as

$$m_i\ddot{\boldsymbol{q}}_{ik} + C_i(\omega_k)\dot{\boldsymbol{q}}_{ik} + \boldsymbol{N}_i(\cdot) + \boldsymbol{F}_i^d = W\left(\boldsymbol{\xi}_{ik}, \dot{\boldsymbol{q}}_{ik}, \boldsymbol{q}_{ik}, \omega_k, \dot{\omega}_k, R_k, \boldsymbol{F}_k\right)\boldsymbol{\Theta}_i. \tag{6.31}$$

where $\boldsymbol{\xi}_{ik} = (\xi_{ik}^x, \xi_{ik}^y, \xi_{ik}^z)^T \in R^3$ is a dummy variable equal to $\ddot{\boldsymbol{q}}_{ik}$, $W(\cdot) \in R^{3\times 4}$ is the regression matrix composed of known functions and defined by

$$W = \begin{bmatrix} \xi_{ik}^x - 2\omega_k\dot{\tilde{y}}_{ik} - \omega_k^2\tilde{x}_{ik} - \dot{\omega}_k\tilde{y}_{ik} + \dfrac{\mu_e\tilde{x}_{ik}}{\|R_k + \boldsymbol{q}_{ik}\|^3} + \dfrac{F_{kx}}{m_k} & 1 & 0 & 0 \\ \xi_{ik}^y + 2\omega_k\dot{\tilde{x}}_{ik} - \omega_k^2\tilde{y}_{ik} + \dot{\omega}_k\tilde{x}_{ik} + \dfrac{\mu_e(\tilde{y}_{ik} + r_k)}{\|R_k + \boldsymbol{q}_{ik}\|^3} - \dfrac{\mu_e}{r_k^2} + \dfrac{F_{ky}}{m_k} & 0 & 1 & 0 \\ \xi_{ik}^z + \dfrac{\mu_e\tilde{z}_{ik}}{\|R_k + \boldsymbol{q}_{ik}\|^3} + \dfrac{F_{kz}}{m_k} & 0 & 0 & 1 \end{bmatrix}, \tag{6.32}$$

and $\boldsymbol{\Theta}_i \in R^4$ is the constant parameter vector defined by

$$\boldsymbol{\Theta}_i = \left(m_i, F_i^{dx}, F_i^{dy}, F_i^{dz}\right)^T \in R^4. \tag{6.33}$$

Then, Equation (6.28) can be equivalently written in the form

$$W\left(\ddot{\boldsymbol{q}}_{ik}, \dot{\boldsymbol{q}}_{ik}, \boldsymbol{q}_{ik}, \omega_k, \dot{\omega}_k, R_k, \boldsymbol{F}_k\right) = \boldsymbol{F}_i, \tag{6.34}$$

indicating that the equation describing the relative motion of two spacecraft has been converted into a form similar that in (5.166). Therefore, the adaptive control strategy developed in Section 5.7 is applicable.

6.2 Adaptive Formation Control

Formation flying depends on accurate relative position control. In the following, we first introduce a leader–follower formation configuration to develop the controller. Only two spacecraft are concerned, called the leader spacecraft and follower spacecraft, respectively. Then, we apply the controller to the case of multiple-spacecraft formation flying.

Consider a desired relative position trajectory vector $\boldsymbol{q}_d = (\boldsymbol{q}_{d1}^T, \cdots, \boldsymbol{q}_{dN}^T)^T \in R^{3N}$, with entry $\boldsymbol{q}_{di} = (q_{di}^x, q_{di}^y, q_{di}^z)^T \in R^3$, $i \in \mathcal{I}$, representing the desired relative position trajectory for the ith spacecraft with respect to the leader spacecraft. Assume that the first two time derivatives of $\boldsymbol{q}_d(t)$ are bounded functions of time t. We are not concerned here about the control problem of the leader spacecraft, or about formation control with multiple leaders. For presentational simplicity, we drop the subscript k in the variables involved in the relative-motion equations (6.28).

For the ith spacecraft ($i \in \mathcal{I}$), the 3-D position tracking error $\tilde{\boldsymbol{q}}_i(t) = (\tilde{q}_i^x, \tilde{q}_i^y, \tilde{q}_i^z)^T \in R^3$ is defined as

$$\tilde{\boldsymbol{q}}_i = \boldsymbol{q}_{ik}(t) - \boldsymbol{q}_d(t), i \in \mathcal{I}, \tag{6.35}$$

where $\boldsymbol{q}_{ik}(t)$ evolves according to (6.28), and the superscripts x, y, and z are added to indicate the variables for different the motion axes. For the ith spacecraft, the three-axis motion synchronization corresponds to

$$\tilde{q}_i^x = \tilde{q}_i^y = \tilde{q}_i^z = 0. \tag{6.36}$$

For an MVS consisting of N spacecraft as followers, motion synchronization or formation flying corresponds to

$$\tilde{\boldsymbol{q}}_1 = \tilde{\boldsymbol{q}}_2 = \cdots = \tilde{\boldsymbol{q}}_N = \boldsymbol{0}_3, \tag{6.37}$$

Since the motion equation considered has been successfully converted into the form (5.166), the control strategy proposed in Section 5.7 is easily applied to either of the above two motion synchronization problems. In the following section, two applications are considered.

6.3 Applications and Simulation Results

6.3.1 Application 1: Leader–Follower Spacecraft Pair

6.3.1.1 Simulation Condition

A leader–follower formation flying configuration and the synchronization of multiple-axis motion are considered in this section. The leader spacecraft follows an elliptical orbit with orbital elements as follows:

- semi-major axis $a = 42241$ km
- eccentricity $e = 0.2$
- mean motion $n = 7.2722 \times 10^{-5}$ rad/s.

The masses of the leader and the follower spacecraft are $m_l = 1550$ kg and $m_f = 410$ kg and the disturbance vector $\boldsymbol{F}_d = [-1.025 \quad 6.248 \quad -2.415]^T \times 10^{-5}$ N.

The desired relative position trajectory is

$$q_{di}^x(t) = 100\sin(4\omega t)[1.0 - \exp(-0.05t^3)] \text{ m}, \tag{6.38}$$

$$q_{di}^y(t) = 100\cos(4\omega t)[1.0 - \exp(-0.05t^3)] \text{ m}, \tag{6.39}$$

$$q_{di}^z(t) = 0. \tag{6.40}$$

The initial conditions are given by

$$\boldsymbol{q}_i(0) = [30 \quad 0 \quad -200]^T \text{ m}, \tag{6.41}$$

$$\dot{\boldsymbol{q}}_i(0) = [0 \quad 0 \quad 0]^T \text{ m/s}, \tag{6.42}$$

$$\hat{\boldsymbol{\Theta}}_i(0) = \text{diag}[0.8 \quad 0.7 \quad 0.7 \quad 0.8]\boldsymbol{\Theta}_i, \tag{6.43}$$

where $\hat{\boldsymbol{\Theta}}_i(t)$ denotes the estimate of $\boldsymbol{\Theta}_i$ at time t, as defined in Section 5.7.

6.3.1.2 Control Parameters

The control and adaptation gains for controller (5.173) are

$$\boldsymbol{K}_D = \text{diag}[0.13 \quad 0.12 \quad 0.09], \boldsymbol{K}_S = \text{diag}[0.03 \quad 0.03 \quad 0.03], \tag{6.44}$$

$$\boldsymbol{\Lambda} = \text{diag}[0.04 \quad 0.04 \quad 0.04], \boldsymbol{B} = \text{diag}[8.0 \quad 8.0 \quad 8.0] \times 10^{-4}, \tag{6.45}$$

$$\Gamma = \text{diag}[900 \quad 28 \quad 28 \quad 9] \times 10^{-5}. \tag{6.46}$$

For multiple-axis motion synchronization, we consider the Type II LSEs with $\mathcal{L}_N$ given by (3.45):

$$\mathcal{L}_3 = \begin{bmatrix} 2 & -1 & -1 \\ -1 & 2 & -1 \\ -1 & -1 & 2 \end{bmatrix}. \tag{6.47}$$

6.3.1.3 Simulation Results and Analysis

Figure 6.1 shows the simulation results of adaptive tracking control of formation flying without a synchronization strategy. Figure 6.2 gives the corresponding simulation results with a synchronization strategy. For the purpose of comparison, the 2-norms of $\tilde{\boldsymbol{q}}(t)$ and $\Xi(t)$ after 5 h are calculated for all simulations and are listed in Table 6.1. It can be seen from the results that although the position tracking error vector $\boldsymbol{e}(t) \rightarrow 0$ can be achieved using an adaptive controller without synchronization strategy, the differences among the position tracking errors of the three axes are obvious: the synchronization errors among different axes are large.

With the adaptive synchronization controller, the synchronization performance is clearly improved. Specifically, the 2-norms of the synchronization errors on the three axes resulting from control without synchronization strategy are 419.2 m, 1442.8 m and 1576.7 m. With a synchronization strategy, the 2-norms of the three-axis synchronization errors have become 88.1 m, 150.7 m and 160.3 m, respectively. Thus, the synchronization errors have been markedly reduced. Moreover, the effort needed to perform each of these control strategies is compared in Table 6.1. The results show that more fuel consumption is needed if synchronization control is applied. For example, to maneuver and maintain the x-axis relative position for 30 h, a fuel consumption of 382.1 N·s is needed, whereas 742.6 N·s is needed when the synchronization strategy is additionally applied.

6.3.2 Application 2: Multiple Spacecraft in Formation

For the multiple spacecraft case, we can apply the synchronization strategy in two ways: internal and external. The internal synchronization error, $\Xi(t)$, is the synchronization error between different axes of one spacecraft. This is the same as that in the leader–follower configuration. The external synchronization error, $\boldsymbol{E}(t)$, denotes the synchronization error between a given axis of all spacecraft. Then, we modify the coupled position error defined in (5.132) as

$$\boldsymbol{E}^*(t) = \tilde{\boldsymbol{q}}(t) + \mathbf{B}'\mathbf{T}_\beta \int_0^t \Xi(\tau)d\tau + \mathbf{A}'\mathbf{T}_E\mathbf{T}_\alpha \int_0^t \boldsymbol{E}(\tau)d\tau \in \mathbb{R}^{mN\times mN}, \tag{6.48}$$

where N denotes the number of spacecraft involved in the formation, m is the number of motion axes of each spacecraft (generally equal to 3), the involved variables are written into block vectors as

$$\boldsymbol{E}^* = [\boldsymbol{E}_1^{*T}, \cdots, \boldsymbol{E}_1^{*T}]^T \in \mathbb{R}^{mN}, \tag{6.49}$$

$$\tilde{\boldsymbol{q}} = [\tilde{\boldsymbol{q}}_1^T, \cdots, \tilde{\boldsymbol{q}}_N^T]^T \in \mathbb{R}^{mN}, \tag{6.50}$$

$$\Xi = [\Xi_1^T, \cdots, \Xi_N^T]^T \in \mathbb{R}^{mN}, \tag{6.51}$$

$$\boldsymbol{E} = [\boldsymbol{E}_1^T, \cdots, \boldsymbol{E}_m^T]^T \in \mathbb{R}^{mN}, \tag{6.52}$$

with

$$\boldsymbol{E}_i^* = [e_{i1}^*, \cdots, e_{im}^*]^T \in \mathbb{R}^m, \tag{6.53}$$

$$\tilde{\boldsymbol{q}}_i = [\tilde{q}_{i1}, \cdots, \tilde{q}_{im}]^T \in \mathbb{R}^m, \tag{6.54}$$

$$\Xi_i = [\xi_{i1}, \cdots, \xi_{im}]^T \in \mathbb{R}^m, \tag{6.55}$$

$$\boldsymbol{E}_i = [e_{i1}, \cdots, e_{iN}]^T \in \mathbb{R}^N, \tag{6.56}$$

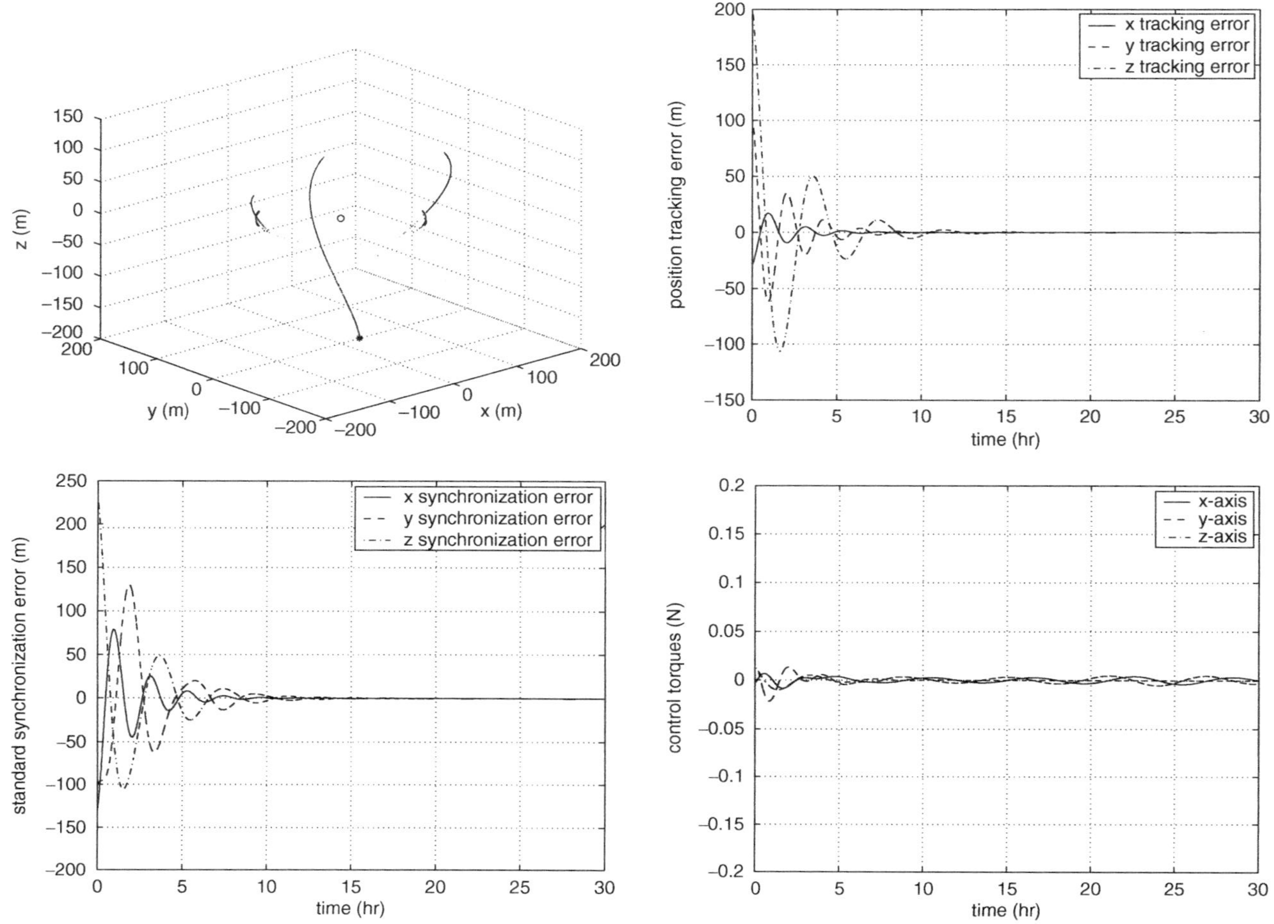

Figure 6.1 Application 1 without synchronization strategy.

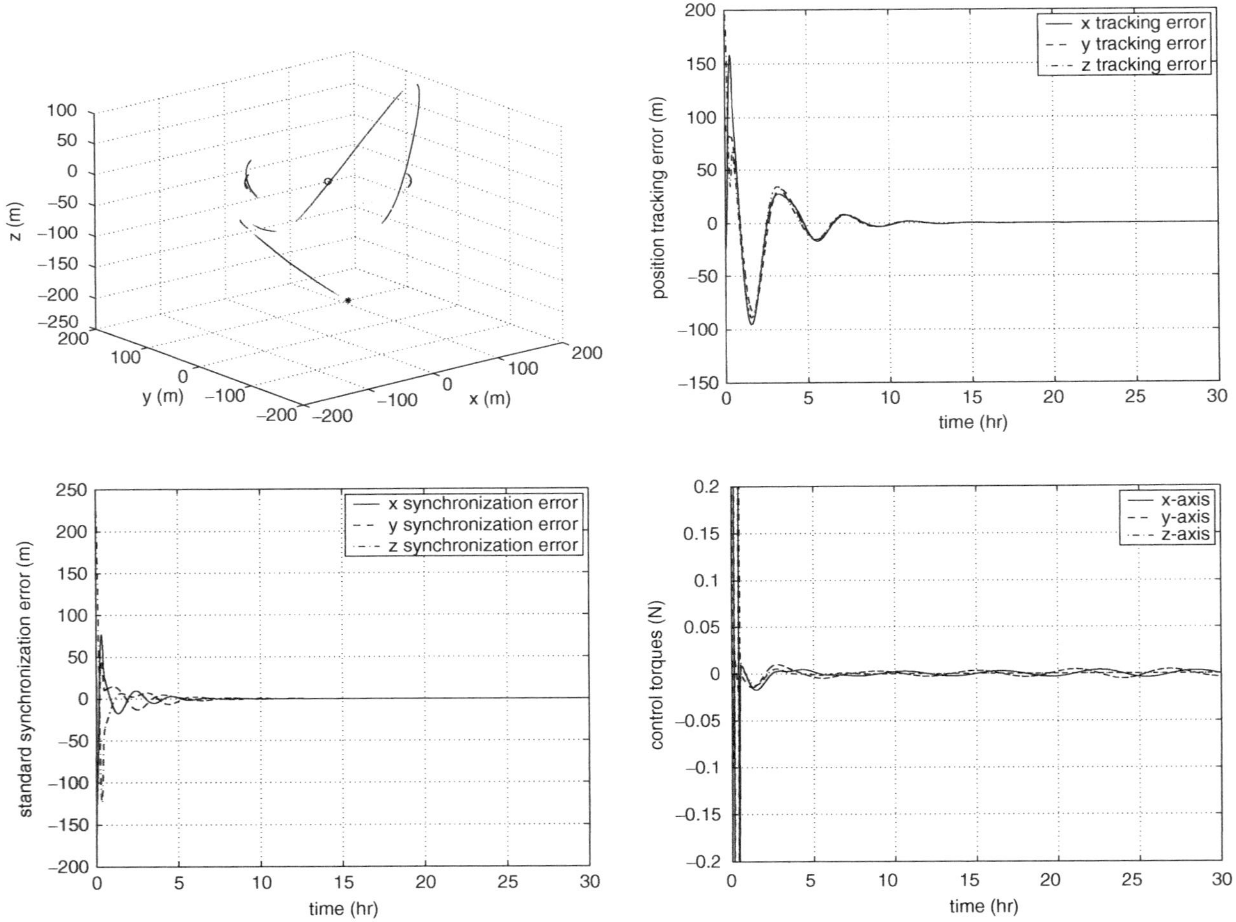

Figure 6.2 Application 1 with synchronization strategy.

Table 6.1 Performance evaluation without and with a synchronization strategy.

Error/control	Units	Without	With
$\|\|\tilde{q}_x\|\|_2$	(m)	78.8	940.8
$\|\|\tilde{q}_y\|\|_2$	(m)	341.5	940.8
$\|\|\tilde{q}_z\|\|_2$	(m)	1540.1	887.3
$\|\|\varepsilon_x\|\|_2$	(m)	419.2	88.1
$\|\|\varepsilon_y\|\|_2$	(m)	1442.8	150.7
$\|\|\varepsilon_z\|\|_2$	(m)	1576.7	160.3
$\int \|u_{xf}\|$	(N·s)	382.1	742.6
$\int \|u_{yf}\|$	(N·s)	498.9	527.9
$\int \|u_{zf}\|$	(N·s)	115.7	424.0

Table 6.2 Parameters of multiple spacecraft in formation flight.

Parameter	Unit	Value (i = 1, 2, 3, 4)
m_{fi}	kg	410, 500, 600, 660
$\boldsymbol{F}_{di}(\times 10^{-5})$	N	[−1.025, 6.248, −2.415], [1.9106, −1.960, −1.517] [−1.925, 4.850, −2.455], [−2.250, 6.850, −3.156]
$\mathbf{R}_{i0}$	m	[150, 10, 20], [−10, −130, −20] [−140, 10, −20], [30, 160, 20]
$\dot{\boldsymbol{R}}_{i0}$	m	[0, 0, 0]
$\mathbf{R}_{if}$	m	[100, 0, 0], [0, −100, 0], [−100, 0, 0], [0, 100, 0]
$\dot{\boldsymbol{R}}_{if}$	m	[0, 0, 0]
$\hat{\boldsymbol{\Theta}}(0)$		$0.7\boldsymbol{\Theta}, 0.85\boldsymbol{\Theta}, 1.15\boldsymbol{\Theta}, 1.3\boldsymbol{\Theta}$
$\mathbf{B}_i$ $(\times 10^{-3})$		[8.0, 8.0, 8.0]
$\mathbf{A}_i$ $(\times 10^{-3})$		[8.0, 8.0, 8.0]

and the gain matrices are written into block matrices as

$$\mathbf{B}' = diag[\mathbf{B}_1^T, \cdots, \mathbf{B}_N^T] \in \mathbb{R}^{mN\times mN}, \tag{6.57}$$

$$\mathbf{A}' = diag[\mathbf{A}_1^T, \cdots, \mathbf{A}_N^T] \in \mathbb{R}^{mN\times mN}, \tag{6.58}$$

$$\mathbf{T}_\beta = diag[\mathbf{T}_{\beta 1}, \cdots, \mathbf{T}_{\beta N}] \in \mathbb{R}^{mN\times mN}, \tag{6.59}$$

$$\mathbf{T}_\alpha = diag[\mathbf{T}_{\alpha 1}, \cdots, \mathbf{T}_{\alpha m}] \in \mathbb{R}^{mN\times mN}, \tag{6.60}$$

with matrices $\mathbf{T}_{\beta i} \in \mathbb{R}^{m\times m}$, $\mathbf{T}_{\alpha i} \in \mathbb{R}^{N\times N}$, and vectors

$$\mathbf{B}_i = [\beta_{i1}, \cdots, \beta_{im}]^T \in \mathbb{R}^m, \tag{6.61}$$

$$\mathbf{A}_i = [\alpha_{i1}, \cdots, \alpha_{im}]^T \in \mathbb{R}^m, \tag{6.62}$$

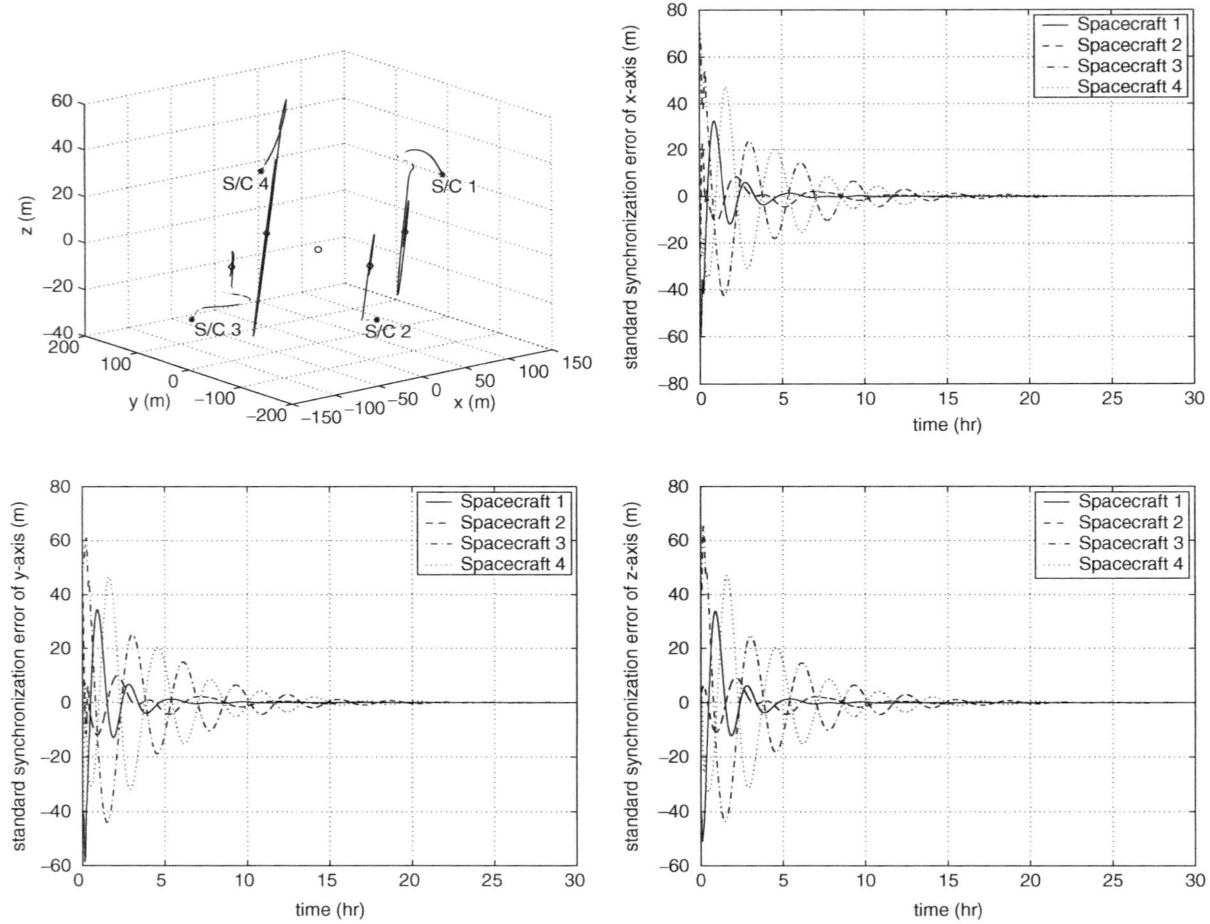

Figure 6.3 Application 2 with only internal synchronization.

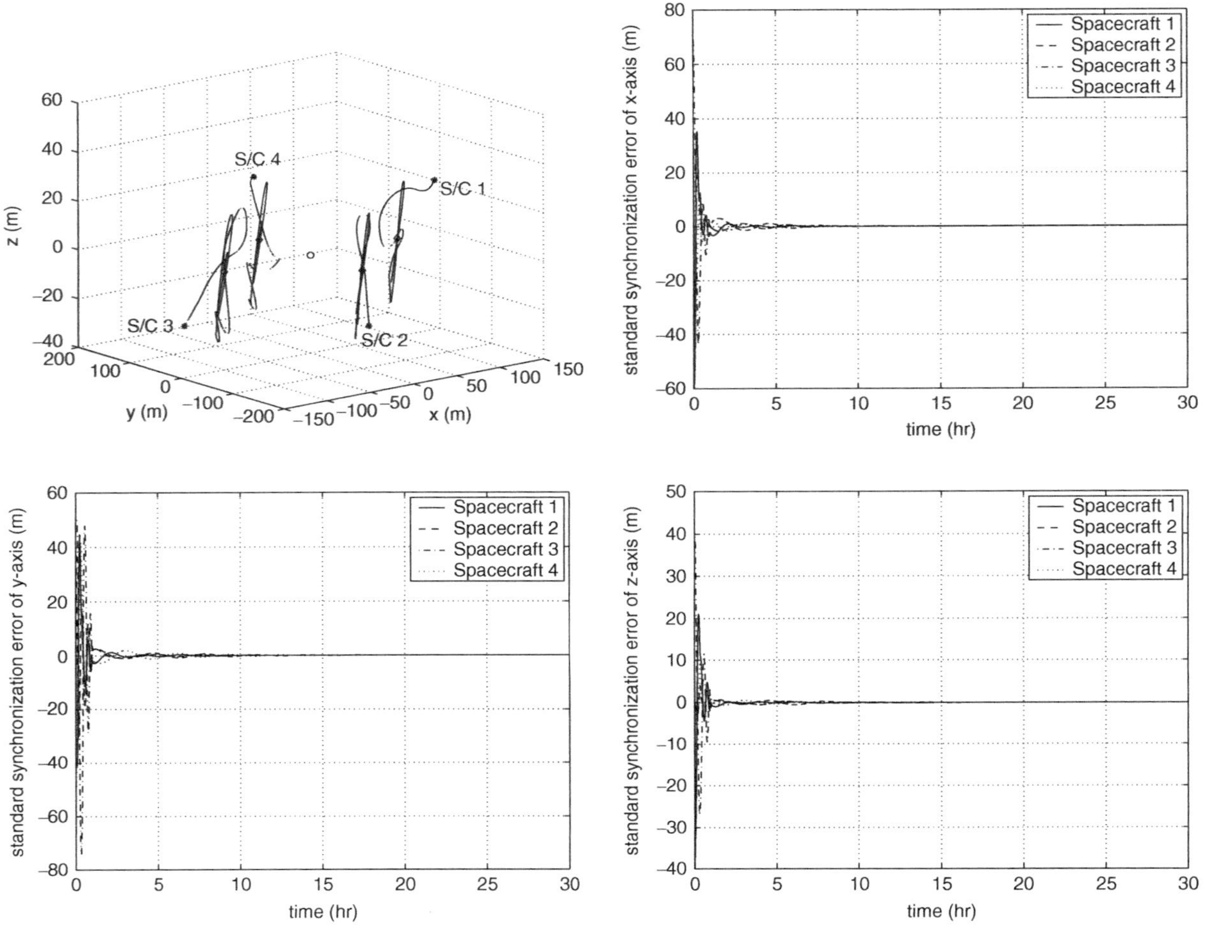

Figure 6.4 Application 2 with both internal and external synchronization.

and another transformation matrix $\mathbf{T}_E$ is

$$\mathbf{T}_E = \begin{cases} 1 & \text{for} \quad ((i-1)m+j,(j-1)N+i); i = 1,\cdots,N, j = 1,\cdots,m \\ 0 & \text{for} \quad \text{others.} \end{cases}$$

In the simulation, we assume four spacecraft are requested to maneuver from their initial relative positions $\mathbf{R}_{i0}$ ($i = 1,\ 2,\ 3,\ 4$) to final positions $\mathbf{R}_{if}$ along the following trajectory and to form a circular formation

$$\mathbf{R}_i^d(t) = \mathbf{R}_{i0} + (\mathbf{R}_{if} - \mathbf{R}_{i0}) \cdot [1 - \exp(C_i t^3)] \tag{6.63}$$

where $C_1 = -0.01$, $C_2 = -0.02$, $C_3 = -0.03$, $C_4 = -0.04$; that is, $N = 4$ and $m = 3$ in the simulation.

Table 6.2 shows the parameters and control gains for MSFF simulation. Other gains are the same as those in Application 1 (the leader–follower configuration). Figure 6.3 gives the simulation results using an internal synchronization strategy only. Figure 6.4 shows the results with both internal and external synchronization strategies. It can be seen from these simulation results that the synchronization errors of these four spacecraft about any given axis are markedly reduced by applying an external synchronization strategy.

6.4 Summary

This chapter describes the application of the adaptive nonlinear synchronization controller developed in Section 5.7 for multiple spacecraft formation fight. With the controller, both position tracking errors and position synchronization errors converge globally to zero even in the presence of uncertain parameters. It can achieve synchronized motion along multiple axes of one spacecraft, and/or any given axis of multiple spacecraft while ensuring the convergence of position tracking errors. Simulations are conducted for the leader–follower configuration and multiple spacecraft formation flight to verify the effectiveness of the controller.

7

Formation Control of Aerial Systems

Formation flight of multiple aircraft has been an active research topic for many years. For the classic leader–follower configuration, when the follower is properly positioned with respect to the leader aircraft, the drag on the follower can be markedly reduced due to the strong wingtip vortices generated by the leader aircraft. Such close-formation flight configurations can lead to reductions in fuel consumption and thus an increase in flight range. Real flight tests on Dryden F/A-18s have illustrated that close formation flight could bring 20% drag reductions and 15–18% fuel savings in the trailing aircraft in certain flight conditions [2, 3].

The high efficiency of close-formation flight relies on accurate relative position control between the follower and the leader aircraft, especially under the effect of coupled aerodynamics. Many control strategies have been proposed for close-formation flight with coupled aerodynamics.

In this chapter, we will introduce how to apply the motion synchronization control strategy proposed in Chapter 5 to multiple fixed-wing aircraft in order to realize close-formation flight. To be specific, we first apply a motion synchronization control strategy to synchronize the relative position tracking motion between multiple follower aircraft. The NASA–Hallock–Burnham vortex profile is used to calculate the vortex-induced forces and moments and the autopilot models of the followers are modified with consideration of the coupled aerodynamics. We then improve the robustness of the control strategy by adopting an uncertainty and disturbance estimator (UDE). Finally, we present simulation results to demonstrate the formation control performance.

7.1 Vortex-induced Aerodynamics

Figure 7.1 shows the schematic diagram of a formation flight. As shown in Figure 7.1a, the formation geometry between the leader and follower aircraft can be described by three relative coordinates: the longitudinal separation x, the lateral separation y, and the vertical separation z. Compared with the dynamics in free flight, the dynamics of aircraft in close-formation flight are much more complicated due to the aerodynamic interaction that arises from the vortex generated by the leader aircraft. As this formation flight phenomenon significantly alters the follower dynamics, its effect has to be sufficiently captured in modeling, so that the controller design will ensure reliable performance of the control system in the real operating environment.

Formation Control of Multiple Autonomous Vehicle Systems, First Edition. Hugh H.T. Liu and Bo Zhu.

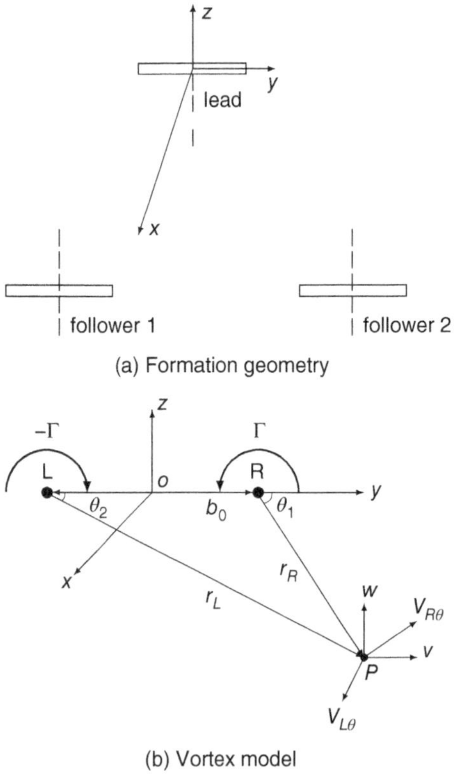

Figure 7.1 The schematic diagram of formation flight.

For close-formation flight, the impact of the longitudinal separation x on the induced forces and moments is much smaller than the lateral separation y and the vertical separation z [4]. Therefore, we neglect the effect of the longitudinal separation x in vortex-induced aerodynamics modeling without loss of generality. Furthermore, we assume that the leader and follower aircraft fly in parallel almost all of the time and there is no attitude difference between them (or the difference is small enough to be tracked quickly).

The tangential velocity $V_\theta(r)$ is frequently used to model a vortex in the rolled-up wake behind an aircraft. Note that the NASA–Hallock–Burnham profile correlates well with experimental data [4]. We here adopt it as follows:

$$V_\theta(r) = \frac{\Gamma}{2\pi r} \frac{r^2}{r^2 + r_c^2}, \tag{7.1}$$

where r is the radius from the vortex center, r_c is the core radius of the vortex, $\Gamma = Mg/\rho V b_0$ is the circulation that describes the vortex strength, Mg is the weight of the aircraft, V is the aircraft velocity, ρ is the air density at flight altitude h, b is the wingspan, and $b_0 = \pi b/4$ is the displacement between the vortex pair.

Figure 7.1b illustrates the tangential velocities $V_{R\theta}$ and $V_{L\theta}$ of the right-hand and the left-hand vortex respectively at point P. These tangential velocities can be decomposed into an upwash w and sidewash v. The vortex-induced upwash w changes the velocity vector of the follower aircraft. This change translates into an increase in the angle of attack and thus the lift. Therefore, the upwash w leads to changes in lift (ΔL), rolling moment (ΔR), and drag (ΔD) on the follower aircraft. On the other hand, the sidewash v at the vertical tail generates a side force (ΔSF). The changes in these force and moment coefficients, ΔC_L, ΔC_R, ΔC_D, and ΔC_{SF}, can be calculated as follows:

$$\Delta C_L = \frac{\Delta L}{(1/2)\rho V^2 A} = \frac{C_{L\alpha}}{VA} \int_0^b c(s) w(y+s,z) \mathrm{d}s \tag{7.2}$$

$$\Delta C_R = \frac{\Delta R}{(1/2)\rho V^2 Ab} = \frac{C_{L\alpha}}{VAb} \int_0^b c(s) w(y+s,z) \left(s - \frac{b}{2}\right) \mathrm{d}s \tag{7.3}$$

$$\Delta C_D = \frac{\Delta D}{(1/2)\rho V^2 A} = \frac{C_L + \Delta C_L}{VA} \int_0^b c(s) w(y+s,z) \mathrm{d}s \tag{7.4}$$

$$\Delta C_{SF} = \frac{\Delta SF}{(1/2)\rho V^2 A} = \frac{C_{vt}}{VA} \int_0^{h_z} c_{\mathrm{tail}}(s) v(y, z+s) \mathrm{d}s \tag{7.5}$$

where the lift curve slope $C_{L\alpha} = 5.67$ is used [4], C_L is the local lift coefficient, $c(s)$ is the chord distribution along the wing, A is the wing area, C_{vt} is the lift curve slope of the vertical tail, h_z is the tail height, and $c_{\mathrm{tail}}(s)$ is a width function of the vertical tail. The upwash $w(y+s,z)$ and sidewash $v(y,z+s)$ have the following expressions:

$$w(y+s,z) = \frac{\Gamma}{2\pi} \left\{ \frac{y+s-b_1}{(y+s-b_1)^2 + z^2 + r_c^2} - \frac{y+s-b_2}{(y+s-b_2)^2 + z^2 + r_c^2} \right\} \tag{7.6}$$

$$v(y,z+s) = \frac{\Gamma}{2\pi} \left\{ \frac{z+s}{(y+\pi b/8)^2 + (z+s)^2 + r_c^2} - \frac{z+s}{(y-\pi b/8)^2 + (z+s)^2 + r_c^2} \right\} \tag{7.7}$$

with $b_1 = (1 + \pi/4)b/2$ and $b_2 = (1 - \pi/4)b/2$.

Consider a pair of F/A-18 class aircraft in formation flight. The induced force and moment coefficients are shown in Figure 7.2. The parameters of the aircraft are given in Table 7.1 and taken from http://www.fas.org/man/dod-101/sys/ac/f-18.htm. The results show that the optimal relative positions for maximal induced lift are $[y_o,\ z_o] = [\pm(1 + \pi/4)b/2,\ 0] = [\pm 10.2\ \mathrm{m},\ 0\ ]\ \mathrm{m}$.

The upwash and sidewash also introduce changes in moments on the follower aircraft, including the yawing moment, pitching moment, and the rolling moment. The rolling moment model is introduced here since it is one factor of stability, especially when a large aircraft followed by a smaller one. If the leader and follower aircraft are the same type, such as F/A-18 aircraft, flight tests show that the induced moments on

Table 7.1 Parameters of F/A-18 class aircraft.

Parameter	Value	Parameter	Value
M, kg	10810	V, m/s	236.0
A, m^2	37.16	h, m	12192
b, m	11.43	ρ, kg/m^3	0.3031
c_R, m	4.04	c_T, m	1.68
AR	3.52	h_z, m	2.7

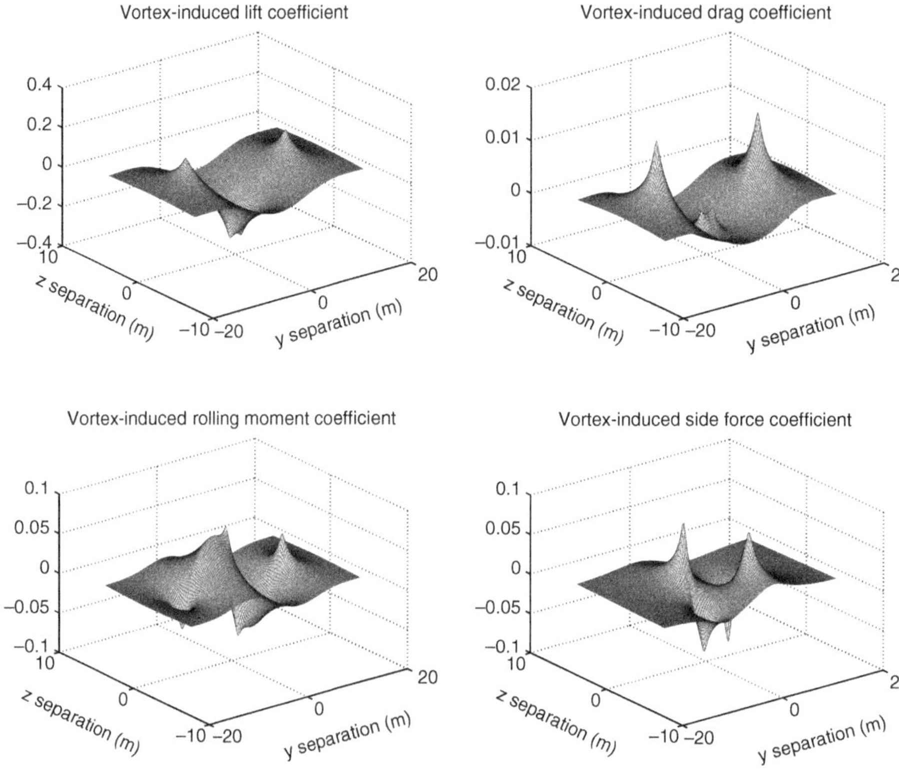

Figure 7.2 The vortex-induced force and moment coefficients.

the follower are easily controllable. We here introduce two typical models, developed using the Biot–Savart law and Kutta–Joukowski theorem, respectively.

7.1.1 Model of the Trailing Vortices of Leader Aircraft

In this section, the trailing vortices of the leader aircraft are modelled using two different approaches. The difference between the two resulting models can be utilized to verify the robustness of controller, as will be shown later.

Lemma 7.1 ***(Biot–Savart law [5])***

As shown in Figure 7.3, the induced velocity at P by the vortex segment $\overline{AB}$ is

$$q = \frac{\Gamma}{4\pi d}(\cos\beta_1 - \cos\beta_2) \tag{7.8}$$

where Γ is the vortex strength and d is the distance from Point P to the vortex segment.

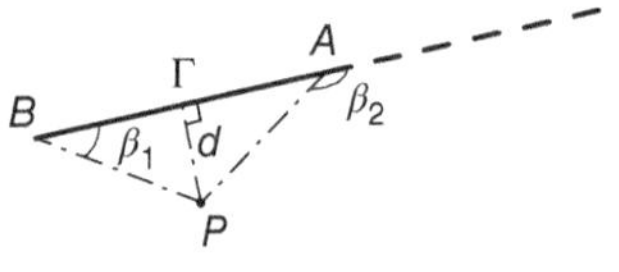

Figure 7.3 Biot–Savart law.

Remark 7.2 Lemma 7.1 is a version of the Biot-Savart law for a particular finite vortex filament. In the context, the vortex filaments shed by the leader aircraft are always assumed to be semi-infinite, which means $\beta_2 = \pi$. Therefore the equation

$$q = \frac{\Gamma}{4\pi d}(1 + \cos \beta_1) \tag{7.9}$$

is commonly used.

Lemma 7.3 ***(Kutta–Joukowski theorem)***
Let L' be the lift per unit span on an airfoil. Then L' has the following relationship with the vortex strength Γ on that airfoil.

$$L' = \rho_\infty V_\infty \Gamma \tag{7.10}$$

where ρ_∞ is the air density and V_∞ denotes the airspeed relative to the airfoil.

7.1.2 Single Horseshoe Vortex Model

The single horseshoe vortex model (SHVM) that will be introduced in this section is also known as the Burnham–Hallock model. In this model, Equation (7.9) will be rewritten as

$$q = \frac{\Gamma}{4\pi} \frac{d}{d^2 + r_c^2}(1 + \cos \beta_1) \tag{7.11}$$

where r_c denotes the vortex core radius. In the single horseshoe vortex model, only a pair of vortex filaments are assumed to be shed at the wingtips of the leader aircraft. The lift distribution along a wing span is assumed to be constant. In light of (7.10), we can get

$$L = bL' = b\rho_\infty V_\infty \Gamma \simeq W \Rightarrow \Gamma \simeq \frac{W}{\rho_\infty V_\infty b} \tag{7.12}$$

where W is the weight of the aircraft. Usually, a reduced wing span $b' = \frac{\pi b}{4}$ is used instead of b in the SHVM to compensate for an ideal elliptical lift distribution along a wing span. Then, (7.12) is rewritten as

$$\Gamma \simeq \frac{W}{\rho_\infty V_\infty b'} \tag{7.13}$$

As shown in Figure 7.4, the induced upwash velocity distribution due to the right-hand vortex filament is

$$w_r(s) = q_r \sin(\lambda) = \frac{\Gamma}{4\pi} \frac{\Delta y + s}{(\Delta y + s)^2 + \Delta z^2 + r_c^2} \left[1 + \frac{\Delta x}{\sqrt{\Delta x^2 + \Delta z^2 + (\Delta y + s)^2}} \right] \tag{7.14}$$

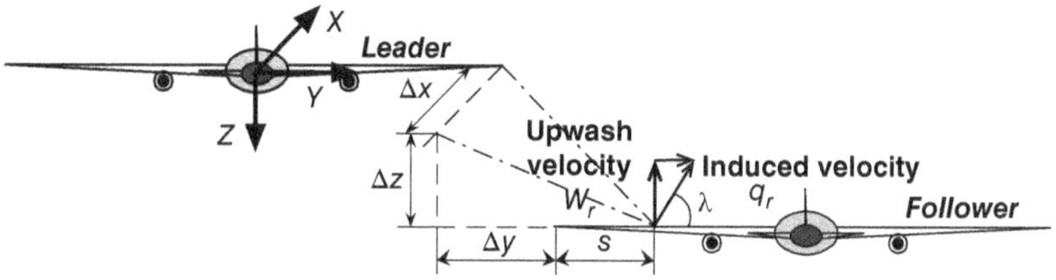

Figure 7.4 Induced velocity on follower by the right-hand vortex filament.

where Γ is from (7.13). Similarly, the induced upwash velocity due to the left-hand vortex filament is calculated as

$$w_l(s) = \frac{\Gamma}{4\pi} \frac{\Delta y + b + s}{(\Delta y + b + s)^2 + \Delta z^2 + r_c^2} \left[1 + \frac{\Delta x}{\sqrt{\Delta x^2 + \Delta z^2 + (\Delta y + b + s)^2}} \right] \tag{7.15}$$

The total induced upwash velocity distribution along the trailing aircraft wing span is

$$w(s) = w_l(s) + w_r(s) \tag{7.16}$$

The total sidewash velocity at $s = \frac{b}{2}$ is computed as:

$$\begin{aligned} u(z) = {} & \frac{\Gamma}{4\pi} \frac{\Delta z + z}{\left(\Delta y + \frac{b}{2}\right)^2 + (\Delta z + z)^2 + r_c^2} \left[1 + \frac{\Delta x}{\sqrt{\Delta x^2 + (\Delta z + z)^2 + \left(\Delta y + \frac{b}{2}\right)^2}} \right] \\ & + \frac{\Gamma}{4\pi} \frac{\Delta z + z}{\left(\Delta y + \frac{3b}{2}\right)^2 + (\Delta z + z)^2 + r_c^2} \left[1 + \frac{\Delta x}{\sqrt{\Delta x^2 + (\Delta z + z)^2 + \left(\Delta y + \frac{3b}{2}\right)^2}} \right] \end{aligned} \tag{7.17}$$

where z is from 0 to the top of the vertical tail at $-h_z$, denoting the distance from a certain point at the vertical tail to the tail bottom.

The induced upwash will increase the angle of attack of the trailing aircraft by $\Delta\alpha$. The aerodynamic forces of the trailing aircraft would be rotated as illustrated in Figure 7.5. The induced lift at a certain point on the trailing aircraft wing is

$$d\Delta L(s) = \frac{1}{2} \rho_\infty V_\infty^2 c(s) C_{l_\alpha}(s) \Delta\alpha(s) ds \Rightarrow \Delta L(s) = \int_0^b d\Delta L(s) ds \tag{7.18}$$

where $c(s)$ denotes the chord at this point, $C_{l_\alpha}(s)$ is the local section lift curve slope, and $\Delta\alpha(s)$ is the induced angle of attack, calculated as

$$\Delta\alpha(s) = \tan^{-1}\left(\frac{w(s)}{V_\infty}\right) \approx \frac{w(s)}{V_\infty} \tag{7.19}$$

The reduction in induced drag is due to the forward tilting of the lift, as shown in Figure 7.5. So the drag reduction is:

$$\Delta D = L' \sin(\Delta\alpha) \approx L' \frac{\overline{w}}{V_\infty} \tag{7.20}$$

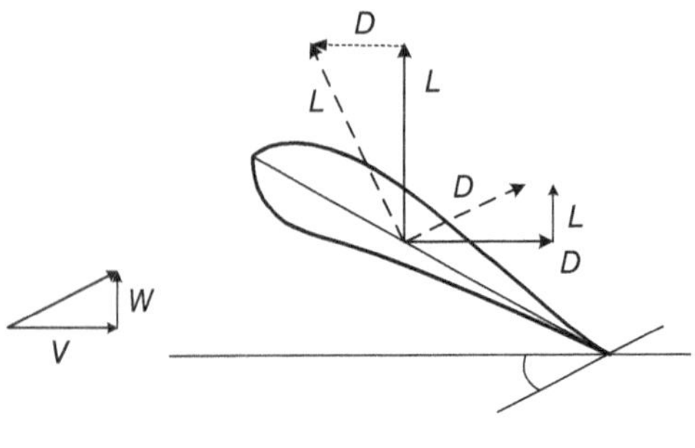

Figure 7.5 Rotation of the aerodynamic forces.

where $\overline{w} = \frac{1}{b}\int_0^b w(s)ds$ represents the averaged upwash velocity along the trailing aircraft wing. The induced side force is [6]:

$$\Delta F_Y = \eta \overline{q} S_{vt} a_{vt} \frac{\overline{u}}{V_\infty} \tag{7.21}$$

where η is the aerodynamic efficiency factor at the tail, S_{vt} is the vertical tail area, a_{vt} is the lift curve slope of the vertical tail, $\overline{q} = \frac{\rho_\infty V_\infty^2}{2}$ is the dynamic pressure, and $\overline{u} = \frac{1}{h_z}\int_0^{h_z} u(z)dz$.

7.1.3 Continuous Vortex Sheet Model

In the continuous vortex sheet model (CVSM), the lift distribution is assumed to vary in the spanwise direction. An elliptical distribution is assumed for the lift distribution in our analysis. According to Lemma 7.3, the vortex strength Γ has an elliptical distribution similar to that shown in Figure 7.6. The elliptical distribution equation is

$$\Gamma(y) = \Gamma_0 \sqrt{1 - \left(\frac{2y}{b}\right)^2}, -\frac{b}{2} \le y \le \frac{b}{2} \tag{7.22}$$

where $\Gamma_0 = \frac{4L}{\rho_\infty V_\infty b\pi}$.

In this model, the induced upwash velocity at a certain point on the trailing aircraft wing (Figure 7.4) is

$$w(s) = \frac{1}{4\pi}\int_{-\frac{b}{2}}^{\frac{b}{2}} \frac{d\Gamma/dy\left(\Delta y + s + \frac{b}{2} - y\right)}{\Delta z^2 + \left(\Delta y + s + \frac{b}{2} - y\right)^2} f_w(s)dy, \tag{7.23}$$

where

$$\frac{d\Gamma}{dy} = -\frac{4\Gamma_0}{b^2}\frac{y}{\sqrt{1-\left(\frac{2y}{b}\right)^2}}, \tag{7.24}$$

$$f_w(s) = 1 + \frac{\Delta x}{\sqrt{\Delta x^2 + \Delta z^2 + \left(\Delta y + s + \frac{b}{2} - y\right)^2}}. \tag{7.25}$$

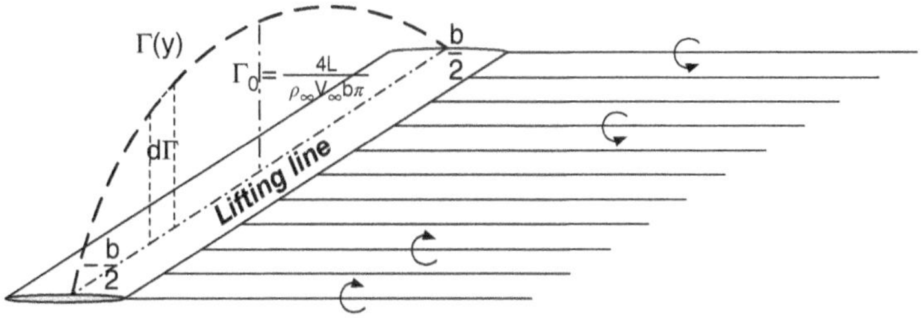

Figure 7.6 Continuous vortex sheet.

The sidewash velocity distribution on the vertical tail of the trailing aircraft is

$$u(z) = \frac{1}{4\pi}\int_{-\frac{b}{2}}^{\frac{b}{2}} \frac{d\Gamma/dy(\Delta z + z)}{(\Delta z + z)^2 + \left(\Delta y + \frac{3b}{2} - y\right)^2} f_u(z)dy, \tag{7.26}$$

where

$$f_u(z) = 1 + \frac{\Delta x}{\sqrt{\Delta x^2 + (\Delta z + z)^2 + \left(\Delta y + \frac{3b}{2} - y\right)^2}}. \tag{7.27}$$

Then, we can obtain the induced lift, drag reduction and induced side force from the CVSM, by substituting (7.23) and (7.26) into each of (7.18), (7.20) and (7.21). Using the F-16 data given by Pachter *et al.* [6], we obtain the curves of ΔC_D and ΔC_L with respect to Δy shown in Figure 7.7. In Figure 7.7a, we compare the predictions of ΔC_D and ΔC_L using the SHVM and CVSM, respectively, where a negative ΔC_D means that formation flight leads to a drag reduction. It can be seen that the predictions by the SHVM and CVSM have similar trends but different values at some locations.

The CVSM was developed under the assumption that the lift distribution along a wing span is elliptical. In this case, the aerodynamic effect should be smaller than the real effect. The SHVM is derived under the assumption of a constant lift distribution of magnitude $\Gamma = \frac{4L}{\pi\rho_\infty V_\infty}$. Hence, the prediction obtained by the SHVM is usually much larger than the real effect.

Since no experimental data are available, the total drag coefficient is not accessible for a free-flight aircraft. However, the induced drag coefficient can be estimated by $C_{D_i} = \frac{C_L^2}{e\pi AR}$, where the efficiency factor $e = 0.74 \sim 0.8$ for many subsonic jet aircraft. For a free-flight F-16 aircraft, the induced drag coefficient is around 0.05. As for ΔC_D, the difference among the estimates using the two models is obvious at some positions, for instance at $\Delta y = -0.1b$.

7.2 Aircraft Autopilot Models

Aircraft in formation are equipped with a flight control system including three autopilots: Mach-hold, heading-hold, and altitude-hold. These are introduced in the paper by Pachter *et al.* [6] and are given by

$$\dot{V} = \frac{1}{\tau_V}(V_c - V) \tag{7.28}$$

$$\dot{\psi} = \frac{1}{\tau_\psi}(\psi_c - \psi) \tag{7.29}$$

$$\ddot{h} = -\left[\frac{1}{\tau_{h_a}} + \frac{1}{\tau_{h_b}}\right]\dot{h} - \frac{1}{\tau_{h_a}\tau_{h_b}}h + \frac{1}{\tau_{h_a}\tau_{h_b}}h_c \tag{7.30}$$

where subscript c is added to variables that are the command signals of the autopilot. ψ is the heading angle, and τ_V, τ_ψ, τ_{ha} and τ_{hb} are the aircraft velocity and heading angle, and two time constants, respectively.

In formation flight, these autopilots can be directly applied to the leader aircraft. For the follower aircraft, the models need to be modified to take into account the vortex-induced aerodynamic forces.

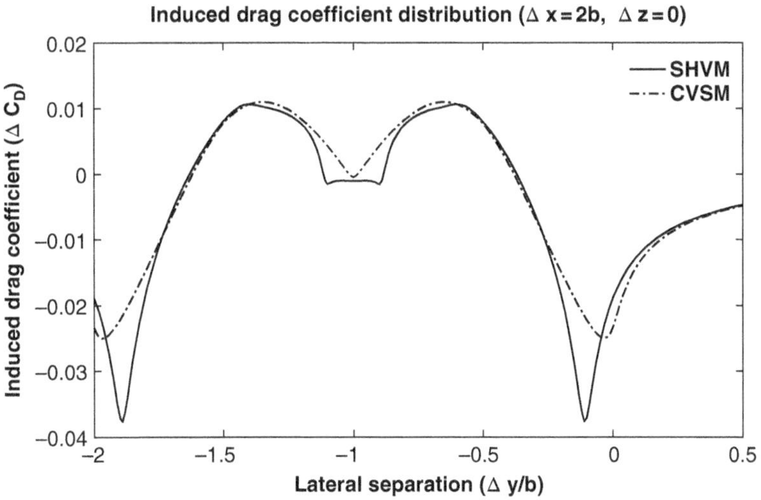

(a) ΔC_D distribution with respect to Δ_y

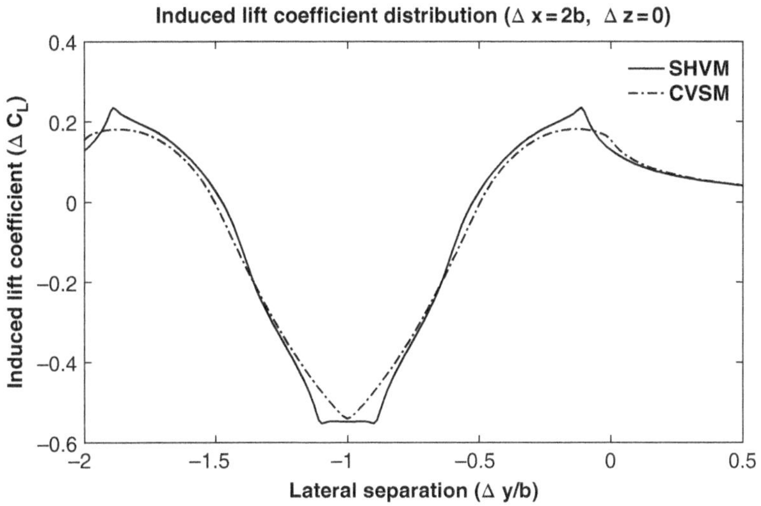

(b) ΔC_L distribution with respect to Δ_y

Figure 7.7 Comparison of the predictions of SHVM and CVSM.

7.2.1 Models for the Follower Aircraft

Modifying (7.28)–(7.30) as in Pachter *et al.* [6] yields the following autopilot models for a follower aircraft:

$$\dot{V}_F = \frac{1}{\tau_V}(V_{Fc} - V_F) + \frac{qA}{M}\Delta C_{DFy}(y - y_d) \tag{7.31}$$

$$\dot{\psi}_F = \frac{1}{\tau_\psi}(\psi_{Fc} - \psi_F) + \frac{qA}{MV}\left[\Delta C_{SFy}(y - y_d) + \Delta C_{SFz}(z - z_d)\right] \tag{7.32}$$

$$\ddot{h}_F = -\left[\frac{1}{\tau_{h_a}} + \frac{1}{\tau_{h_b}}\right]\dot{h}_F - \frac{1}{\tau_{h_a}\tau_{h_b}}h_F + \frac{1}{\tau_{h_a}\tau_{h_b}}h_{Fc} + \frac{qA}{M}\Delta C_{LFy}(y - y_d) \tag{7.33}$$

where subscript F denotes the follower aircraft, $q = \rho V^2/2$ is the dynamic pressure, and ΔC_{DFy}, ΔC_{SFy}, ΔC_{SFz}, ΔC_{LFy} are the stability derivatives evaluated at the optimal relative position $[y_o, z_o]$.

As in Pachter *et al.* [6], the induced moments are not incorporated into the design of the outer-loop formation-hold autopilots, and the formation control task is to construct commands V_{Fc}, ψ_{Fc} and h_{Fc} for system (7.31)–(7.33).

7.2.2 Kinematics for Close-formation Flight

The kinematics between the follower and leader aircraft are governed by:

$$\dot{x} = -y\dot{\psi}_F - V_F + V_L \cos e_\psi \tag{7.34}$$

$$\dot{y} = x\dot{\psi}_F + V_L \sin e_\psi, \tag{7.35}$$

where $e_\psi = \psi_F - \psi_L$ is the heading angle error. Substituting equations (7.31)–(7.33) into the kinematic equations, we can obtain the following six-dimensional non-linear equations for the follower aircraft

$$\begin{cases} \dot{x} &= -\dfrac{y}{\tau_\psi}(\psi_{Fc} - \psi_F) - \frac{qA}{MV}[\Delta C_{SFy}(y - y_d) + \Delta C_{SFz}(z - z_d)]y \\ & \quad -V_F + V_L\cos(\psi_F - \psi_L) \\ \dot{y} &= \dfrac{x}{\tau_\psi}(\psi_{Fc} - \psi_F) + \frac{qA}{MV}[\Delta C_{SFy}(y - y_d) + \Delta C_{SFz}(z - z_d)]x \\ & \quad +V_L\sin(\psi_F - \psi_L) \\ \dot{V}_F &= \dfrac{1}{\tau_V}(V_{Fc} - V_F) + \frac{qA}{M}\Delta C_{DFy}(y - y_d) \\ \dot{\psi}_F &= \dfrac{1}{\tau_\psi}(\psi_{Fc} - \psi_F) + \frac{qA}{MV}[\Delta C_{SFy}(y - y_d) + \Delta C_{SFz}(z - z_d)] \\ \dot{Z} &= \xi \\ \dot{\xi} &= -\left[\dfrac{1}{\tau_{h_a}} + \dfrac{1}{\tau_{h_b}}\right]\xi - \dfrac{1}{\tau_{h_a}\tau_{h_b}}z + \dfrac{1}{\tau_{h_a}\tau_{h_b}}h_{Fc} + \dfrac{qA}{M}\Delta C_{LFy}(y - y_d) \\ & \quad -\dfrac{1}{\tau_{h_a}\tau_{h_b}}h_{Lc} \end{cases}$$

where the signals associated with the leader aircraft, V_L, ψ_L and h_{Lc}, are considered as disturbances.

7.3 Controller Design

7.3.1 Linear Proportional-integral Controller

A triangular-formation flight configuration, with one leader and two followers – one behind and to the right, the other behind and to the left – is considered. The control objective is two-fold:

- maintain optimal relative positions between the follower and the leader aircraft in order to obtain the maximal induced lift, even in the presence of maneuvers by the leader
- regulate the responses of two follower aircraft to achieve synchronized motion tracking.

A linear proportional integral (PI) controller is employed for the tracking control of the follower aircraft [7]. The controller for the x/y channel contains a linear mixer on the x/y error signals and proportional plus integral action. The z channel controller is a standard PI controller driven by the tracking error:

$$V_{Fc_i} = K_{xp_i} E_{x_i} + K_{xi_i} \int_0^t E_{x_i} \mathrm{d}t, \tag{7.36}$$

$$\psi_{Fc_i} = K_{yp_i} E_{y_i} + K_{yi_i} \int_0^t E_{y_i} \mathrm{d}t, \tag{7.37}$$

$$h_{Fc_i} = K_{zp_i} e_{z_i} + K_{zi_i} \int_0^t e_{z_i} \mathrm{d}t + h_{0_i}, \tag{7.38}$$

with

$$E_{x_i} = k_{x_i} e_{x_i} + k_{V_i} e_{V_i} \qquad E_{y_i} = k_{y_i} e_{y_i} + k_{\psi_i} e_{\psi_i}, \tag{7.39}$$

where i denotes the ith follower aircraft, $e_{x_i} = x_i - x_{di}$, $e_{y_i} = y_i - y_{di}$, and $e_{z_i} = z_i - z_{di}$ are the relative position tracking errors, $e_{\psi_i} = \psi_{F_i} - \psi_L$ is the heading angle tracking error, $e_{V_i} = V_{F_i} - V_L$ is the velocity tracking error, K_{xp_i}, K_{xi_i}, K_{yp_i}, K_{yi_i}, K_{zp_i}, K_{zi_i}, k_{x_i}, k_{V_i}, k_{y_i}, and k_{ψ_i} are control gains, and h_{0_i} is the initial flight altitude.

The cross-coupling concept introduced in Section 5.6 is used to synchronize the relative position tracking motion of the two follower aircraft. Consider the Type II position synchronization errors as follows:

$$\varepsilon_{x_1} = e_{x_1} - e_{x_2} \qquad \varepsilon_{x_2} = e_{x_2} - e_{x_1} \tag{7.40}$$

$$\varepsilon_{y_1} = e_{y_1} - e_{y_2} \qquad \varepsilon_{y_2} = e_{y_2} - e_{y_1} \tag{7.41}$$

$$\varepsilon_{z_1} = e_{z_1} - e_{z_2} \qquad \varepsilon_{z_2} = e_{z_2} - e_{z_1} \tag{7.42}$$

Then, the coupled position errors are formed so as to include both the position tracking errors and the position synchronization errors:

$$e^*_{x_i} = e_{x_i} + \beta_{x_i} \varepsilon_{x_i}, \tag{7.43}$$

$$e^*_{y_i} = e_{y_i} + \beta_{y_i} \varepsilon_{y_i}, \tag{7.44}$$

$$e^*_{z_i} = e_{z_i} + \beta_{z_i} \varepsilon_{z_i}. \tag{7.45}$$

where β_{x_i}, β_{y_i}, and β_{z_i} are positive synchronization gains for the x, y, and z channels of the ith follower. Therefore, the generalized errors for x and y channels are

$$E^*_{x_i} = k_{x_i} e^*_{x_i} + k_{V_i} e_{V_i} \qquad E^*_{y_i} = k_{y_i} e^*_{y_i} + k_{\psi_i} e_{\psi_i}. \tag{7.46}$$

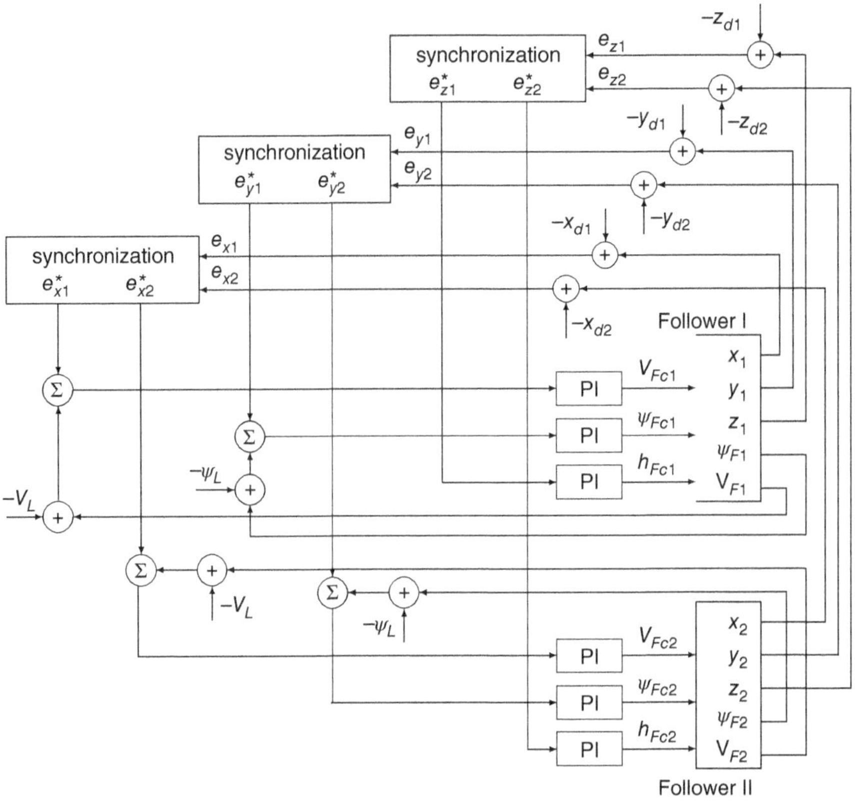

Figure 7.8 The structure of the formation flight controller.

Substituting E_{x_i}, E_{y_i}, and e_{z_i} into (7.36)–(7.38) with $E^*_{x_i}$, $E^*_{y_i}$, $e^*_{z_i}$, respectively, yields the synchronized tracking controllers

$$V_{Fc_i} = K_{xp_i}E^*_{x_i} + K_{xi_i}\int_0^t E^*_{x_i}\,\mathrm{d}t, \tag{7.47}$$

$$\psi_{Fc_i} = K_{yp_i}E^*_{y_i} + K_{yi_i}\int_0^t E^*_{y_i}\,\mathrm{d}t, \tag{7.48}$$

$$h_{Fc_i} = K_{zp_i}e^*_{z_i} + K_{zi_i}\int_0^t e^*_{z_i}\,\mathrm{d}t + h_{0_i}. \tag{7.49}$$

The overall structure of the proposed formation flight controller is illustrated in Figure 7.8. The position tracking errors of the two follower aircraft are fed to synchronization blocks to generate the coupled position errors, which are mixed by the heading angle or velocity tracking errors to form the final error signals for the PI controller.

7.3.2 UDE-based Formation-flight Controller

The proposed formation control is shown in Figure 7.9. The principal formation controller aims to stabilize the formation. The constant compensation is introduced to reject the slowly-varying component of the formation aerodynamic effects. The other

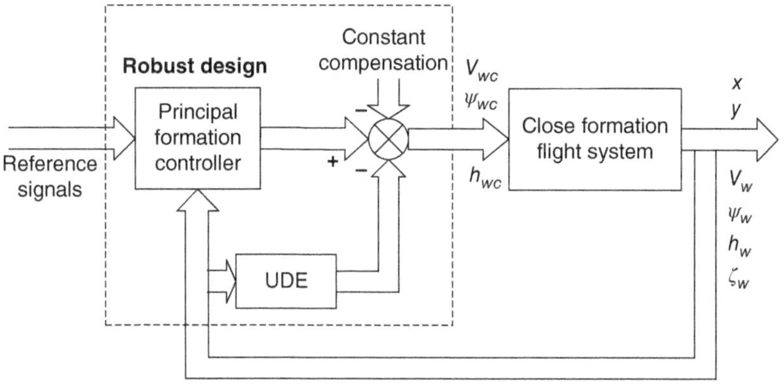

Figure 7.9 Robust control configuration for close-formation flight.

unpredictable uncertain component are compensated by the signal generated by a UDE. The basic idea behind the design of UDE is to choose a suitable filter to estimate the uncertainties and disturbances using system states and control inputs [8]. The filter may be chosen according to the frequency-domain characteristics of the uncertainties or disturbances. For instance, if the uncertainties to be estimated vary slowly with respect to time (i.e. they are a low-frequency signal), a low-pass filter can be chosen. Applications of UDE-based control can be found in many publications in the literature [9–11].

7.3.2.1 Formation Flight Controller Design

A nonlinear formation flight controller is developed based on the following nominal formation flight dynamics:

$$\begin{cases} \dot{x} &= V_l \cos\psi_e + y\left[-\frac{1}{\tau_\psi}(\psi_w - \psi_{w_c})\right] - V_w \\ \dot{y} &= V_l \sin\psi_e - x\left[-\frac{1}{\tau_\psi}(\psi_w - \psi_{w_c})\right] \\ \dot{V}_w &= -\frac{1}{\tau_V}(V_w - V_{w_c}) \\ \dot{\psi}_w &= -\frac{1}{\tau_\psi}(\psi_w - \psi_{w_c}) \\ \dot{h}_w &= \zeta_w \\ \dot{\zeta}_w &= -\left(\frac{1}{\tau_{h_a}} + \frac{1}{\tau_{h_b}}\right)\zeta_w - \frac{1}{\tau_{h_a}\tau_{h_b}}h_w + \frac{1}{\tau_{h_a}}h_{w_c} \end{cases} \tag{7.50}$$

The heading angle dynamics are much faster than the translational dynamics. For this reason, the longitudinal controller is designed by assuming that $-\frac{1}{\tau_\psi}(\psi_w - \psi_{w_c})$ is equal to zero, yielding the following equations:

$$\dot{x} = V_l - V_w$$
$$\dot{V}_w = -\frac{1}{\tau_V}(V_w - V_{w_{c0}})$$

Clearly, if $V_w = V_l + K_1(x - x_r)$ with $K_1 > 0$, the trajectories of equation $\dot{x} = V_l - V_w$ will asymptotically track a constant reference x_r. So, we aim to design a controller $V_{w_{c0}}$ to render $V_w \to V_l + K_1(x - x_r)$. To this end, we consider the Lyapunov function candidate

$$\mathbb{V}_1 = \frac{[V_w - V_l - K_1(x - x_r)]^2}{2} \geq 0.$$

Thus,

$$\dot{\mathbb{V}}_1 = [V_w - V_l - K_1(x - x_r)]\left[-\frac{1}{\tau_V}(V_w - V_{w_c}) - K_1(V_l - V_w)\right].$$

The longitudinal controller

$$V_{w_{c0}} = \tau_V(K_1 + K_2)(V_l - V_w) + \tau_V K_1 K_2(x - x_r) + V_w, K_2 > 0$$

would render

$$\dot{\mathbb{V}}_1 = -K_2(V_w - V_l - K_1(x - x_r))^2 \leq 0,$$

which means V_w will converge to $V_l + K_1(x - x_r)$ and finally $\dot{x} = V_l - V_w$ would be stabilized.

The lateral controller is also designed by a Lyapunov function approach. In particular, we choose $\mathbb{V}_2 = \frac{(y-y_r)^2}{2}$. Then,

$$\dot{\mathbb{V}}_2 = (y - y_r)\left[V_l \sin\psi_e + \frac{x}{\tau_\psi}(\psi_w - \psi_{w_c})\right].$$

The lateral controller

$$\psi_{w_{c0}} = \psi_w + \frac{\tau_\psi}{x}\left[V_l \sin\psi_e + K_y(y - y_r)\right], K_y > 0 \tag{7.51}$$

would render

$$\dot{\mathbb{V}}_2 = -K_y(y - y_r)^2 \leq 0.$$

Similarly, the altitude controller can be designed. It is easy to check that the following altitude controller

$$h_{w_{c0}} = \tau_{ha}K_{h1}K_{h2}(h_l - h_w) + \left[\left(1 + \frac{\tau_{ha}}{\tau_{hb}}\right) - \tau_{ha}(K_{h1} + K_{h2})\right]\zeta + \frac{1}{\tau_{hb}}h_w \tag{7.52}$$

is effective, where $K_{h1} > 0$ and $K_{h2} > 0$ are constant parameters.

7.3.2.2 Uncertainty and Disturbance Estimator

The formation aerodynamics effects are compensated by the constant compensation signals and the outputs of the UDE. The constant compensation signals, D_0, F_{Y_0}, and L_0, are predicted using SHVM. The other unpredictable aerodynamic effects can be expressed as a function of the aircraft states and control inputs. In the following, a UDE is designed to estimate those effects using the aircraft states and control inputs.

By incorporating the constant compensation terms, we obtain the formation controller

$$\begin{cases} V_{w_{c1}} &= V_{w_{c0}} - \dfrac{D_0}{m}\tau_V \\ \psi_{w_{c1}} &= \psi_{w_{c0}} - \dfrac{F_{Y_0}}{mV_w}\tau_\psi \\ h_{w_{c1}} &= h_{w_{c0}} - \dfrac{L_0}{m}\tau_{h_a} \end{cases} \tag{7.53}$$

Considering the above expressions for (7.50), we obtain

$$\begin{cases} \dot{x} &= V_l \cos\psi_e + y\left[-\dfrac{1}{\tau_\psi}(\psi_w - \psi_{w_{c0}}) + \dfrac{\delta_Y}{mV_w}\right] - V_w \\ \dot{y} &= V_l \sin\psi_e - x\left[-\dfrac{1}{\tau_\psi}(\psi_w - \psi_{w_{c0}}) + \dfrac{\delta_Y}{mV_w}\right] \\ \dot{V}_w &= -\dfrac{1}{\tau_V}(V_w - V_{w_{c0}}) + \dfrac{\delta_D}{m} \\ \dot{\psi}_w &= -\dfrac{1}{\tau_\psi}(\psi_w - \psi_{w_{c0}}) + \dfrac{\delta_Y}{mV_w} \\ \dot{h}_w &= \zeta_w \\ \dot{\zeta}_w &= -\left(\dfrac{1}{\tau_{h_a}} + \dfrac{1}{\tau_{h_b}}\right)\zeta_w - \dfrac{1}{\tau_{h_a}\tau_{h_b}}h_w + \dfrac{1}{\tau_{h_a}}h_{w_{c0}} + \dfrac{\delta_L}{m} \end{cases} \tag{7.54}$$

where $\delta_D = \Delta D - D_0$, $\delta_Y = \Delta F_Y - F_{Y_0}$, and $\delta_L = \Delta L - L_0$ are residual aerodynamic effects, respectively.

A UDE-based formation flight controller has the form

$$\begin{cases} V_{w_c} &= V_{w_{c1}} - \dfrac{\hat{\delta}_D}{m}\tau_V \\ \psi_{w_c} &= \psi_{w_{c1}} - \dfrac{\hat{\delta}_Y}{mV_w}\tau_\psi \\ h_{w_c} &= h_{w_{c1}} - \dfrac{\hat{\delta}_L}{m}\tau_{h_a} \end{cases} \tag{7.55}$$

where $\hat{\delta}_D$, $\hat{\delta}_Y$, and $\hat{\delta}_L$ denote the estimates of δ_D, δ_Y, and δ_L, respectively. Since the aerodynamic effects have slow dynamics with respect to time, we may consider a group of first-order low-pass filters to give

$$\begin{cases} \dot{\hat{\delta}}_D &= \dfrac{1}{T_D}(\delta_D - \hat{\delta}_D) \\ \dot{\hat{\delta}}_Y &= \dfrac{1}{T_Y}(\delta_Y - \hat{\delta}_Y) \\ \dot{\hat{\delta}}_L &= \dfrac{1}{T_L}(\delta_L - \hat{\delta}_L) \end{cases} \tag{7.56}$$

where T_D, T_Y, and T_L are time constants.

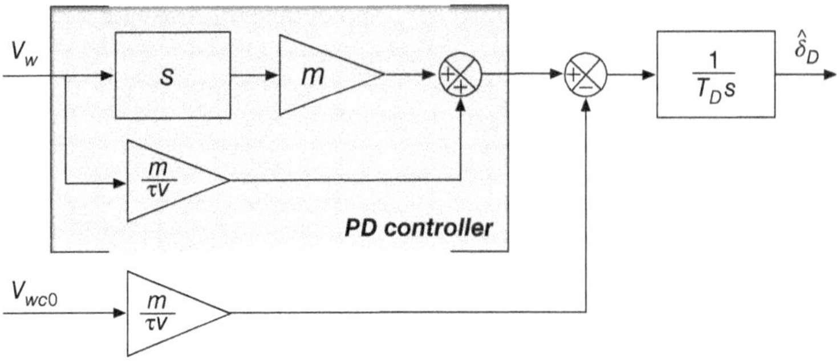

Figure 7.10 Uncertainty and disturbance estimator for δ_D.

Generally, δ_D, δ_Y, and δ_L can be expressed as functions of the aircraft states and control inputs. In particular, we derive from (7.54)–(7.56) that

$$\begin{cases} \dot{\hat{\delta}}_D = \dfrac{m}{T_D}\left(\dot{V}_w + \dfrac{1}{\tau_V}(V_w - V_{w_{c0}})\right) \\ \dot{\hat{\delta}}_Y = \dfrac{mV_w}{T_Y}\left(\dot{\psi}_w + \dfrac{1}{\tau_\psi}(\psi_w - \psi_{w_{c0}})\right) \\ \dot{\hat{\delta}}_L = \dfrac{m}{T_L}\left(\dot{\zeta}_w + \left(\dfrac{1}{\tau_{h_a}} + \dfrac{1}{\tau_{h_b}}\right)\zeta_w + \dfrac{1}{\tau_{h_a}\tau_{h_b}}h_w - \dfrac{1}{\tau_{h_a}}h_{w_{c0}}\right). \end{cases} \tag{7.57}$$

Although some state derivatives appear in (7.57), it is not necessary to measure these derivatives for computing the control law, since the control law may be approximately realized by a mechanism shown in Figure 7.10. Finally, the overall robust formation flight controller is

$$\begin{cases} V_{w_c} = \tau_V(K_1 + K_2)(V_l - V_w) + \tau_V K_1 K_2 (x - x_r) + V_w - \dfrac{D_0}{m}\tau_V - \dfrac{\hat{\delta}_D}{m}\tau_V \\ \psi_{w_c} = \psi_w + \dfrac{\tau_\psi}{x}[V_l \sin\psi_e + K_y(y - y_r)] - \dfrac{Y_0}{mV_w}\tau_\psi - \dfrac{\hat{\delta}_Y}{mV_w}\tau_\psi \\ h_{w_c} = \tau_{ha}K_{h1}K_{h2}(h_l - h_w) + \left[\left(1 + \dfrac{\tau_{ha}}{\tau_{hb}}\right) - \tau_{ha}(K_{h1} + K_{h2})\right]\zeta \\ \qquad + \dfrac{1}{\tau_{hb}}h_w - \dfrac{L_0}{m}\tau_{h_a} - \dfrac{\hat{\delta}_L}{m}\tau_{h_a} \\ \dot{\hat{\delta}}_D = \dfrac{m}{T_D}\left(\dot{V}_w + \dfrac{1}{\tau_V}(V_w - V_{w_{c0}})\right) \\ \dot{\hat{\delta}}_Y = \dfrac{mV_w}{T_Y}\left(\dot{\psi}_w + \dfrac{1}{\tau_\psi}(\psi_w - \psi_{w_{c0}})\right) \\ \dot{\hat{\delta}}_L = \dfrac{m}{T_L}\left(\dot{\zeta}_w + \left(\dfrac{1}{\tau_{h_a}} + \dfrac{1}{\tau_{h_b}}\right)\zeta_w + \dfrac{1}{\tau_{h_a}\tau_{h_b}}h_w - \dfrac{1}{\tau_{h_a}}h_{w_{c0}}\right). \end{cases} \tag{7.58}$$

7.4 Simulation Results

7.4.1 Simulation Results for Controller 1

Simulations are performed on the triangular close-formation flight of three F/A-18 aircraft. The positions $[x_o, y_o, z_o]$=[50.0 m, ±10.2 m, 0 m] are chosen as optimal for induced lift and are therefore the relative positions that the two follower aircraft should maintain. The longitudinal separation is chosen for safety. The vortex-induced stability derivatives and the controller gains are given in Tables 7.2 and 7.3, respectively. The time constants for Follower I are chosen to be $\tau_V = 6.0$ s, $\tau_\psi = 1.0$ s, $\tau_{ha} = 0.5$ s, $\tau_{hb} = 4.1$ s, and 95% of those values are applied to Follower II.

At forst, the two follower aircraft fly in the optimal relative positions with respect to the leader. At 10 s, the leader begins to execute maneuvers:

1. the heading angle changes from 0 to 0.524 rad (30°) at a rate of 0.0524 rad/s;
2. the flight altitude changes from 12,192 m to 13,192 m at a vertical velocity of 100 m/s.

Figure 7.11 shows the simulation results without a synchronization strategy; that is, $\beta_{x_i}, \beta_{y_i}, \beta_{z_i} = 0$. Asymptotic tracking of the three relative positions and the aircraft velocity is achieved, while the obvious differences between the tracking errors of two followers can be seen. Figure 7.12 illustrates the simulation results with the synchronization strategy; that is, $\beta_{x_i}, \beta_{y_i}, \beta_{z_i} = 1$. In this case, both followers achieve asymptotic relative position tracking. In addition, the differences between the relative position tracking errors of the two followers are greatly reduced. In other words, the

Table 7.2 Vortex-induced stability derivatives.

Derivative	Follower I	Follower II
ΔC_{LFy}	0.0276	−0.0276
ΔC_{DFy}	0.0033	0.0033
ΔC_{SFy}	0.000849	0.000849
ΔC_{SFz}	0.0018	−0.0018

Table 7.3 Controller gains for the follower aircraft.

Parameter	Value	Parameter	Value
K_{xp_i}, /s	3.2	K_{xi_i}, /s^2	0.3
K_{yp_i}, /m	5.5	K_{yi_i}, /ms	0.4
K_{zp_i}	3.0	K_{zi_i}, /s	0.25
k_{x_i}	−3.0	k_{V_i}, s	6.4
k_{y_i}	−1.0	k_{ψ_i}	2.5

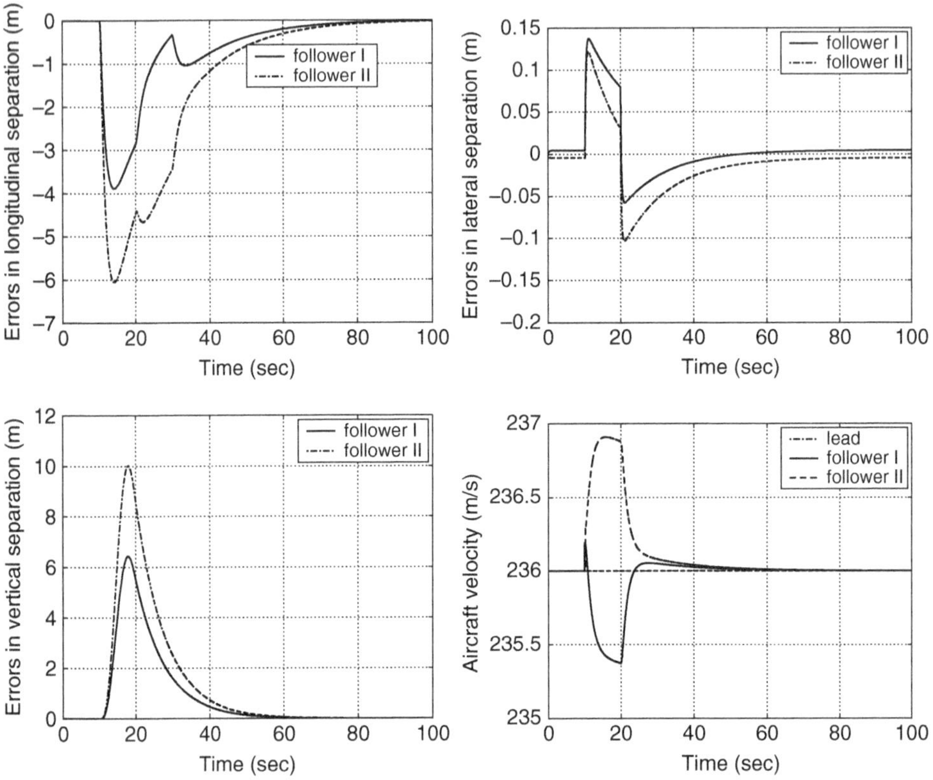

Figure 7.11 Formation flight control without synchronization.

two followers form a more "synchronized" pattern of close-formation flight with the leader aircraft.

7.4.2 Simulation Results for Controller 2

In this section, the proposed formation controller is verified for an example of close-formation flight of two F-16 aircraft. The F-16 aircraft is represented by the simple aircraft/autopilot model (7.28)–(7.30), while the formation kinematics are nonlinear and shown by (7.34)–(7.35). The principal data of the simple F-16 aircraft/autopilot model are given in Table 7.4 (see Pachter *et al.* [6] for more detailed aerodynamic data for the F-16). The formation aerodynamics are described by the SHVM and CVSM, respectively. In one simulation case, the CVSM is deemed to describe the "real" formation aerodynamics. The complete formation controller with compensation terms given by the SHVM will be verified for the case where the formation aerodynamics are described by the CVSM.

Assume that the leader aircraft maintains level and straight flight with

$$V_l = 825 \text{ ft/s},\ \psi_l = 30°,\ h_l = 45000 \text{ ft}.$$

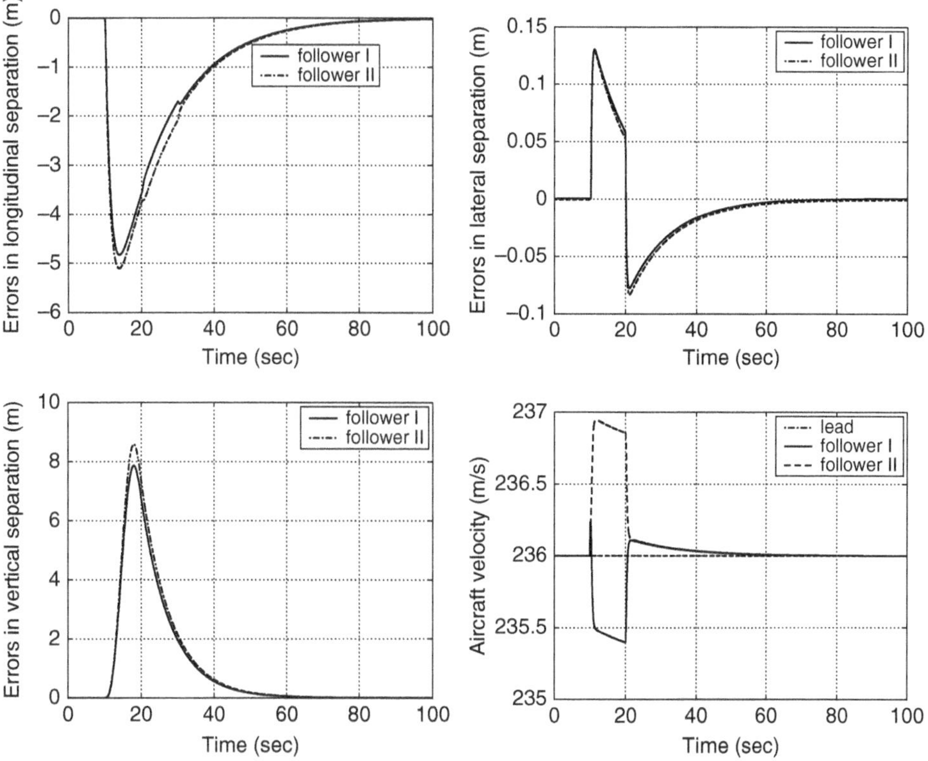

Figure 7.12 Formation flight control with synchronization.

The initial state of the trailing aircraft are

$$V_w = 800 \text{ ft/s}, \ \psi_w = 15°, \ h_w = 45500 \text{ ft}.$$

The initial formation geometry is defined by $x_0 = 100$ ft, $y_0 = 80$ ft, and the desired formation geometry is $x_r = 90$ ft, $y_r = \frac{\pi b}{4}$ ft and $z_r = 0$. Note that $z_r = 0$ means the trailing aircraft and the leader aircraft will be at the same altitude; that is, $h_w = h_l$. Actually, z is always equal to $h_l - h_w$. The constant compensation terms in (7.58) will be predicted by the SHVM at the steady state of formation flight with $(x, y, z) = \left(90, \frac{\pi b}{4}, 0\right)$. The parameters of formation controller and UDE are given in Table 7.5. Numerical simulations were carried out for the three different cases illustrated in Figure 7.13.

Case 1 In Case 1, the performance of the UDE-based controller is verified by comparing the simulation results with and without UDE, as shown in Figures 7.14 and 7.15. The formation aerodynamics are described by the SHVM. The constant compensation terms are predicted by the SHVM for the situation where $x = 90$ ft, $y = \frac{\pi b}{4}$ ft and $z = 0$. It appears that the principal formation controller can stabilize the nonlinear formation system, but this controller is not robust enough. The steady-state error in longitudinal response is around 8ft, while the steady error in altitude response is around 12ft. A steady-state error of 12ft in altitude may cause a great loss of

Table 7.4 F-16 aircraft/autopilot model parameters.

Parameter	Value
Velocity time constant τ_V (s)	5
Heading time constant τ_ψ (s)	0.75
Altitude time constant h_a (s)	0.3075
Altitude time constant h_b (s)	3.85
Gross mass m (*slugs*)	776.4
Wing span b (*ft*)	30

Table 7.5 Formation controller and UDE parameters.

Parameter	Value	Parameter	Value
K_1	0.1	K_2	0.1
K_y	0.2	K_{h1}	0.5
K_{h2}	0.3	T_D	0.2
T_Y	0.2	T_L	0.5

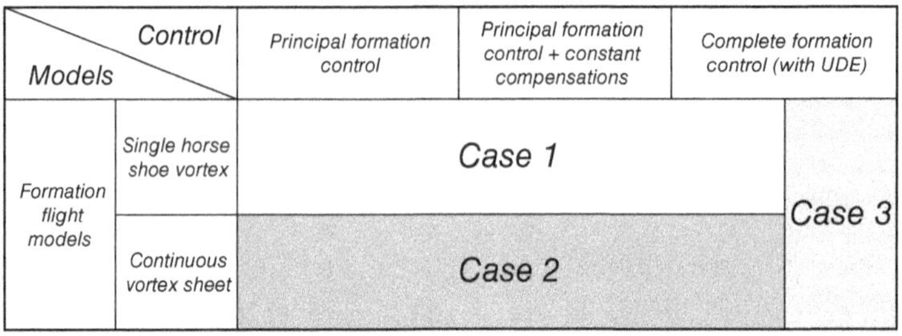

Figure 7.13 Three different simulation cases.

aerodynamic efficiency, since the error is much larger than 10%. Note that the side force ΔF_Y is much smaller than ΔD and ΔL, and that ΔF_Y is divided by mV in the formation model. As a result, the effect of side force ΔF_Y is much less obvious than that of ΔD or ΔL.

The simulation results become worse if constant aerodynamic compensation terms are incorporated. This is because the constant aerodynamic compensation terms only make sense if they closely match the effects of constant disturbance. If this is not the case, compensation using a constant signal might lead to worse results. In the simulation, the principal formation controller fails to achieve reference tracking. Therefore, we conclude that the "real" aerodynamic effects differ markedly from a constant. Worse responses also indicate that the formation controller with only constant compensation does not work better for close-formation flight. However, as shown in Figure 7.15b, the

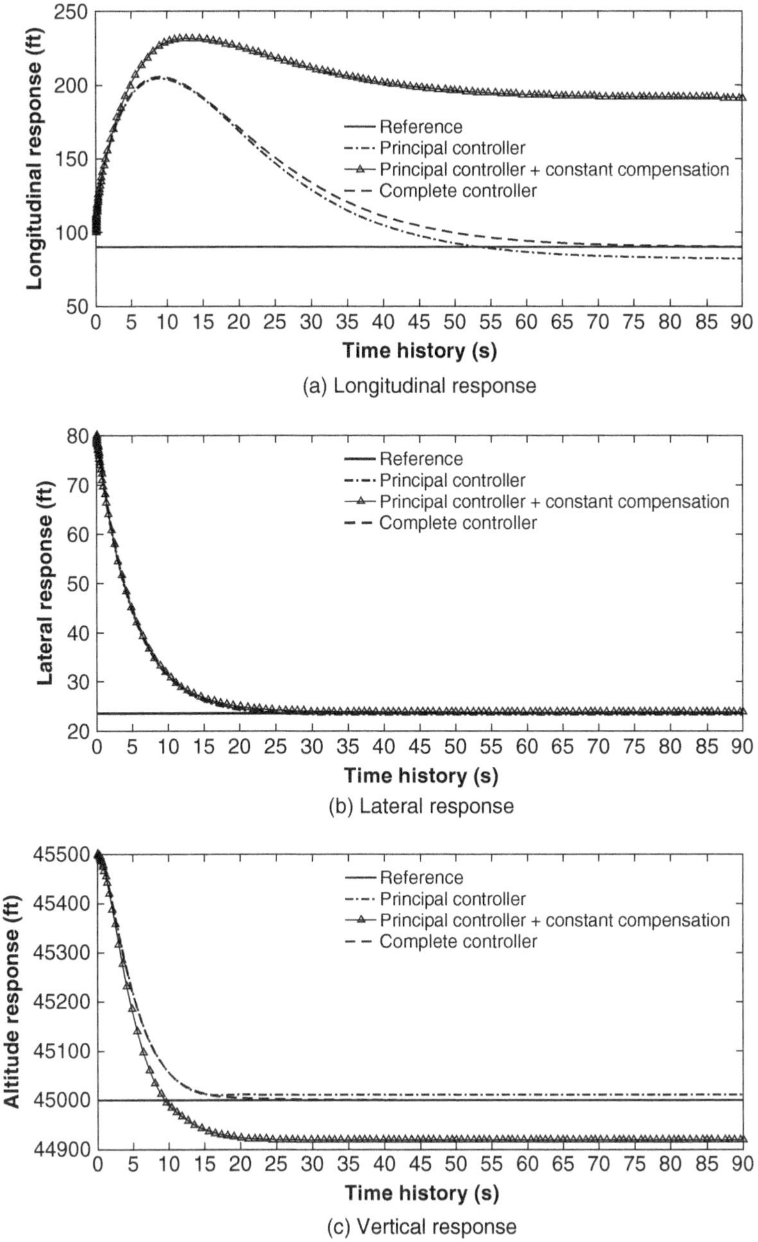

Figure 7.14 Results for the SHVM: Part 1.

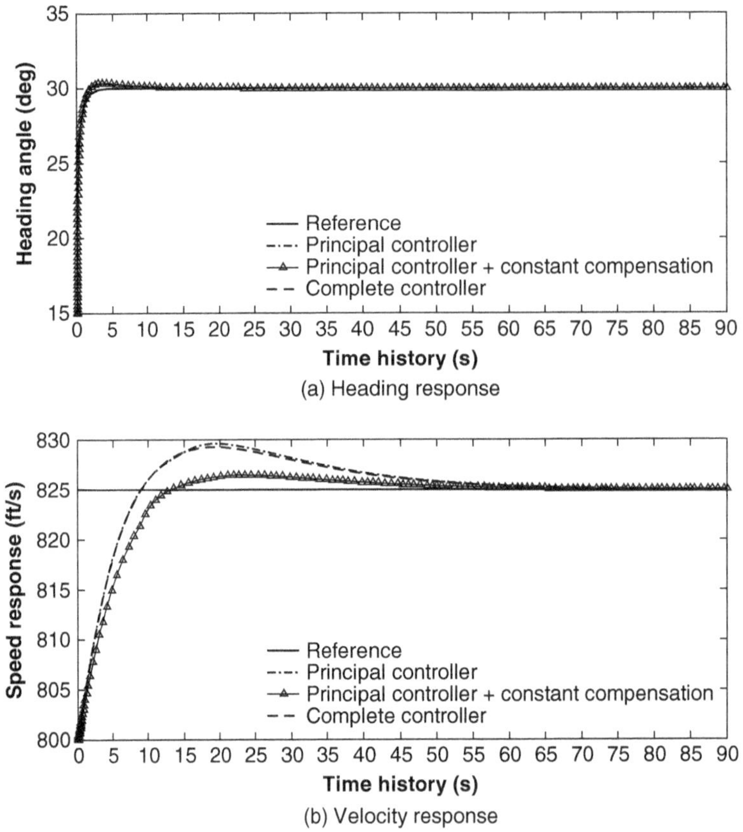

Figure 7.15 Results for the SHVM: Part 2.

speed tracking performance is somewhat different, owing to the constant compensation terms. By additional use of UDE, the steady-state errors are eliminated and a more precise close formation is achieved, indicating that UDE helps to improve the robustness of the principal formation controller.

Case 2 In the second case, the formation aerodynamics are described by the CVSM, while the constant aerodynamic terms are calculated by the SHVM. The controllers used in this case are the same as those in Case 1. The performances of all three controllers are illustrated in Figures 7.16 and 7.17. Only the principal controller can still stabilize the system but steady-state errors still exist. After incorporating constant compensation terms, the performances become worse. However, steady-state errors are eliminated by the UDE-based controller, even though different aerodynamic models are employed. This indicates that the UDE generates an accurate estimation of the formation aerodynamics.

Case 3 In the final case, the UDE-based formation-flight controller will be applied to formation models in which the formation aerodynamics are described by the SHVM

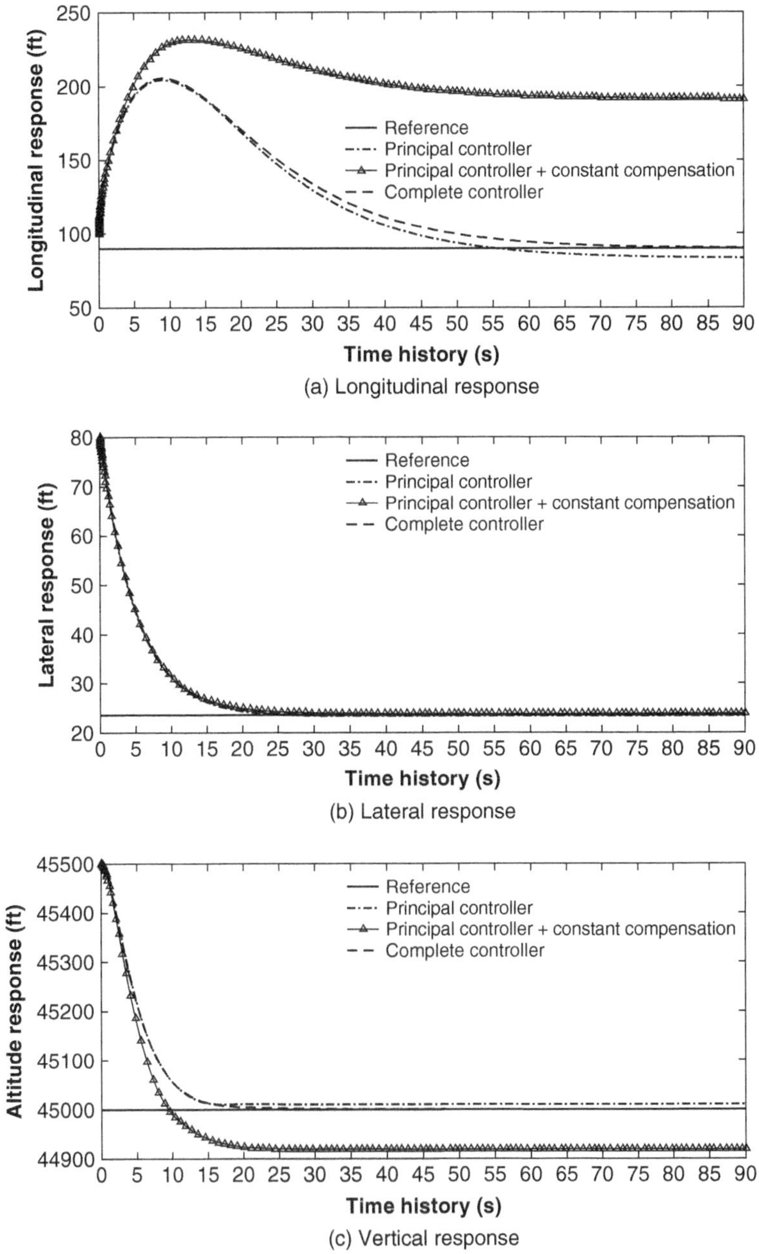

Figure 7.16 Results for the CVSM: Part 1.

and the CVSM. The simulation results are illustrated in Figures 7.18 and 7.19. Assume the "real" formation aerodynamics are described by the CVSM. The constant aerodynamic compensation terms are still calculated using the SHVM. The differences of the two models are apparent, especially in the reference formation state. The results show

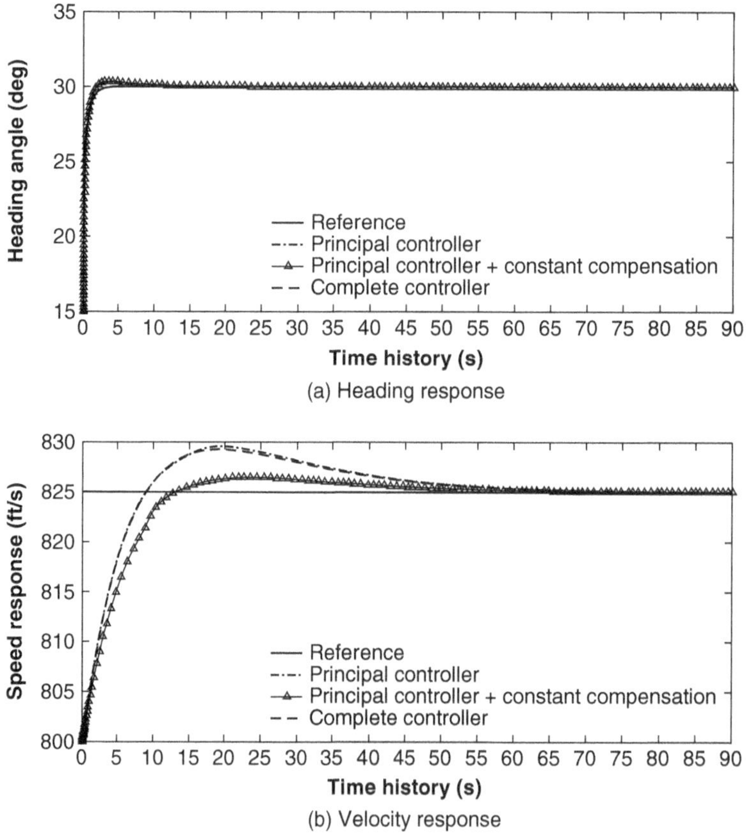

Figure 7.17 Results for the CVSM: Part 2.

that the aerodynamic effects can be well compensated by constant aerodynamic compensation terms and UDE output. Therefore, the formation flight system will have the same simulation results even if formation aerodynamic effects are modelled using two different approaches.

7.5 Summary

In this chapter, we first introduced a linear PI controller for the follower aircraft to track their optimal relative positions with respect to the leader aircraft. Synchronized motion between two follower aircraft was achieved using the cross-coupling concept introduced in Section 5.6. This synchronization strategy was combined with a linear PI controller to form an outer-loop synchronized tracking controller. We presented simulation results for three fixed-wing aircraft in triangular formation flight, demonstrating the effectiveness and performance improvement of the proposed synchronization strategy.

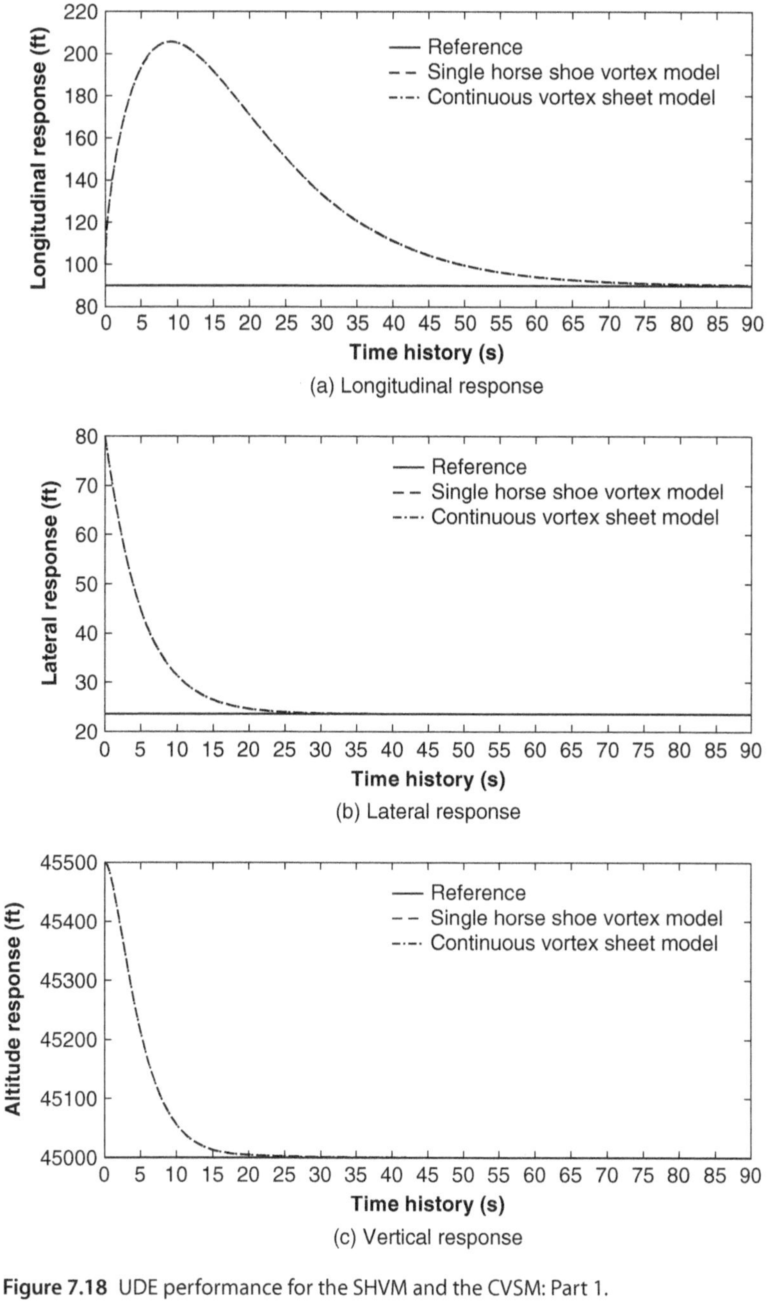

Figure 7.18 UDE performance for the SHVM and the CVSM: Part 1.

In addition, a UDE-based robust controller was proposed for close-formation flight. The formation aerodynamics are described by two different models: SHVM and CVSM. The former was used to calculate the constant aerodynamic compensation terms, while the latter was used in one simulation case to verify the robustness of the proposed

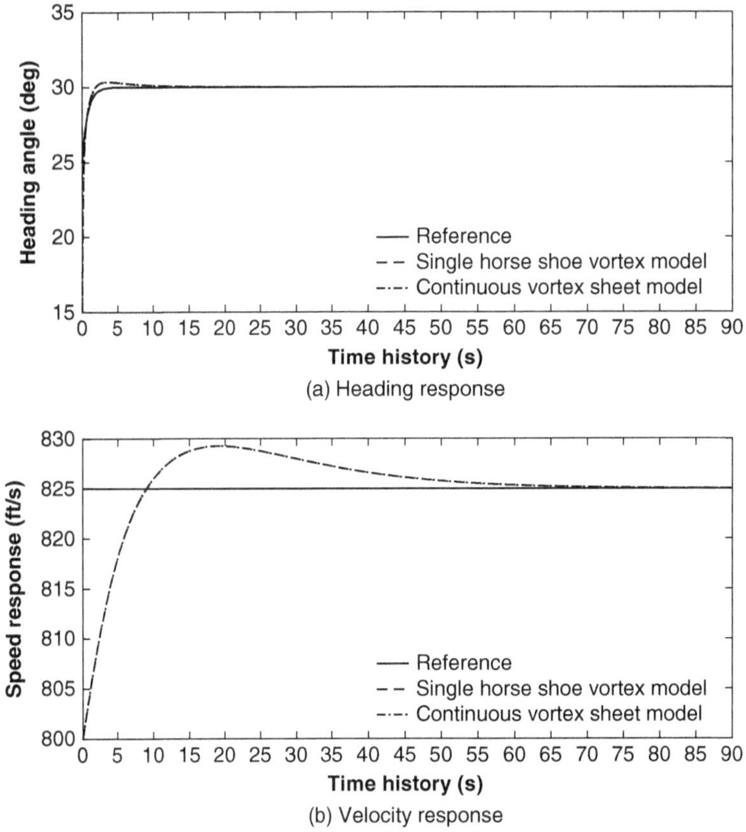

(a) Heading response

(b) Velocity response

Figure 7.19 UDE performance for the SHVM and the CVSM: Part 2.

controller. In the design, a principal nonlinear formation flight controller was developed based on the nominal formation model without considering formation aerodynamics, and then robustness improvement was achieved by incorporating constant compensation terms and a UDE.

8

Formation Control of Robotic Systems

This chapter focuses on the sensing and control aspects of MVS formations. Two formation-control laws – synchronization and passivity – are applied to conventional leader–follower formation control in simulation, and their performances are compared. Additionally, a low-cost omnivision sensor and concomitant visual tracking algorithms are developed and tested in formation-keeping experiments.

8.1 Introduction

Over the past decade, there has been a growing interest in the control of autonomous multi-agent robotic systems. Compared to single-agent systems, these are more robust to individual agent failure, can be reconfigured with greater ease, and have more flexible functionality [12]. It is often expected that a group of robots will maintain a predefined formation during task execution. This is done using formation control, driving each robot in a group to a desired reference position and orientation relative to the others [12].

Formation-keeping has many applications. For example, unmanned aerial vehicles (UAVs) can reduce drag and improve fuel efficiency by flying in a V-shape [13]. When carrying out ground surveillance or search-and-rescue missions, sparse formations are needed to maximize combined UAV sensor coverage [14]. On the ground, formation control has many applications in unmanned ground vehicle (UGV) convoys and patrols [15]. In space, a group of spacecraft equipped with interferometers and traveling in formation can achieve greater resolution than a single large space telescope in imaging distant objects [12]. In all these applications, formation control is essential in making an MVS perform better than a single-vehicle counterpart.

While constant formations can be easy to analyse and implement, many real-life applications rely on time-varying formations. For example, a group of micro air vehicles (MAVs) can adopt a tight formation to enter buildings when executing an urban reconnaissance mission. After completing the indoor component of their mission, they can return to a sparse formation to maximize sensor coverage. Quick and efficient transitions between formations are crucial for the success of many unmanned MVSs.

In this chapter, the subscript i denotes the ith follower and L the leader, the superscript d denotes a desired value, and the angular sign convention assumes that angles measured counterclockwise from their baselines are positive. The acronym UGV is used to denote an unmanned ground vehicle; RGB to denote red, green, and blue; HSV to denote hue, saturation, and value. The variables and parameters are explained in Tables 8.1 and 8.2.

Formation Control of Multiple Autonomous Vehicle Systems, First Edition. Hugh H.T. Liu and Bo Zhu.

Table 8.1 Some of the symbols and parameters used.

Term	Description
ϵ_{xi} ϵ_{yi}	Position synchronization errors
β_{xi} β_{yi}	Synchronization gains
e^*_{xi} e^*_{yi}	Coupled position errors
k_{xi} k_{yi} $k_{\beta i}$ $k_{\upsilon i}$	Weighting factors
β_i	ith follower's relative heading angle
υ_i	ith follower's speed
υ_L	Leader's speed

Table 8.2 Some of the variables used.

Variable(s)	Description	Defined in
s	Pixel distance from camera optical axis to UGV base centroid on the distorted image	Fig. 8.13
β	Relative heading angle between the leader and a follower	Fig. 8.13
r	Separation distance between the centers of a pair of UGVs	Fig. 8.15
(x, y)	UGV's position in global coordinates	
(x^d_L, y^d_L)	Desired position in the leader's reference frame	Fig. 8.15
(e_x, e_y)	x and y-direction tracking errors in the leader's reference frame	Fig. 8.15
e	Magnitude of the x and y-direction tracking errors	Eq. (8.13)
υ	Linear velocity	
ω	Angular velocity	
θ	Position angle of the leader relative to the follower's local y-axis	Fig. 8.15
ϕ	UGV's heading angle relative to the global x-axis	
(E_x, E_y)	x and y-channel errors	Eqs (8.10), (8.11)
(β_x, β_y)	x and y-direction synchronization gains	Eqs (8.10), (8.11)
(k_x, k_y)	Weighting factors for the x and y-direction coupled position errors	Eqs (8.10), (8.11)
k_β	Weighting factor for relative heading angle	Eqs (8.10), (8.11)
k_υ	Weighting factor for relative speed	Eqs (8.10), (8.11)
K_P, K_I, K_D	Proportional, integral and derivative gains	
k, κ	Linear and torsional spring constants	Eqs (8.21), (8.23)
F, F_k, F_T	Force, spring force, torsional spring force	Eqs (8.21), (8.23)
(F_x, F_y)	x and y-direction force components in global coordinates	Eqs (8.24), (8.26)
τ	Torque OR a dummy variable for integration	Eq. (8.22)
γ	Position angle of a follower relative to the leader's local negative y-axis	Eqs (8.19), (8.20)
e_F	Magnitude of resultant force	Eq. (8.28)
e_ϕ	Difference between resultant force heading and follower's current heading	Eq. (8.29)

8.2 Visual Tracking

Techniques for obtaining relative position and orientation information of UGVs using omnivision cameras are discussed in this section. Visual tracking involves identifying the object of interest in captured images and relating its image location to a real-world position through projective geometry. After this is done, relative positioning data between pairs of UGVs can be used for feedback in formation control laws.

8.2.1 Imaging Hardware

In vision-based leader–follower formation control, followers are equipped with cameras that continuously image their surroundings, enabling them to localize themselves relative to the leader. Because most commercially available cameras have fields of view of under 50°, a forward-facing camera on a follower UGV is not able to see the leader when the pair's relative heading angle is large. To solve this problem, an omnidirectional vision system can be implemented. These have a hemispherical mirror above an upward-facing camera (Figure 8.1). Here, Ikea STABIL stainless steel double-boiler inserts (Figure 8.2) were used as hemispherical mirrors due to their highly reflective outer surfaces. For prototyping purposes, cooking utensils are preferable to commercially available omnidirectional vision systems because of the lower cost. Imaging was performed using Logitech C510 high definition webcams (Figure 8.2).

The test cases conducted only involved formations where the followers were behind the leader. As such, the followers only needed a 180° forward field of view to detect the leader. Mounting the camera so that its optical axis was aligned with the hemispherical

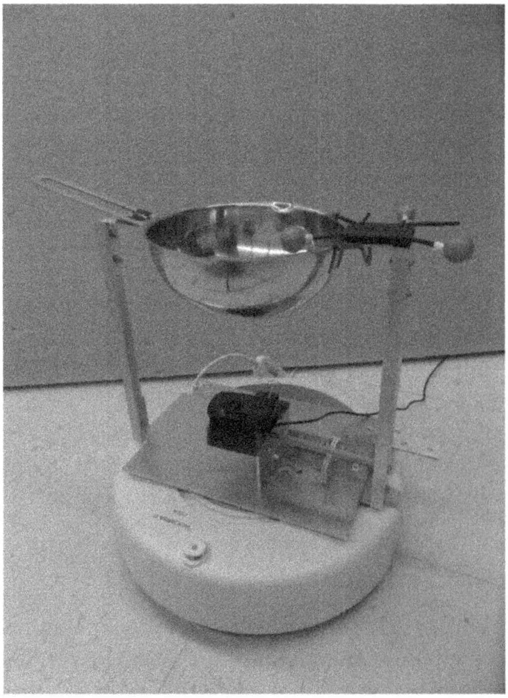

Figure 8.1 An omnivision system with a camera.

Figure 8.2 Equipment used: (left) Ikea STABIL stainless steel double-boiler insert; (right) Logitech C510 high definition webcam.

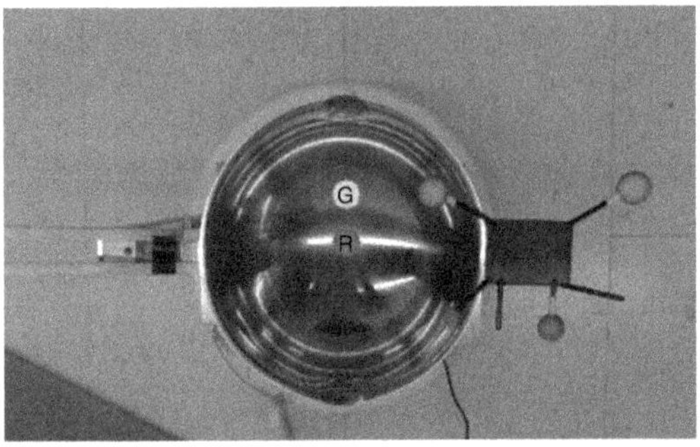

Figure 8.3 Overhead view of a follower UGV.

mirror's axis of symmetry would have produced a 360° field of view and wasted approximately half of the resultant image pixels. Instead, the camera was mounted slightly forward of the hemispherical mirror's axis of symmetry (Figure 8.3), where the R dot marks the hemispherical mirror's axis of symmetry, while the G dot marks the camera's optical axis. This resulted in a more efficient use of image pixels to cover the required 180° forward field of view, as will be seen later on, in Figure 8.6.

The leader UGV was equipped with easily recognizable color markers to aid the follower UGVs' color threshold image processing, which is described in Section 8.2.3. Since the experiments in this project only involved formations where the leader was ahead of the followers, color markers were mounted on the leader's rearward half (features on the leader's forward half are usually hidden from a follower's view). The rearward portion of the leader's base was covered with blue tape, while three tennis balls, each of a different color, were mounted above it (Figure 8.4). These markers were brightly colored for easy detection within captured images, since they are small compared to the blue base. A black cover for the leader's electronics increased the color contrast of the spherical markers with their immediate surroundings.

8.2.2 Image Distortion

In order to extract localization information from an image, it is necessary to establish a mathematical relationship between image points and real world points. While a great advantage of an omnidirectional vision system is its increased field of view, a

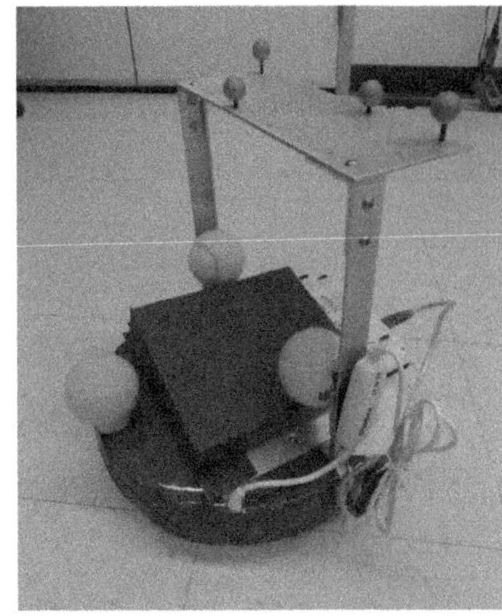

Figure 8.4 The leader UGV is outfitted for easy visual recognition

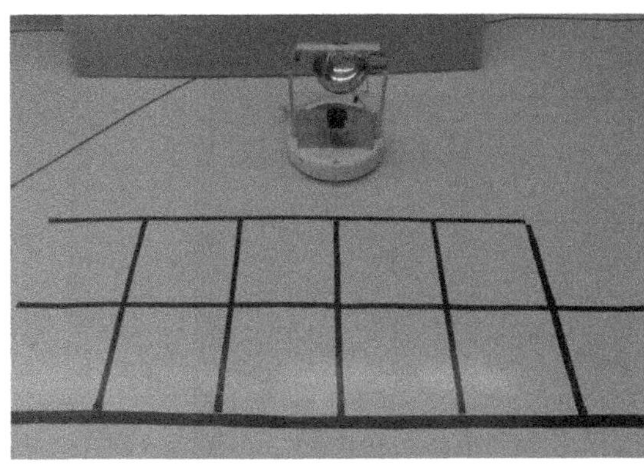

Figure 8.5 A follower UGV placed at the origin of the RWC system.

significant disadvantage is that optical analysis using projective geometry is complicated. A complete theoretical analysis of this project's imaging setup would involve three-dimensional light-ray tracing from distant objects to the hemispherical mirror and eventually to the camera's image plane. Because of irregularities in the manufacture and mounting of the double-boiler insert relative to the camera, radial symmetry cannot be assumed to reduce the three-dimensional problem to a two-dimensional one. In light of the difficulties related to formulating a theoretical relationship between image points and real world points according to a pinhole camera model, another method is used.

The alternative method used in this project eliminates the geometric optics model in favour of creating a direct mapping from image points to real world points. First, a follower UGV is placed beside a clearly visible calibration grid of known dimensions, but at the origin of the real world coordinate (RWC) system alongside the calibration grid, as shown in Figure 8.5. The pixel coordinates of all grid corners are then recorded from the follower UGV's raw camera image (Figure 8.6). Next, pixel coordinates undergo a transformation from image axes (anchored at the upper-left corner of the image) to camera axes (anchored at the camera's optical axis), as shown in Figure 8.6. Finally, nonlinear multivariable linear fits are performed to create a pair of mappings from the RWCs centered on the UGV to the camera-axes pixel coordinates (CAPC), as shown in Figure 8.7.

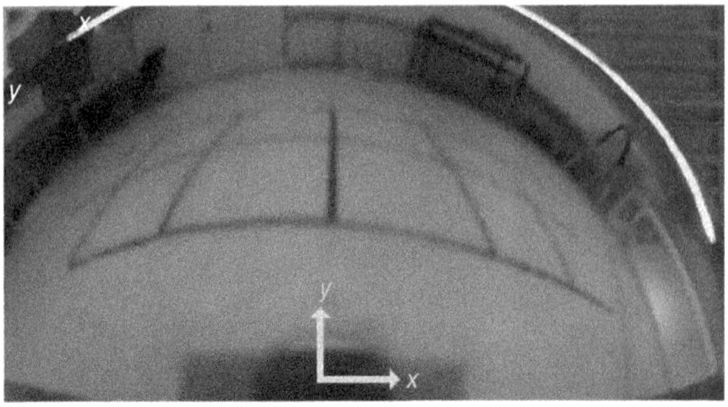

Figure 8.6 Image axes and camera axes depicted in a raw image.

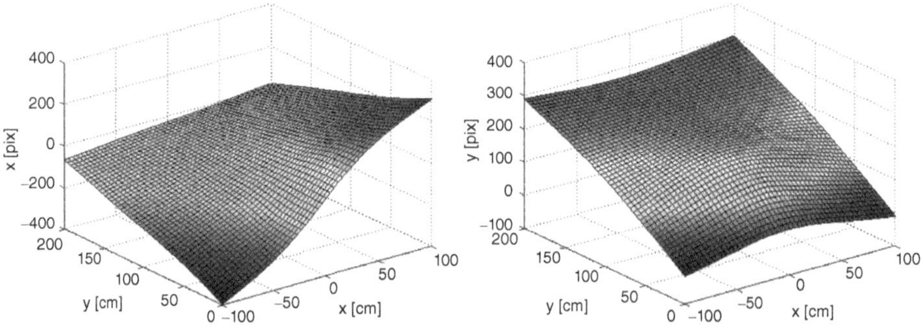

Figure 8.7 Mappings from RWC to CAPC in pixels: left, x; right, y.

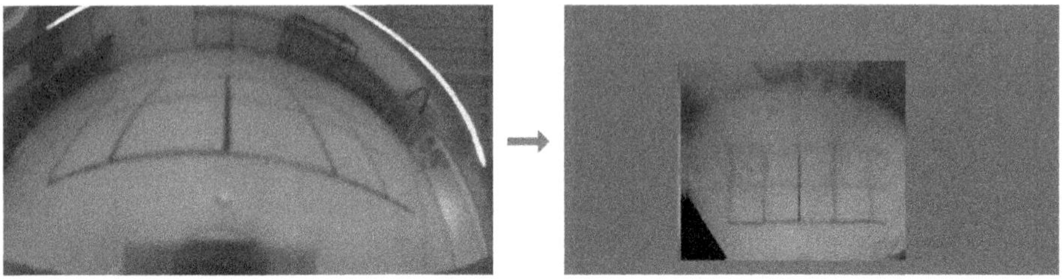

Figure 8.8 The mapped pixels.

The calibration grid allows for easy determination of the RWC of the grid corners, for which the pixel coordinates are known.

In this project, a raw image size of 544×288 pixels was used to achieve a compromise between capturing detailed features and reducing the computational cost. When a raw image enters the image processing pipeline, the origin of the camera axes is taken to be the centroid of the black region in the bottom third of the image (the black Logitech C510 HD webcam). After applying the transformation from image axes to camera axes, each pixel in the distorted image is mapped (Figure 8.8) from the raw image using bilinear interpolation of lookup tables that discretize the mappings in Figure 8.7. To reduce the computational cost, only the pixels of the distorted image that correspond to the area immediately in front of the follower UGV are mapped from the raw image.

As depicted in Figure 8.8, the overall action of the mapping is to transform the curvilinear grid image to a rectilinear grid image, where the right-hand image shows the results of applying the mapping in Figure 8.7 to the left-hand image (gray areas are not mapped to reduce computational cost). This provides a top-down perspective on the region immediately in front of the follower UGV (roughly a meter ahead and half a meter to either side), which is where the leader is expected to be. The primary advantage of the top-down perspective is that the pixel distances between objects and the follower UGV are linearly proportional to the corresponding real world distance, a fact which is exploited in Section 8.2.5.

8.2.3 Color Thresholding

For simplicity, this project relies on color thresholding to identify uniquely colored objects of interest within an image. With this method, finding an object of interest only requires finding the region of an image with its associated color. Because the experiments for this project were performed in an indoor environment with relatively constant lighting conditions, color thresholding was sufficient for followers to consistently identify the leader. However, it is recognized that practical applications of vision-based formation control will require more sophisticated image processing techniques.

The main advantage of color thresholding is that it is computationally cheaper than other commonly used image processing techniques such as edge, corner, or blob detection. For experiments, a fast image processing stage is needed to provide feedback control inputs at a high enough rate to maintain formation stability.

Computers perceive images as arrays of numbers that represent pixel color values. In most digital imaging systems, color is represented by the RGB model, which assigns a triplet of integers (each between 0 and 255) that quantify the relative levels of red, green, and blue that make up a given color. In this project, color thresholding takes place after converting RGB images to the HSV (hue, saturation, value) color space in order to reduce the variability introduced by ambient lighting conditions. The color markers shown in Figure 8.4 were chosen to have distinct HSV ranges to facilitate HSV color thresholding. RGB and false-color HSV images of a distorted image from a follower UGV's video stream are shown in Figure 8.9.

Thresholding is a process where an image is divided into object pixels and background pixels. In HSV color thresholding, object pixels are pixels whose HSV values fall within prespecified HSV thresholds (Table 8.3), while all other pixels are background pixels. The resulting binary image features white object pixels and black colored pixels (Figure 8.10).

8.2.4 Noise Rejection

Under ideal conditions, HSV color thresholding would result in the perfect detection of color marker regions in all input images. In practice, false negatives and false positives arise due to the presence of similarly colored background objects, abnormal lighting conditions, or other factors.

A false negative occurs when a region of the image that belongs to a color marker is not identified as such. This type of error commonly arises when part of a spherical

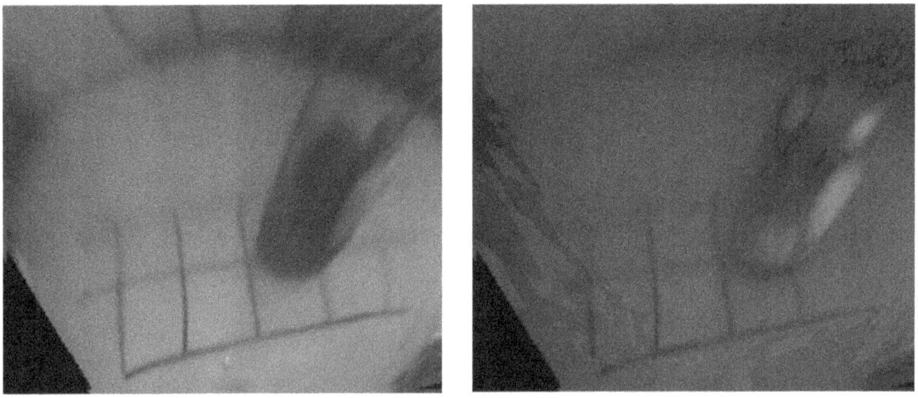

Figure 8.9 The leader robot as seen by a follower robot: left, in RGB space; right, in HSV (false color).

Table 8.3 HSV ranges for each colored region on the leader UGV.

Colour	H range	S range	V range
Blue	[108,126]	[22,121]	[74,110]
Pink	[166,180]	[54,133]	[112,220]
Green	[24,39]	[44,169]	[75,110]
Orange	[0,13]	[62,168]	[93,131]

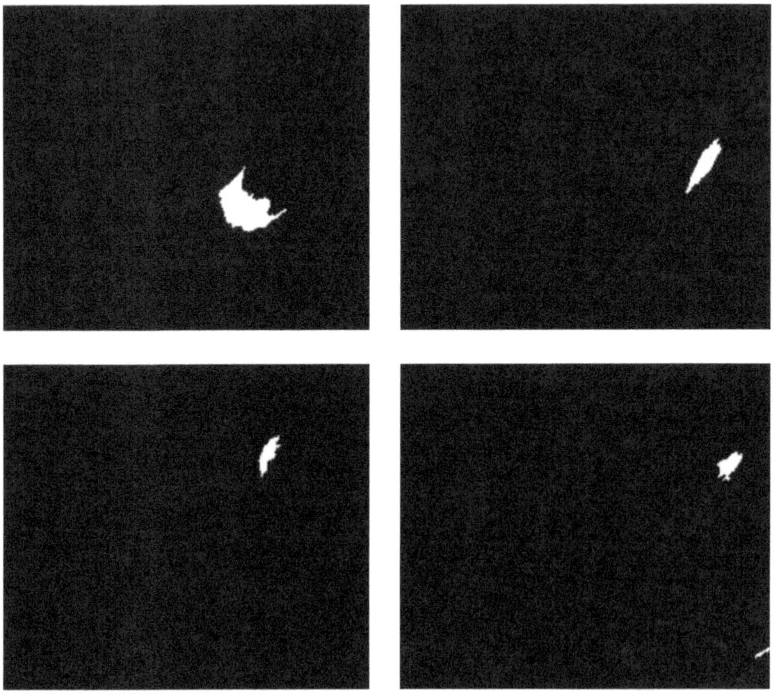

Figure 8.10 Object and background pixels following HSV thresholding.

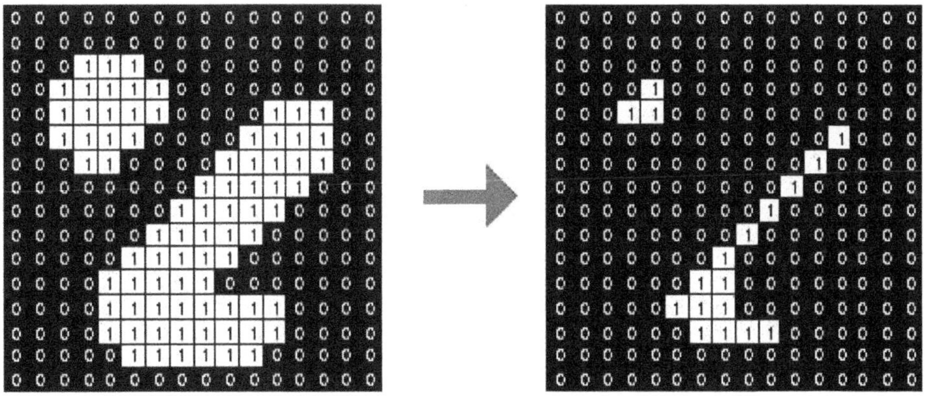

Figure 8.11 An example of erosion applied to a binary image.

marker is covered by a shadow. False negatives do not impact data extraction as long a color marker is not completely undetected in an image.

A false positive occurs when a region of the image is identified as belonging to a color marker when in fact it does not. For example, a blue chair close to a follower UGV might be erroneously identified as part of the leader's base. These types of errors are primarily reduced by modifying the testing environment so that it has no color regions with HSV values overlapping those of the leader.

When modifying the test environment, it is difficult to completely eliminate all background objects with conflicting HSV values. Therefore, tiny groups of pixels representing background objects (usually around the image edges) are identified as part of the object region. These extraneous pixels constitute image noise, which can be reduced or even eliminated by erosion. Erosion is a morphological operation on a thresholded binary image that causes the background regions to encroach on the object regions (Figure 8.11). Since falsely detected object regions that constitute image noise tend to be much smaller than the true object region, applying the erosion operator eliminates the former while only shrinking the latter. In experiments, this technique was found to be more reliable when the leader was relatively close to the follower. When separation distance increases, the leader's image size decreases relative to the image sizes of background objects. Consequently, the true object regions and falsely detected object regions are of comparable size. Applying the erosion operator to such images either fails to eliminate falsely detected object regions or eliminates the true object regions along with them.

8.2.5 Data Extraction

Colour thresholding produces separate binary images for the different color regions on the leader, from which information about the separation distance, position angle, and relative heading angle between the leader and follower UGVs must be extracted. The first step in this process involves identifying the locations of color regions in an image. These are taken to be the centroids of object regions in their respective binary images.

Figure 8.12 depicts the centroids and object regions of all four color markers, where the centroids of the different color regions resulting from HSV thresholding are marked.

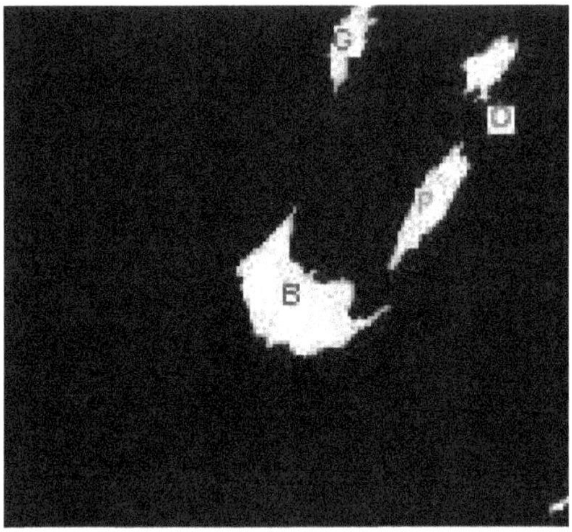

Figure 8.12 The centroids of the different color regions.

While the blue (B), green (G) and pink (P) centroids are roughly in the center of their respective object regions, the orange (O) centroid lies in the background pixel region below the white blob representing the orange marker. This is because a falsely detected object region was included in the centroid calculations. Despite the failure of the erosion operator to eliminate the falsely detected object region, the calculated orange centroid is still relatively close to its actual image position due to the larger size of the true object region. Good results can be obtained provided the leader is close enough to the follower to ensure that the true object region dwarfs all falsely detected object regions.

Using the linear proportionality between image distances and real world distances, the real world separation distance is obtained by scaling the image distance from the blue centroid to the follower camera axis (Figure 8.13, left). The scalar conversion factor for the imaging setup used in this project was 0.8 cm/pixel. Due to the top-down perspective of the distorted image, position angle θ can be directly obtained from the image. The disturbance s is linearly proportional to the separation distance.

Relative heading angle, β, is determined using the three luminous spherical markers. Being higher above the ground than the UGV base, they appear further away from the

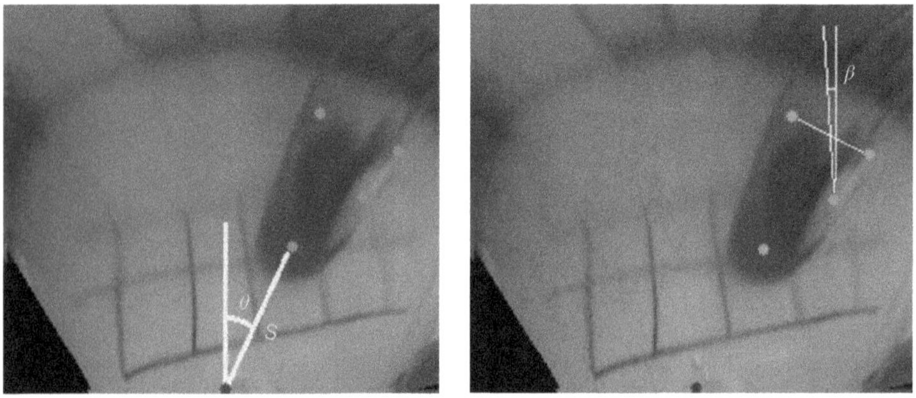

Figure 8.13 The geometry involved: left, s and θ; right, β.

camera's optical axis in the distorted image. This fact precludes the ground position of the neon markers from being simply read off the image. However, the relative positions of the three neon markers are preserved reasonably well when the leader is immediately in front of the follower. The line joining the pink centroid with the midpoint of the green and orange centroids is roughly parallel to the leader's heading vector as seen from the follower. The angle between this line and the vertical is taken to be the relative heading angle (Figure 8.13, right).

8.3 Synchronization Control

This section describes synchronization formation control for a group of three UGVs. First, the principles of synchronization control and the definitions of the geometric formation parameters are reviewed. Next, a control architecture for this formation of three UGVs is examined. Finally, simulation results are presented.

8.3.1 Synchronization

The goal of synchronization formation control is to converge each robotic agent's position towards its desired position while maintaining a given kinematic constraint. This is achieved by simultaneously driving individual robotic agents' tracking errors and synchronization errors to zero. Tracking error is the difference between a robotic agent's current position and desired position, while synchronization error is a weighted difference between the tracking errors of two robotic agents. Synchronization control improves upon other control laws by incorporating information from other robotic agents into a given robotic agent's feedback loop (Figure 8.14).

Synchronization is implemented within a leader–follower framework. Conventional leader–follower formations feature a single robotic agent (the leader) that determines the desired motion of the other robotic agents (the followers). Followers travel along paths that keep them at the desired positions relative to the leader, which travels along an arbitrary predefined path. Most implementations of this control strategy to date have relied on followers' individual tracking errors for feedback control. Augmenting leader–follower formation control with synchronization incorporates tracking errors from the other followers into each follower's control loop. Here, in simulations and

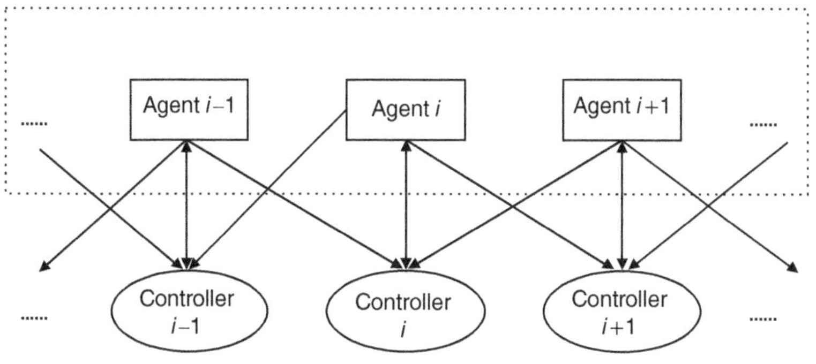

Figure 8.14 Schematic of the synchronization control scheme.

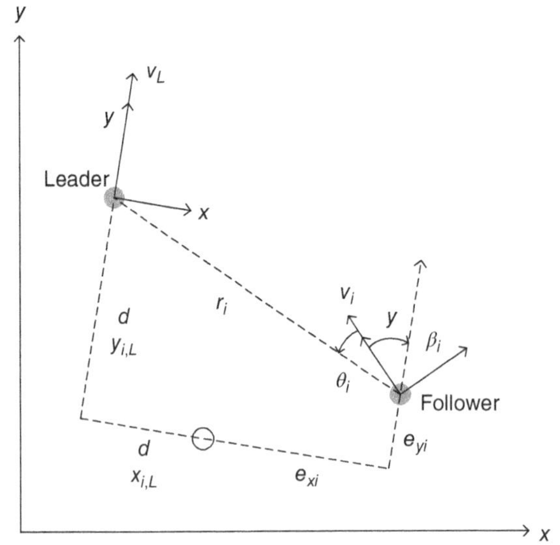

Figure 8.15 Geometric formation parameters.

Table 8.4 Classification of parameters in Figure 8.15.

Measurement	Variables	Reference frame
Sensor inputs	r_i θ_i β_i	Follower i
Desired position	$x^d_{i,L}$ $y^d_{i,L}$	Leader
Tracking errors	e_{xi} e_{yi}	Leader

experiments, this novel formation-control scheme is compared with a conventional leader–follower control scheme.

8.3.2 Formation Parameters

The formation-control parameters used in this project are illustrated in Figure 8.15 and classified in Table 8.4. Both the leader and follower UGVs have local reference frames attached to their centers, with the y-axes aligned with their heading directions. The separation distance r_i is the Cartesian distance from the ith follower to the leader. The position angle θ_i is measured from the ith follower's local y-axis to a line joining the centers of the ith follower and the leader. The relative heading angle β_i is measured from the ith follower's local y-axis to the leader's heading vector.

In a conventional leader–follower formation, the desired position of each follower is specified relative to the leader's local reference frame. The desired position of the ith follower is denoted by an empty circle in Figure 8.15 and its coordinates relative to the leader's local axes are $(x^d_{i,L}, y^d_{i,L})$. In a time-varying formation, the desired position coordinates are functions of t. The tracking errors are defined with respect to the leader's local axes as

$$e_{xi} = x_{i,L} - x^d_{i,L}, \tag{8.1}$$

$$e_{yi} = y_{i,L} - y^d_{i,L}. \tag{8.2}$$

With sensor input variables, (8.1) and (8.2) are equivalent to

$$e_{xi} = r_i \sin(\theta_i - \beta_i) - x^d_{i,L}, \tag{8.3}$$

$$e_{yi} = -r_i \cos(\theta_i - \beta_i) - y^d_{i,L}, \tag{8.4}$$

respectively.

8.3.3 Architecture

A high-level overview of the control architecture implemented in both simulations and experiments is provided in Figure 8.14. The details of the sensor block are described in Section 8.2. Arrows between blocks show data transmissions. The key difference between conventional leader–follower formation control and the approach followed here is that there is mutual exchange of tracking error information between the two followers.

UGVs were modelled with unicycle dynamics for simulations, with linear and angular velocities v and ω as control inputs. In global coordinates, unicycle robot motion can be modelled by the equations

$$\dot{x} = v \cos \phi, \tag{8.5}$$

$$\dot{x} = v \cos \phi, \tag{8.6}$$

$$\dot{\phi} = \omega. \tag{8.7}$$

The UGV's global coordinates are given by (x, y), while ϕ denotes the heading angle of the UGV relative to the global x-axis. The right-hand sign convention is used, assigning positive values to angles counterclockwise of the global x-axis.

8.3.4 Control Law

Note that the controller is required to drive tracking errors of the followers to zero while ensuring that they move synchronously. If PI controllers are used to determine the linear and angular velocity inputs from the x and y-channel errors, we then have

$$v_i = K_{P,vi} E_{yi} + K_{I,vi} \int_0^t E_{yi} d\tau, \tag{8.8}$$

$$\omega_i = K_{P,\omega i} E_{xi} + K_{I,\omega i} \int_0^t E_{xi} d\tau. \tag{8.9}$$

The ith follower's x-channel error E_{xi} and y-channel error E_{yi} are defined as follows:

$$E_{xi} = \underbrace{k_{xi}[e_{xi} + \beta_{xi}\overbrace{(e_{xi} - e_{xj})}^{\epsilon_{xi}}]}_{e^*_{xi}} + k_{\beta i}\beta_i, \tag{8.10}$$

$$E_{yi} = \underbrace{k_{yi}[e_{yi} + \beta_{yi}\overbrace{(e_{yi} - e_{yj})}^{\epsilon_{yi}}]}_{e^*_{yi}} + k_{vi}(v_i - v_L). \tag{8.11}$$

The parameters in (8.10) and (8.11) are identified in Table 8.5. The x and y-channel errors are constructed so that driving both to zero results in a follower being at its

Table 8.5 Synchronization parameters for the *i*th follower.

Term	Description
ϵ_{xi} ϵ_{yi}	Position synchronization errors
β_{xi} β_{yi}	Synchronization gains
e^*_{xi} e^*_{yi}	Coupled position errors
k_{xi} k_{yi} $k_{\beta i}$ k_{vi}	Weighting factors

Table 8.6 PI controller gains used in simulations and experiments.

Parameter	Units	Value
$K_{P,vi}$	s^{-1}	−0.5
$K_{I,vi}$	s^{-2}	−0.3
$K_{P,\omega i}$	$m^{-1}s^{-1}$	1.2
$K_{I,\omega i}$	$m^{-1}s^{-2}$	0.4

desired location relative to the leader, heading in the same direction, and moving at the same speed. Within Equations 8.10 and 8.11, the synchronization gains β_{xi} and β_{yi} can be adjusted to change the relative importance of individual position tracking and movement synchronization.

8.3.5 Simulations

The movement of a group of three unicycle robots (one leader and two followers) under a synchronization formation control law was simulated in Matlab Simulink. Multiple test cases were run to determine the performance of this control law under different conditions. The Simulink model directly implemented the proposed control architecture, except that actual tracking errors were fed into the control block instead of image sensor estimates. Actual tracking errors were calculated using the following matrix equation:

$$\begin{bmatrix} e_{xi} \\ e_{yi} \end{bmatrix} = \begin{bmatrix} \sin\phi_L & -\cos\phi_L \\ \cos\phi_L & \sin\phi_L \end{bmatrix} \begin{bmatrix} x_i - x_L \\ y_i - y_L \end{bmatrix} - \begin{bmatrix} x^d_{i,L} \\ y^d_{i,L} \end{bmatrix} \tag{8.12}$$

In the following simulations, all weighting factors were set to unity. The gains of the PI controllers for the followers were set according to Table 8.6. Synchronization gains were set to zero for simulations without synchronization, and unity for simulations with synchronization.

Table 8.7 Circular trajectory test case formation parameters.

	Variable	Units	Value
Desired position	$x^d_{1,L}$	m	−0.5
	$y^d_{1,L}$	m	−0.5
	$x^d_{2,L}$	m	0.5
	$y^d_{2,L}$	m	−0.5
Initial position (at $t = 0$)	$x_{1,L}$	m	−0.8
	$y_{1,L}$	m	−0.8
	$x_{2,L}$	m	0.8
	$y_{2,L}$	m	−0.8
	β_1	rad	$-\frac{\pi}{6}$
	β_2	rad	$-\frac{\pi}{6}$

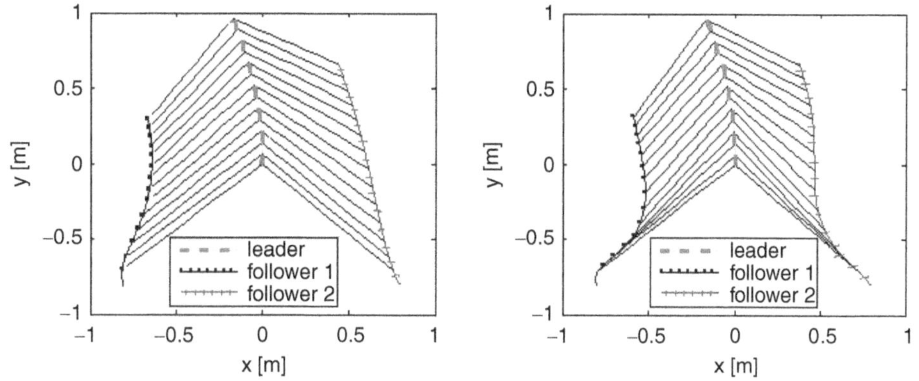

Figure 8.16 Formation trajectories: left, without synchronization; right, with synchronization.

8.3.5.1 Constant Formation along Circular Trajectory

Table 8.7 shows the parameters of a simulation of a constant leader–follower formation moving along a circular trajectory, where the velocities of the leader are $v_L = 0.11$ m/s and $\omega_L = 0.04$ rad/s. The effect of incorporating synchronization into a conventional leader–follower formation controller is shown in Figure 8.16, which depicts the first 10 s of a simulation under Table 8.7 conditions. In the two parts of the figure, the UGV trajectories are as indicated in the key, while the black V-shaped lines show the triangular formation's shape at different time points. Figure 8.17 shows that synchronization greatly reduces the difference between the tracking error magnitudes (Equation (8.13)) of the two followers.

$$e_i = \sqrt{e^2_{xi} + e^2_{yi}}. \tag{8.13}$$

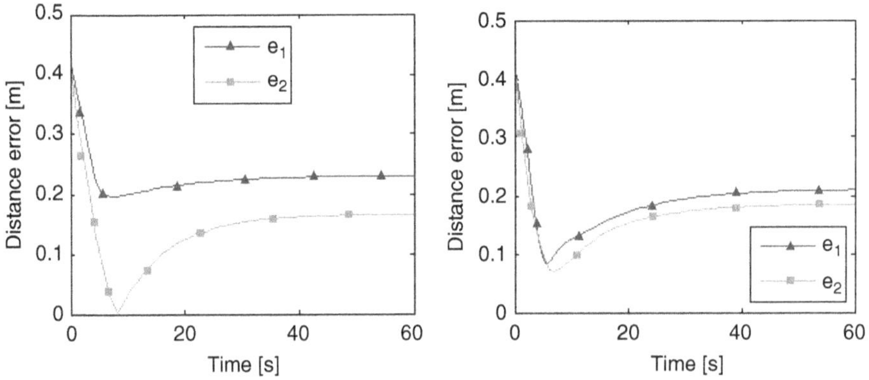

Figure 8.17 Tracking error: left, without synchronization; right, with synchronization.

Under conventional leader–follower control, the separation distance between the two followers r_{12} converges to its desired value r_{12}^d slowly (Figure 8.18). Furthermore, this only occurs as a consequence of r_{1L} and r_{2L} being driven to their desired values r_{1L}^d and r_{2L}^d:

$$r_{12}^d = \sqrt{(x_{1,L}^d - x_{2,L}^d)^2 + (y_{1,L}^d - y_{2,L}^d)^2}, \tag{8.14}$$

$$r_{1L}^d = \sqrt{(x_{1,L}^d)^2 + (y_{1,L}^d)^2}, \tag{8.15}$$

$$r_{2L}^d = \sqrt{(x_{2,L}^d)^2 + (y_{2,L}^d)^2}. \tag{8.16}$$

When synchronization is implemented, all three separation distances are driven to their desired values simultaneously (Figure 8.18). Figure 8.18 does not show a dashed line for r_{1L}^d because $r_{1L}^d = r_{2L}^d$ under the simulation parameters listed in Table 8.7. While the effect of synchronization is most noticeable in the faster convergence of r_{12} to r_{12}^d, it is also responsible for reducing the difference in separation errors $\left||r_{1L} - r_{1L}^d| - |r_{2L} - r_{2L}^d|\right|$. Figure 8.19 shows that synchronization increases the overshoot of follower position angles but does not affect their asymptotic values. The overall effect of synchronization control is to help a leader–follower formation attain its desired geometry

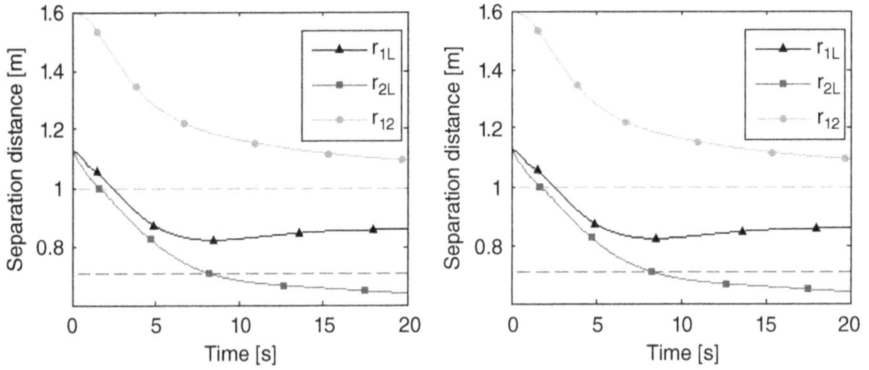

Figure 8.18 Leader–follower distances: left, without synchronization; right, with synchronization.

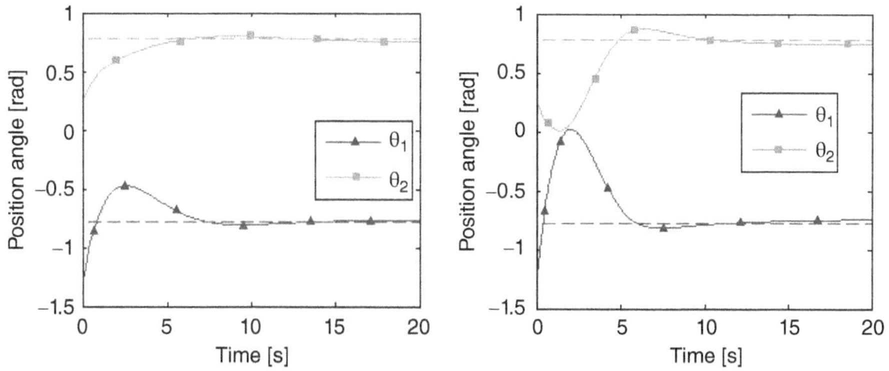

Figure 8.19 Follower angular position: left, without synchronization; right, with synchronization.

more quickly, by enforcing geometric constraints between followers in addition to follower–leader pairs.

8.3.5.2 Time-varying Formation along Linear Trajectory

This test case recreates the formation depicted in Figure 8.20, in which a leader and two followers transition from formation A to formation C via formation B along a linear trajectory. The desired position coordinates successively take on the values for formations A, B, and C, as listed in Table 8.8, with linear transitions between formations. At the beginning of the simulation, the UGVs are in formation A, with relative heading angles of zero between the followers and leader. The desired positions remain in the A, B, and C configurations for 10 s, 15 s, and 15 s respectively, with 5 s transition periods between successive pairs of configurations (Figure 8.21). The leader moves at a linear velocity of 0.11 m/s and an angular velocity of 0 rad/s.

Figure 8.22 illustrates the effect of synchronization on the trajectories of a time-varying leader–follower formation. Figure 8.23 shows that synchronization reduces the follower tracking error magnitudes. Unlike in the case of a constant

Figure 8.20 Time-varying formation along a linear trajectory.

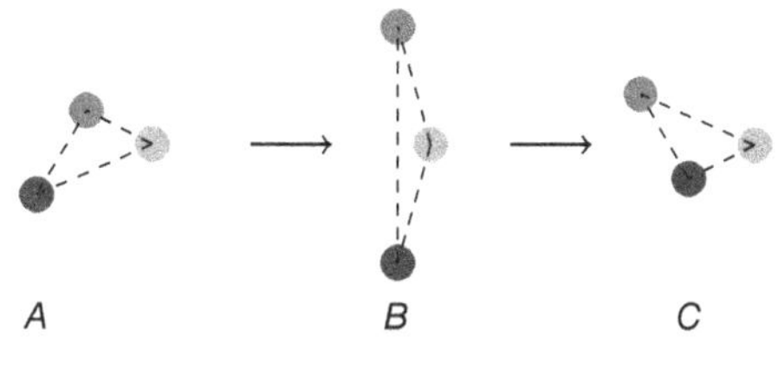

Table 8.8 Details of formations A, B, and C.

Formation	$x^d_{1,L}$ [m]	$y^d_{1,L}$ [m]	$x^d_{2,L}$ [m]	$y^d_{2,L}$ [m]
A	−0.2	−0.4	0.3	−0.7
B	−0.7	−0.2	0.7	−0.2
C	−0.3	−0.7	0.2	−0.4

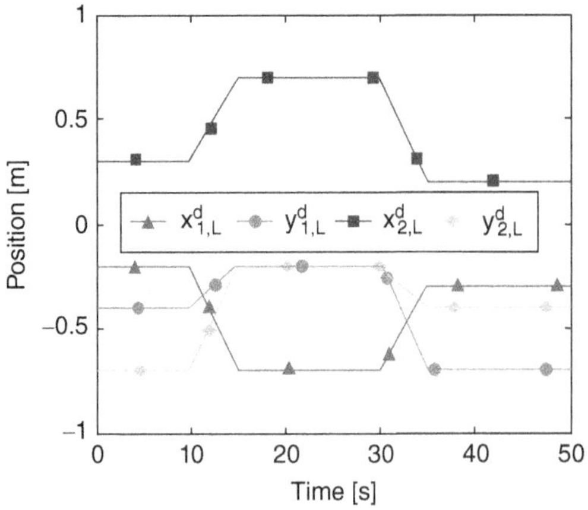

Figure 8.21 Desired position coordinates as functions of time.

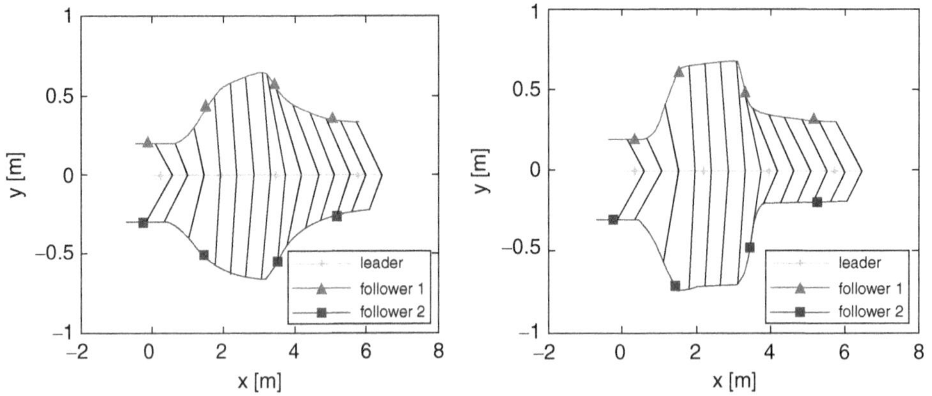

Figure 8.22 Formation trajectories: left, without synchronization; right, with synchronization.

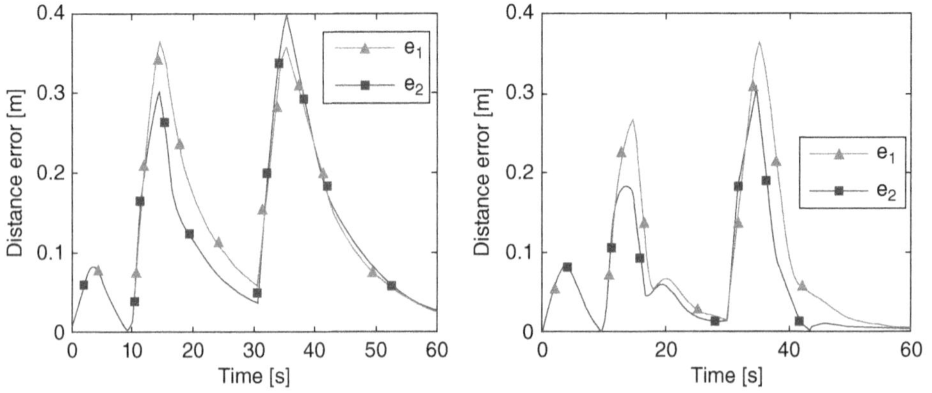

Figure 8.23 Follower tracking error: left, without synchronization; right, with synchronization.

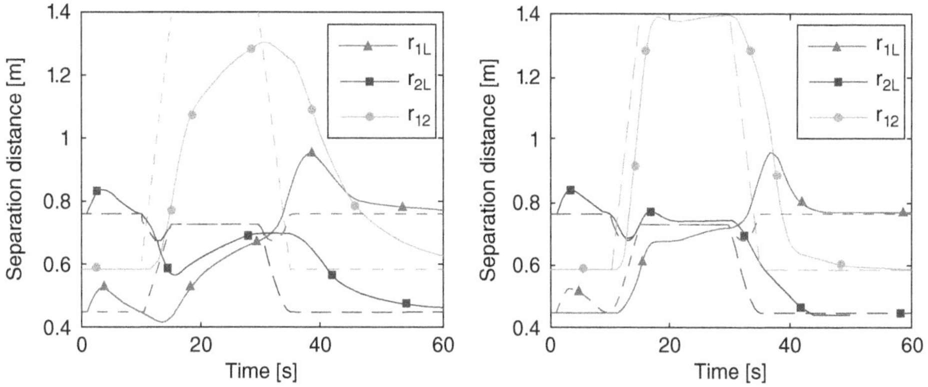

Figure 8.24 Leader–follower distances: left, without synchronization; right, with synchronization.

formation traveling along a circular trajectory, the difference between tracking error magnitudes has increased with synchronization.

From visual inspection of the trajectories and formation shapes in Figure 8.22, it is evident that the followers transition between successive formations much faster when synchronization is implemented. Under conventional leader–follower control, the separation distances between pairs of UGVs lag significantly behind their desired values (Figure 8.24, left). None of the pairwise separation distances for formation *B* have been reached by the time the UGVs start to transition to formation *C*. This stands in sharp contrast to the right-hand part of Figure 8.24, where synchronization is introduced. With synchronization, the pairwise separation distances converge to their desired values quickly each time a formation transition occurs.

Figure 8.25 depicts the follower position angles under a time-varying formation. The major difference between the two parts of the figure is that the right-hand graph shows a higher degree of overshoot, especially after the transition from formation *B* to formation *C*. However, this position angle overshoot is the result of the followers adjusting $x_{1,L}$ and $x_{2,L}$ more quickly than in the case without synchronization. It must also be noted that

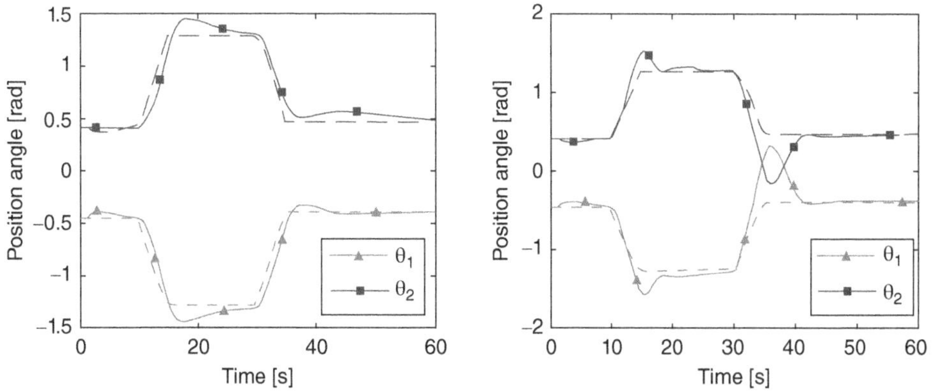

Figure 8.25 Follower angular position: left, without synchronization; right, with synchronization.

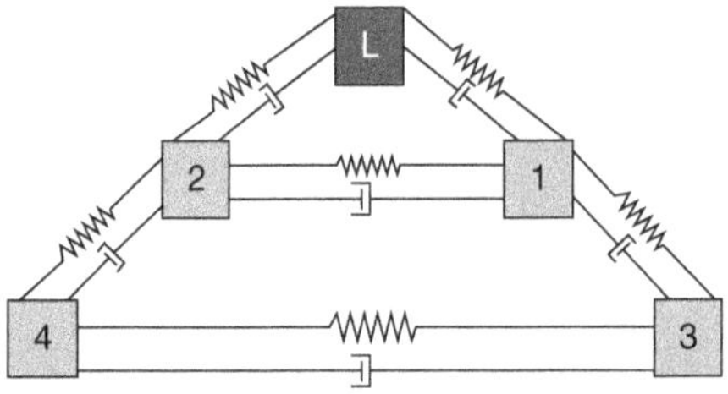

Figure 8.26 A group of agents connected with springs and dampers.

after each transition period the follower position angles track their desired values more closely when synchronization is implemented.

As in the constant formation circular trajectory test case, the time-varying formation linear trajectory test case shows that synchronization decreases the time required for a leader–follower formation to attain its desired geometry.

8.4 Passivity Control

This section describes a theoretical implementation of passivity formation control in a group of three UGVs. First, the principles of passivity control and the definitions of formation parameters are reviewed. Next, a controller for a formation of three UGVs is described in detail. Finally, simulation results for this controller are presented.

8.4.1 Passivity

Passivity theory models dynamics systems as a network of interacting energy transformation devices. A passivity-based controller shapes energy flows between system elements in a way that brings about desired movements [16]. Robotic agents that break formation are modeled as energy disturbances to be damped as the formation is brought back together [17]. For formation control applications, a common approach is to model robotic agents as a system of masses connected by springs and dampers (Figure 8.26).

In this chapter, passivity is implemented within a leader–follower framework for comparison with the previously analysed synchronization controller.

8.4.2 Formation Parameters

As in Section 8.3.3, UGVs are modelled by unicycles within two-dimensional Cartesian space. A unicycle's state is the triplet (x, y, ϕ) that describes position and heading. The relationships between the involved variables are laid out in Equations 8.17 and 8.18, where 1, 2, or L can be substituted for i and j. The angular positions are defined in Equations 8.19 and 8.20, and measured counter-clockwise from the direction opposite the leader's heading.

$$r_{ij} = \sqrt{(x_i - x_j)^2 + (y_i - y_j)^2}, \tag{8.17}$$

$$\phi_{ij} = \arctan \frac{y_i - y_j}{x_i - x_j}, \tag{8.18}$$

$$\gamma_{L1} = \arctan\left[\frac{(x_1 - x_L)\sin\phi_L - (y_1 - y_L)\cos\phi_L}{-(x_1 - x_L)\cos\phi_L - (y_1 - y_L)\sin\phi_L}\right], \tag{8.19}$$

$$\gamma_{L2} = \arctan\left[\frac{(x_2 - x_L)\sin\phi_L - (y_2 - y_L)\cos\phi_L}{-(x_2 - x_L)\cos\phi_L - (y_2 - y_L)\sin\phi_L}\right]. \tag{8.20}$$

8.4.3 Control Law

The objective of the passivity controller is to converge followers to their desired positions by damping out position disturbances. In order to achieve this, a group of three UGVs is modeled as a spring–mass system whose natural equilibrium state is the desired formation. Proportional integral controllers are used to determine linear and angular velocity inputs based on heading errors and vector sums of the spring forces.

The leader–follower framework for passivity control adopted in this project means that only follower UGVs are controlled; the leader tracks an arbitrary path. In this arrangement, the followers' motion is influenced by spring forces but the leader's motion is not. Physically, this can be imagined as a spring–mass system where the leader's mass is much greater than any of the followers' masses.

For control purposes, both linear (Equation (8.21)) and torsional (Equation (8.23)) spring forces are used to compute the resultant force on a follower. The sign convention used in this project is that a positive linear spring force $F_{K,ij}$ is directed from the ith UGV toward the jth UGV for $i \in \{1,2\}, j \in \{1,2,L\}, i \neq j$. Linear spring forces are always parallel to the lines connecting the centers of a given pair of UGVs. The sign convention for torques (Equation (8.22)) is that positive torques obey the right-hand rule with respect to the global Cartesian coordinate system.

$$F_{K,ij} = k_{ij}(r_{ji} - r_{ji}^d), \tag{8.21}$$

$$\tau_{iL} = \kappa_{iL}(\gamma_{Li}^d - \gamma_{Li}), \tag{8.22}$$

$$F_{T,iL} = \frac{\kappa_{iL}}{r_{iL}}(\gamma_{Li}^d - \gamma_{Li}). \tag{8.23}$$

For a group of three vehicles, the resultant force on each follower is the vector sum of two linear spring forces and one torsional spring force; that is,

$$\mathbf{F_1} = \mathbf{F_{k,1L}} + \mathbf{F_{k,12}} + \mathbf{F_{T,1L}} \rightarrow \tag{8.24}$$

$$\begin{bmatrix} F_{x1} \\ F_{y1} \end{bmatrix} = \begin{bmatrix} F_{k,1L}\cos\phi_{L1} + F_{k,12}\cos\phi_{21} + F_{T,1L}\sin\phi_{L1} \\ F_{k,1L}\sin\phi_{L1} + F_{k,12}\sin\phi_{21} - F_{T,1L}\cos\phi_{L1} \end{bmatrix} \tag{8.25}$$

$$\mathbf{F_2} = \mathbf{F_{k,2L}} + \mathbf{F_{k,21}} + \mathbf{F_{T,2L}} \rightarrow \tag{8.26}$$

$$\begin{bmatrix} F_{x2} \\ F_{y2} \end{bmatrix} = \begin{bmatrix} F_{k,2L}\cos\phi_{L2} + F_{k,12}\cos\phi_{12} + F_{T,2L}\sin\phi_{L2} \\ F_{k,2L}\sin\phi_{L2} + F_{k,12}\sin\phi_{12} - F_{T,2L}\cos\phi_{L2} \end{bmatrix} \tag{8.27}$$

The force "error" for a follower is the magnitude of the resultant force applied to it (Equation (8.28)). A follower's angular error consists of the difference between the resultant force's heading angle and its own heading angle. Thus, we have

$$e_{Fi} = ||\mathbf{F_i}|| = \sqrt{F_{xi}^2 + F_{yi}^2}, \tag{8.28}$$

$$e_{\phi i} = \arctan\frac{F_{yi}}{F_{xi}} - \phi_i. \tag{8.29}$$

Using PID control, linear velocity inputs are determined from force "errors" while angular velocity inputs are determined from angular errors:

$$v_i = K_{P,vi} e_{Fi} + K_{I,vi} \int_0^t e_{Fi} d\tau + K_{D,vi} \frac{d}{dt} e_{Fi}, \tag{8.30}$$

$$\omega_i = K_{P,\omega i} e_{\phi i} + K_{I,\omega i} \int_0^t e_{\phi i} d\tau + K_{D,\omega i} \frac{d}{dt} e_{\phi i}. \tag{8.31}$$

In passivity control, the followers do not need to know the leader's velocity. Each follower makes use of position information from other followers by facilitating communication of leader-relative tracking errors between followers. Each follower directly gathers position information from other followers via sensing and uses it to compute spring forces.

8.4.4 Simulation

The movement of three unicycle robots (one leader and two followers) under a passivity formation control law was simulated in Matlab Simulink. PID controller gains were set according to Table 8.9. All spring constants were set to unity. The passivity control test case is characterized by the parameters listed in Table 8.10. Although the followers' desired positions are the same as those in Section 8.3.5, both their initial positions and the leader's angular velocity are changed. In particular, $\omega_L = 0.02$ rad/s and the initial positions are as shown in Table 8.10.

The left-hand part of Figure 8.27 depicts the formation's circular trajectory, while the right-hand part shows the initial portion of that trajectory in greater detail. Like a physical spring–mass system, the follower resultant force magnitudes e_{F1} and e_{F2} gradually decay to zero after a sharp initial drop (Figure 8.28). In the first 50 s of the simulation, the follower angular errors oscillate as an underdamped second-order system (Figure 8.29). After the oscillations are damped out, the angular errors rise instead of decaying to zero as a consequence of the counterclockwise formation trajectory.

Since passivity models a group of UGVs as a spring–mass system, it is useful to characterize their deviation from the desired formation in terms of elastic potential energy. Elastic potential energy is useful for describing the state of a multi-vehicle formation because it incorporates information about all vehicles' deviations from their desired positions in a single scalar value. Figure 8.30 shows a steep decline in total

Table 8.9 PID controller gains used in simulations.

Parameter	Units	Value
$K_{P,vi}$	s kg^{-1}	2.0
$K_{I,vi}$	kg^{-1}	0.016
$K_{D,vi}$	s^{2}kg^{-1}	0
$K_{P,\omega i}$	s^{-1}	0.075
$K_{I,\omega i}$	s^{-2}	−0.0004
$K_{D,\omega i}$	–	0.02

Table 8.10 Circular trajectory test case formation parameters.

	Variable	Units	Value
Desired position	r_{L1}^d	m	$\frac{1}{\sqrt{2}}$
	γ_{L1}^d	rad	$-\frac{\pi}{4}$
	r_{L2}^d	m	$\frac{1}{\sqrt{2}}$
	γ_{L2}^d	rad	$\frac{\pi}{4}$
Initial position (at $t = 0$)	$x_{1,L}$	m	−0.51
	$y_{1,L}$	m	−1
	$x_{2,L}$	m	0.51
	$y_{2,L}$	m	−0.6
	β_1	rad	0
	β_2	rad	0

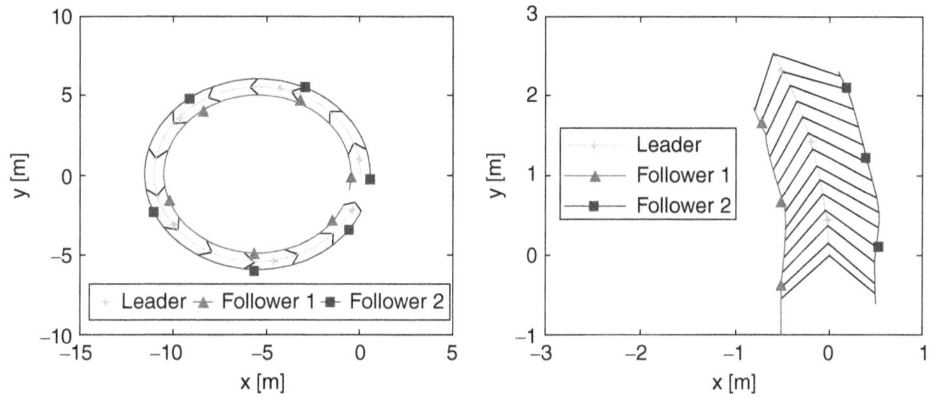

Figure 8.27 Passivity formation trajectories: left, $(0 < t < 300$ s); right, $(0 < t < 25$ s).

Figure 8.28 Follower resultant force magnitudes.

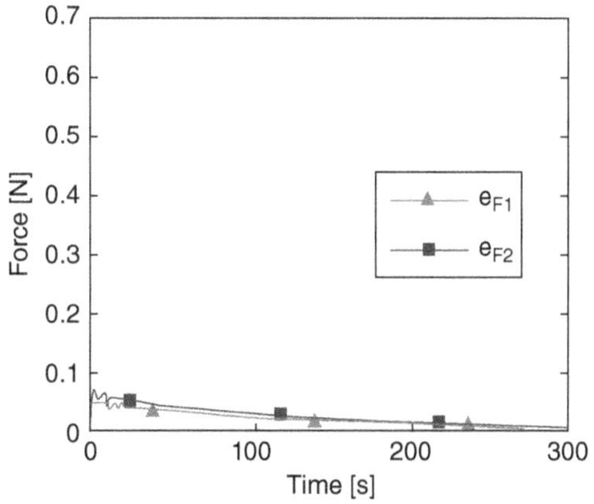

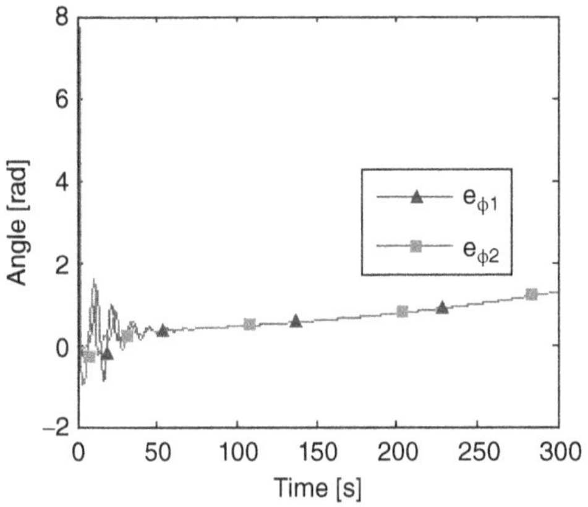

Figure 8.29 Follower angular errors ($0 < t < 300$ s).

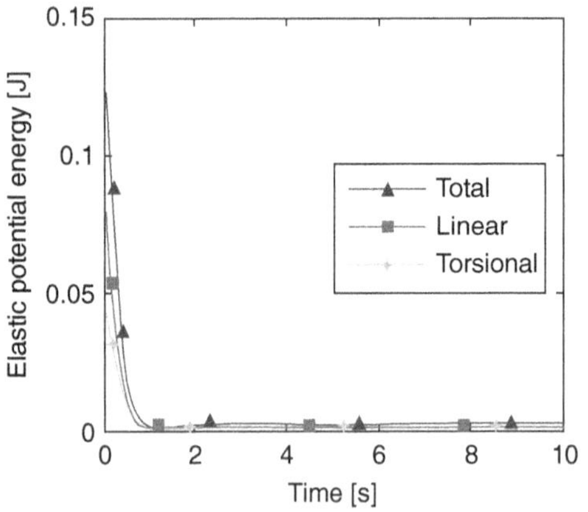

Figure 8.30 Formation spring energy.

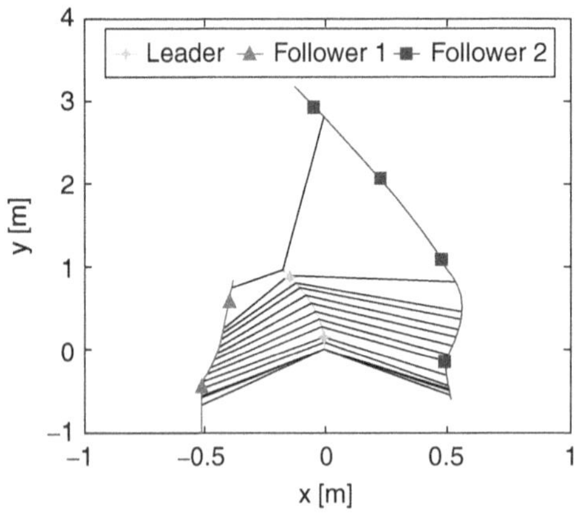

Figure 8.31 Passivity controller instability.

elastic potential energy within the first few seconds of the simulation, which attests to the ability of the passivity controller to quickly damp out initial position disturbances.

Figure 8.31 depicts a simulation run under the conditions of Table 8.10 with the single exception that $\omega_L = 0.04$ m/s (twice its previous value). While this new set of conditions is nearly identical to that of Section 8.3.5, the passivity controller breaks down shortly after 10 s. The higher curvature of the leader's path changes pairwise separation distances dramatically, leading to overshoots in spring forces and control inputs. This indicates that the synchronization controller has a much greater tolerance for leader path curvature than the passivity controller. Additionally, other simulations indicate that the passivity controller is only stable when the followers start out close to their leader-relative desired positions. These facts suggest that the passivity controller is not as effective as the synchronization controller for the formation control of UGVs.

8.5 Experiments

8.5.1 Setup

This project used Roomba robots manufactured by the iRobot Corporation as UGV platforms for vision-based formation control experiments (Figure 8.32). Roomba robots were chosen because they are commonly used in the robotics research community, which increases the amount of documentation available for troubleshooting. Three Roombas along with accessories such as USB cables, chargers, and metal mounting brackets were used in this project. Additionally, custom-made metal mounting brackets provided a convenient superstructure on which to mount imaging equipment.

The three UGVs used in this project's experiments were each controlled by a different computer. The two followers were controlled by a desktop and laptop because their image-processing routines required greater computing power. On the other hand, a netbook was sufficient for the computing needs of the leader, which was not required to do image processing. All three computers were wirelessly connected to a local area network without other computers. Further details of the three computers are provided in Table 8.11.

In the experiments, all three computers ran various versions of Ubuntu. Image processing code for the follower UGVs was written in C++ using OpenCV 2.4.0 image processing libraries. The UGVs were driven by C++ code using ROS (Robot Operating System), an open source package for robotics applications. The Fuerte release of ROS was used on all three computers to avoid wireless communications problems that might arise if different releases of ROS were used. The Optitrack system (Figure 8.33) was used for ground truth measurements of UGV positions. Real-time position data was collected using three tripod-mounted infrared cameras (Figure 8.34) that tracked each UGV's unique infrared marker pattern (Figure 8.35) across the test area.

Figure 8.32 iRobot Create (Roomba) robot.

Table 8.11 Computers used in this project.

Processor	RAM (GB)	Ubuntu version
Intel i7-2600 (3.4 GHz)	8	12.04 LTS
Intel Core 2 Duo P8700 (2.53 GHz)	3	12.04 LTS
Dual Core Intel Atom N270 (1.6 GHz)	1	10.04 LTS

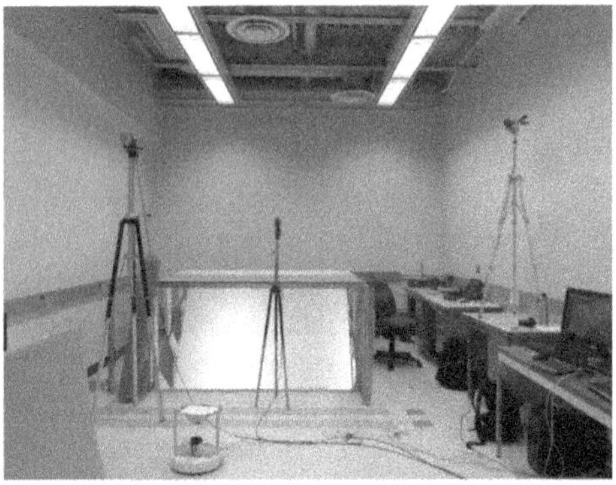

Figure 8.33 Optitrack infrared camera setup.

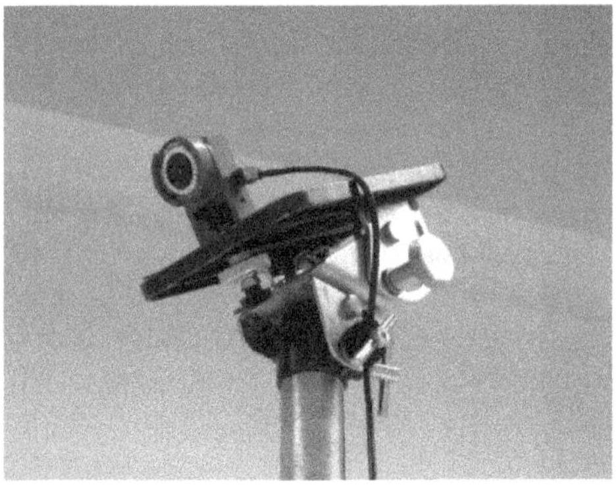

Figure 8.34 Close-up of an infrared camera.

8.5.2 Results

The experimental results are presented here to validate the synchronization control law. In all experiments, both of the followers' PI controller gains were set according to Table 8.6. Due to network latency problems, which erratically delayed communication of tracking errors between the followers, each follower was told to assume that the other's tracking error was zero. As a consequence, the followers were subject to pure leader–follower control rather than synchronization control. However, the experiments were still valuable for testing the image processing algorithms described in Section 8.2.

8.5.2.1 Constant Formation Along Circular Trajectory

This experimental scenario was patterned after the simulations in Section 8.3.5. Experimental formation parameters are given in Table 8.12, while the ground truth formation

(a) Leader

(b) Follower 1

(c) Follower 2

Figure 8.35 Optitrack infrared marker patterns on UGVs.

trajectories and shapes are depicted in Figure 8.36. The velocities of the leader were $v_L = 0.11$ m/s and $\omega_L = -0.002$ rad/s.

In Figure 8.36, the followers' paths show consistent offsets from their desired paths, suggesting systemic errors in the image processing algorithm. One likely cause for these offsets was the imperfect alignment of the Optitrack infrared marker centroids (used as ground truth positions) and camera axes on the follower UGVs. Despite this, the two followers were able to reach what their image sensors perceived to be the desired positions and turn with the leader. Figure 8.37 shows the x and y-direction tracking errors for both followers. In this and other experiments, the magnitudes of tracking errors of the follower on the outside curve of a circular trajectory of the leader were smaller than those of the follower on the inside curve.

8.5.2.2 Time-varying Formation along Linear Trajectory

This experimental scenario was patterned after the simulations done in Section 8.3.5, with an additional simplification that formation changes only affect the followers' relative positions in the direction parallel to the leader's path. Formation parameters are given in Table 8.13, while a graph of time-varying desired position coordinates is shown in Figure 8.38.

The black lines in Figure 8.39 clearly depict the transition in formation as the followers swap y-positions relative to the leader. In this test case the y-direction tracking

Table 8.12 Circular trajectory experiment formation parameters.

	Variable	Units	Value
Desired position	$x^d_{1,L}$	m	−0.3
	$y^d_{1,L}$	m	−0.4
	$x^d_{2,L}$	m	0.3
	$y^d_{2,L}$	m	−0.4
Initial position (at $t = 0$)	$x_{1,L}$	m	−0.42
	$y_{1,L}$	m	−0.56
	$x_{2,L}$	m	0.41
	$y_{2,L}$	m	−0.68
	β_1	rad	0
	β_2	rad	0

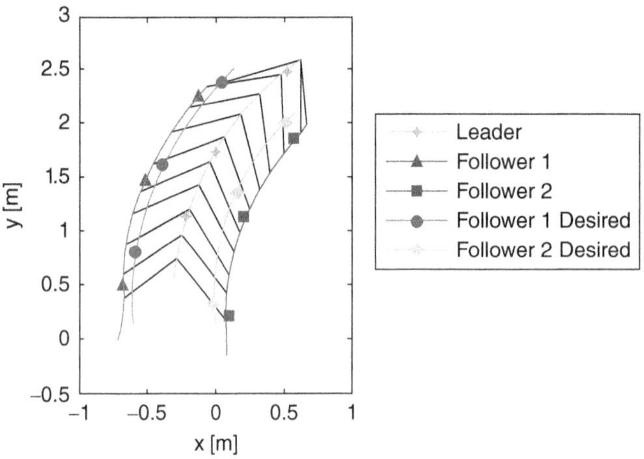

Figure 8.36 Ground truth formation trajectories and shapes.

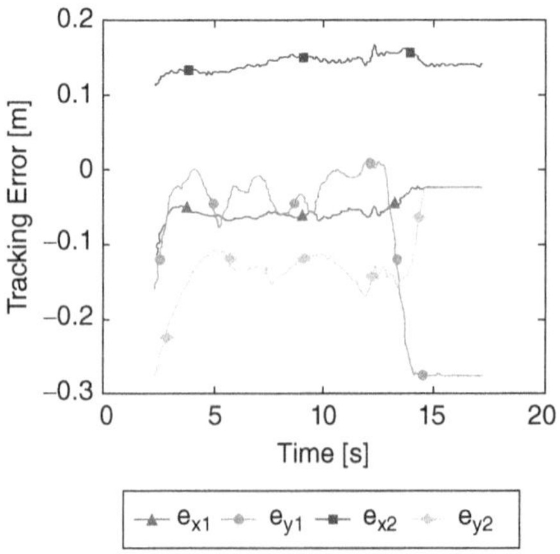

Figure 8.37 Ground truth tracking errors.

Table 8.13 Linear trajectory experiment formation parameters.

	Variable	Units	Value
Leader	v_L	m/s	0.11
	ω_L	rad/s	0
Initial position (at $t = 0$)	$x_{1,L}$	m	−0.35
	$y_{1,L}$	m	−0.48
	$x_{2,L}$	m	0.3
	$y_{2,L}$	m	−0.71
	β_1	rad	0
	β_2	rad	0

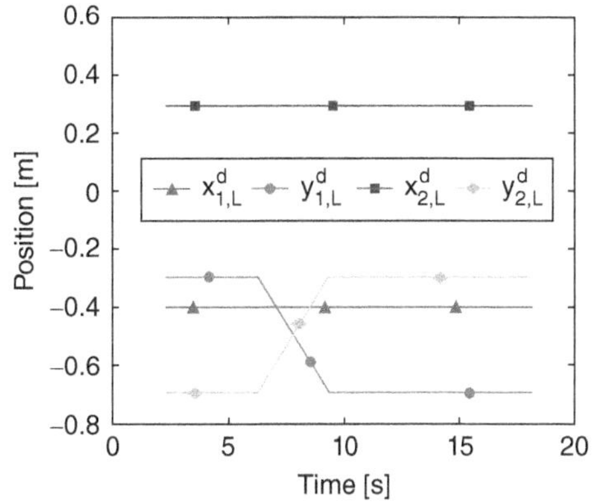

Figure 8.38 Desired formation coordinates as functions of time.

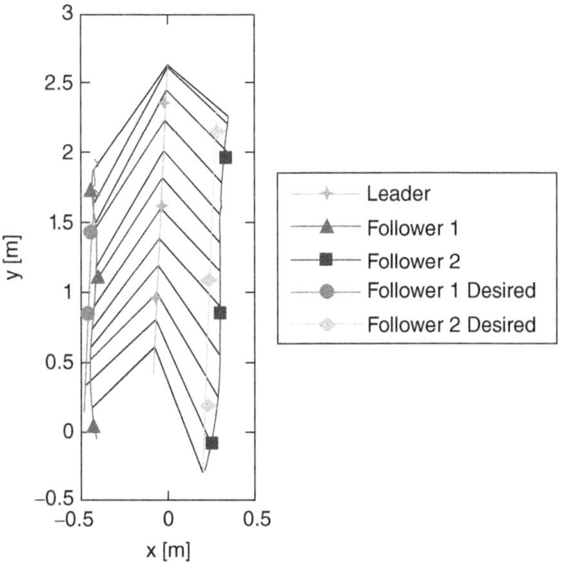

Figure 8.39 Ground truth formation trajectories and shapes.

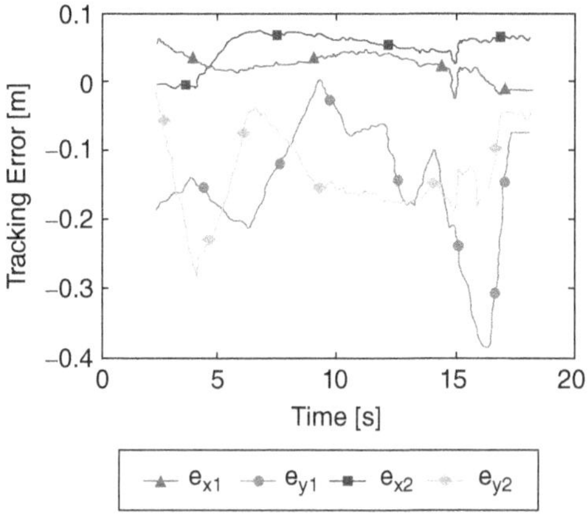

Figure 8.40 Ground truth tracking errors.

errors were larger than the x-direction tracking errors for both followers (as shown in Figure 8.40) because formation changes solely involved $y^d_{1,L}$ and $y^d_{2,L}$.

8.6 Summary

This chapter examined the performance of synchronization and passivity-based formation controllers through simulations and experiments. First, a geometric model for two-dimensional formation control of UGVs was developed. Next, control laws based on both synchronization and passivity were developed within a leader–follower framework. A PID controller was used to provide linear and angular velocity inputs to the follower UGVs, while the leader UGV traveled on a predefined path. Both control laws were implemented and tested in Matlab Simulink for a group consisting of a leader and two followers.

Simulation results showed that, compared to conventional leader–follower control, synchronization decreases the time needed for follower UGVs starting out from arbitrary positions to move into their desired locations relative to the leader. Because synchronization facilitates the exchange of tracking error information between follower UGVs, it ensures that pairwise separation distances between followers converge to their desired values, something that is not explicitly addressed by conventional leader–follower control. This information exchange and coupling ensures that differences between follower tracking errors are very small, leading to synchronized motion. The benefits of synchronization over conventional leader–follower control are prominently highlighted in simulations of time-varying formations, where the former method has much faster convergence rates.

Simulations of passivity control indicated that the range of stable follower starting positions relative to the leader was narrower than that for synchronization control. In particular, the passivity controller was sensitive to changes in follower positions perpendicular to the leader's path. Instability often resulted when followers' paths exhibited angular overshoots or the leader's path curvature was too great. Despite these issues, the passivity formation controller is capable of maintaining followers' leader-relative positions on arbitrary trajectories provided that they start close enough to those desired positions. Furthermore, an advantage of passivity control is that a single scalar value analogous to total elastic potential energy can be used to characterize how close an entire formation is to its corresponding desired formation.

Visual tracking algorithms were developed to experimentally implement synchronization control. Omnivision systems consisting of webcams and hemispherical mirrors were mounted on Roombas to give images with a wide field of view. After distortion, HSV color thresholding, and erosion, color region centroids were computed to determine the locations of tracked objects of interest. Separation distances, position angles, and relative heading angles between leader–follower pairs were calculated from each image and fed into a control law.

Due to network latency problems, conventional leader–follower control was experimentally implemented on the Roombas instead of synchronization control. Ground truth results from test runs showed that aside from systemic position offsets caused by imaging errors, the follower UGVs were able to successfully track their desired positions in both constant and time-varying formations.

This work relied on lookup tables to approximate the mapping between image points and real-world locations. Although functional and computationally cheap, lookup tables lack accuracy and engineering elegance. One alternate solution is to derive mapping functions from a geometric model of the imaging system. Using a pinhole camera model and ray-tracing laws, geometric relationships between object distances and image coordinates can be constructed. Any mathematical derivation would have to take into account the imperfect curvature of the hemispherical mirror, offset of the camera from the mirror's axis of symmetry, and measurement errors associated with modeling the imaging system's geometry.

The simulation results have showed on a case-by-case basis whether or not a controller ensured formation stability. It would be advantageous to derive the conditions for which the followers are theoretically guaranteed to track the leader. Typical conditions would include limiting ranges for the linear and angular velocities of the leader along its predefined path, as well as the followers' initial separation distances and heading angles. Lyapunov stability analysis would be a good choice for carrying out this type of investigation.

We do not feature experimental implementations of synchronization control due to network latency problems. The sending and receiving of tracking errors caused erratic delays in the transmission of driving commands to follower UGVs, undermining the ability of all three UGVs to move simultaneously. Even if there was sufficient bandwidth for all information to be transmitted to the required receivers, the finite amount of time required for transmission means that each follower can only receive historical information about other followers' tracking errors. One way to deal with this limitation is to have a global clock keeping time for all UGVs in the formation. The global clock can be used to time stamp each outgoing tracking error, allowing for listening followers to reconstruct a time history of other followers' tracking errors. Then, an algorithm for estimating other followers' current tracking errors based on sparse information from their past tracking errors could be developed. This algorithm could be compared to simulations in which all followers have perfect, instantaneous knowledge of each other's tracking errors for benchmarking purposes.

Bibliography

1 S.-J. Chung, U. Ahsun, and J.-J. E. Slotine, "Application of synchronization to formation flying spacecraft: Lagrangian approach," *Journal of Guidance, Control, and Dynamics*, vol. 32, no. 2, pp. 512–526, 2009.
2 R. Ray, B. Cobleigh, M. J. Vachon, and C. S. John, "Flight test techniques used to evaluate performance benifits during formation flight," in *Proceedings of AIAA Atmospheric Flight Mechanics Conference and Exhibit*, no. 4492, (Monterey, California), AIAA, 2002.
3 M. J. Vachon, R. Ray, K. Walsh, and K. Ennix, "F/A-18 aircraft performance benefits measured during the autonomous formation flight project," in *Proceedings of AIAA Atmospheric Flight Mechanics Conference and Exhibit*, no. 4491, (Monterey, California), AIAA, 2002.
4 P. Binetti, K. B. Ariyur, M. Krstic, and F. Bernelli, "Formation flight optimization using extremum seeking feedback," *Journal of Guidance, Control, and Dynamics*, vol. 26, no. 1, pp. 132–142, 2003.
5 J. D. Anderson, *Fundamentals of Aerodynamics*. McGraw-Hill, 2nd ed., 1991.
6 M. Pachter, J. J. D'Azzo, and A. W. Proud, "Tight formation flight control," *Journal of Guidance, Control, and Dynamics*, vol. 24, no. 2, pp. 246–254, 2001.
7 M. Pachter, J. D'Azzo, and M. Veth, "Proportional and integral control of nonlinear systems," *International Journal of Control*, vol. 64, no. 4, pp. 679–692, 1996.
8 Q.-C. Zhong and D. Rees, "Control of uncertain LTI systems based on an uncertainty and disturbance estimator," *Journal of Dynamic Systems, Measurement, and Control*, vol. 126, no. 4, pp. 34–44, 2004.
9 S. E. Talole and S. Phadke, "Robust input-output linearisation using uncertainty and disturbance estimation," *International Journal of Control*, vol. 82, no. 10, pp. 1794–1803, 2009.
10 B. Zhu, C. Meng, and G. Hu, "Robust consensus tracking of double-integrator dynamics by bounded distributed control," *International Journal of Robust and Nonlinear Control*, vol. 26, no. 7, pp. 1489–1511, 2016.
11 S. Gadelovits, Q.-C. Zhong, V. Kadirkamanathan, and A. Kuperman, "UDE-based controller equipped with a multi-band-stop filter to improve the voltage quality of inverters," *IEEE Transactions on Industrial Electronics*, vol. 64, no. 9, pp. 7433–7443, 2017.
12 H. Bai, M. Arcak, and J. Wen, *Cooperative Control Design: A Systematic, Passivity-based Approach*. Springer Science & Business Media, 2011.

Formation Control of Multiple Autonomous Vehicle Systems, First Edition. Hugh H.T. Liu and Bo Zhu.

13 C. J. Schumacher and R. Kumar, "Adaptive control of UAVs in close-coupled formation flight," in *Proceedings of the 2000 American Control Conference*, vol. 2, pp. 849–853, IEEE, 2000.
14 B. D. Anderson, B. Fidan, C. Yu, and D. Walle, "UAV formation control: theory and application," in *Recent Advances in Learning and Control*, pp. 15–33, Springer, 2008.
15 T. Kopfstedt, M. Menges, and S. Bullmer, "Methods for control of UGV convoys in unexplored and partially unpaved terrain," *IFAC Proceedings Volumes*, vol. 43, no. 23, pp. 149–154, 2010.
16 R. Ortega, A. J. V. Der Schaft, I. M. Y. Mareels, and B. Maschke, "Putting energy back in control," *IEEE Control Systems Magazine*, vol. 21, no. 2, pp. 18–33, 2001.
17 C. A. Rabbath, N. Léchevin, and J. Apkarian, "Experiments with a passivity-based formation control system for teams of small robotic drones," in *Proceedings of AIAA Guidance, Navigation, Control Conference*, pp. 1–23, 2011.

Part IV

Formation Control: Laboratory

9

Experiments on 3DOF Desktop Helicopters

In this chapter, we will first introduce an experimental apparatus, which consists of four Quanser Inc.'s desktop "helicopters", and then show how the control concepts and approaches introduced in Chapter 6 are applied to this experimental apparatus. In particular, we will present several laboratory cases and show the experimental results obtained.

9.1 Description of the Experimental Setup

As a case study to illustrate the main control concepts and results presented in this book, an experimental apparatus is set up as a possible laboratory exercise. The so-called desktop three-degree-of-freedom "helicopters" (3DOF-Heli) used are shown in Figure 9.1: as can be seen, they are a desktop model of a helicopter rather than a helicopter *per se*. Four of them serve as a platform for conducting experiments on multiple vehicles.

Before the details of the experimental setup are introduced, one may wonder how representative this laboratory equipment can be. How realistic are they in terms of reflecting real-world applications? Of course, they are not commercial systems, yet, one may find in the 3DOF-Heli some resemblances to real-world applications. It is not difficult to understand where the "helicopter" name comes from when one sees a 3DOF-Heli "fly" like a real two-rotor helicopter (which is also called as a tandem rotor helicopter, with one rotor in the front and the other at the back of the vehicle) such as:

- the Piasecki HRP Rescuer, designed by Frank Piasecki and built by Piasecki Helicopter;
- the Boeing CH-47 Chinook, an American twin-engine, tandem rotor heavy-lift helicopter, the primary roles of which are troop movements, artillery placement and battlefield resupply;
- the Boeing Model 360, an experimental medium-lift tandem rotor cargo helicopter developed privately by Boeing to demonstrate advanced helicopter technology.

In addition, the 3DOF-Heli's lifting device, driven by DC motors, provides an excellent representation of the second-order systems that are often found in space, aerial and robotics applications. The DC motor offers plenty of opportunities to address the nonlinearities, uncertainties or measurement errors involved.

The apparatus is well designed, easy to operate, and there are many examples of experiments and simulations that the reader can find in the literature to enrich their

Formation Control of Multiple Autonomous Vehicle Systems, First Edition. Hugh H.T. Liu and Bo Zhu.

Figure 9.1 The setup of four 3DOF-Helis.

study. The 3DOF-Heli equipment is supplied by Quanser Consulting Inc.,[1] a well regarded company in the control engineering education community.

There are four 3DOF-Helis at the Flight Systems and Control (FSC) Laboratory of the University of Toronto. They are used for studying cooperative control of MVSs, multivehicle formation flight control, and advanced controller development. The comprehensive experimental configuration and signal flow chart is shown in Figure 9.2, where all four helicopters are involved. Experimental configurations with fewer helicopters are easily obtained.

As shown in Figure 9.2, four helicopters (H1 to H4), ten universal power modules (UPMs), two Q8 terminal boards and a PC are included in this configuration. Helicopters H2 and H4 are equipped with active disturbance systems (ADSs), whereas the H1 and H3 are not. The elevation and pitch angles and the position signals of the ADSs are measured using encoders installed on the helicopters and transmitted through the Q8 terminal boards to two Q8 data acquisition cards (DACs), which are installed on the PC. The resolution of each encoder is 4096 counts per revolution, which yields a resolution of 0.0879° for the angular position measurement. For H1 (or H3), two UPMs are used to supply power to the front and back DC motors (propellers), respectively. For H2 (or H4), an additional UPM is used for the ADS motor. One of the two Q8 terminal boards is used for H1 and H2, and the other is for H3 and H4. Each terminal board is connected to one Q8 DAC. In each experiment, the sampling time is fixed to be 0.001 s, and the implemented control laws are constructed using Simulink blocks, compiled and downloaded to Q8 DACs and run in real-time therein. The real-time experimental data from the Q8 DAC are sent to a PC through a PCI computer bus. For the full-state feedback, the elevation and pitch angular rates are generated using second-order derivative filters provided by Quanser Consulting Inc. [1].

1 One may find more about the equipment and showcases from Quanser's website: www.quanser.com.

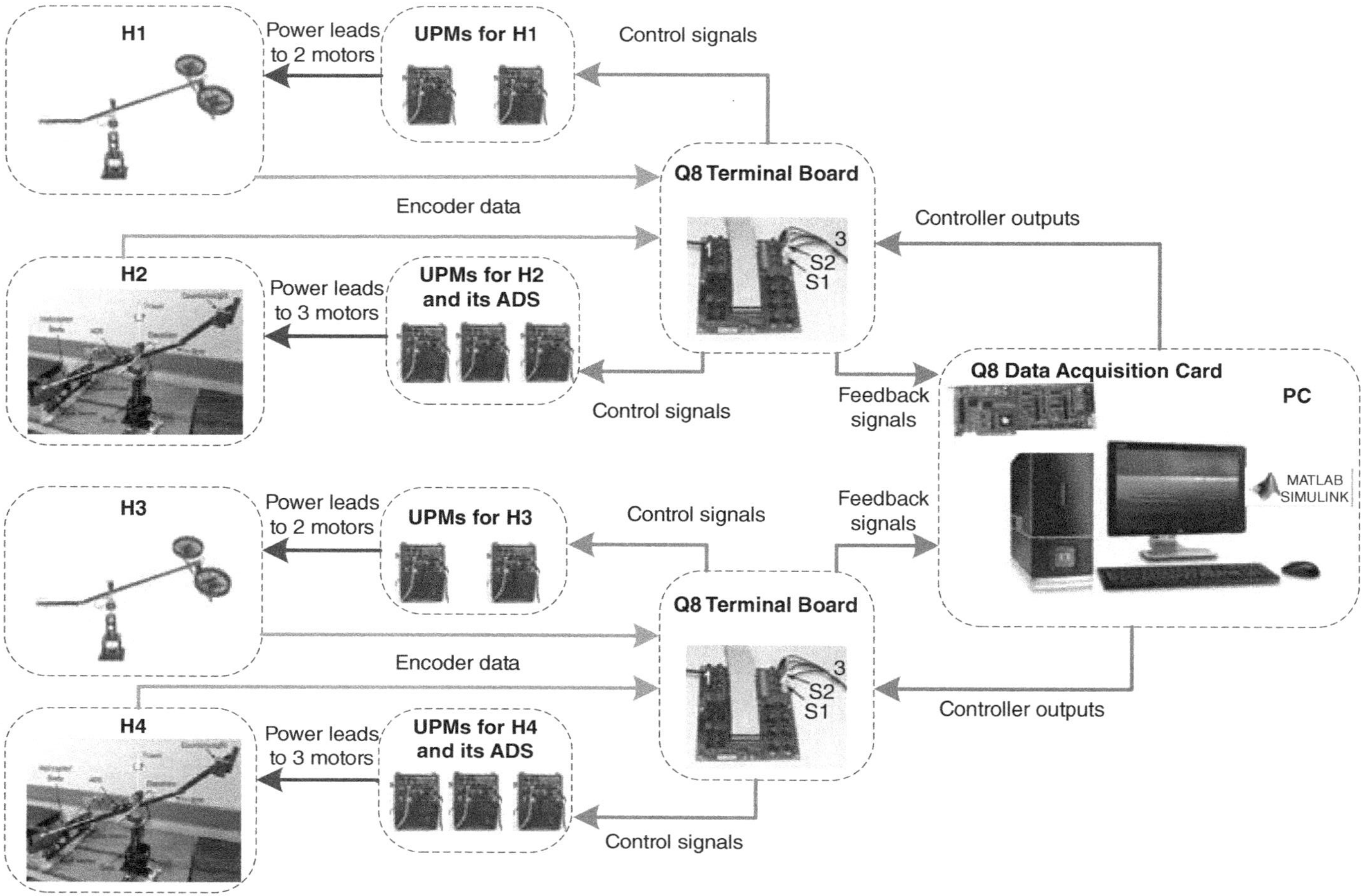

Figure 9.2 The hardware configuration of the experimental setup.

9.2 Mathematical Models

9.2.1 Nonlinear 3DOF Model

To obtain the nonlinear 3-DOF motion model of a helicopter, we first introduce its hardware components. Photos of a single helicopter are shown in Figures 9.3–9.5. Two DC motors are used to drive the propellers, which in turn generate control forces to regulate the 3-DOF motions of the helicopter. These motors are defined as the front and the back motors respectively. The body frame is suspended from an instrumented joint, which connects the centre of the body frame and the end of a long arm, and is free to pitch about its centre. The arm is installed on the base through a 2-DOF instrumented joint, which allows the helicopter body to elevate and travel. A counterweight is located at the other end of the arm such that the effective mass of the helicopter can be adjusted to a suitable value. In addition, an optional ADS can be installed on the arm. This includes an ADS motor that can drive an adjustable weight moving along the arm. The ADS motor is controlled independently, and therefore can act as an external disturbance or model uncertainty. The elevation motion can be generated by applying a positive voltage to each motor, and positive pitch by applying

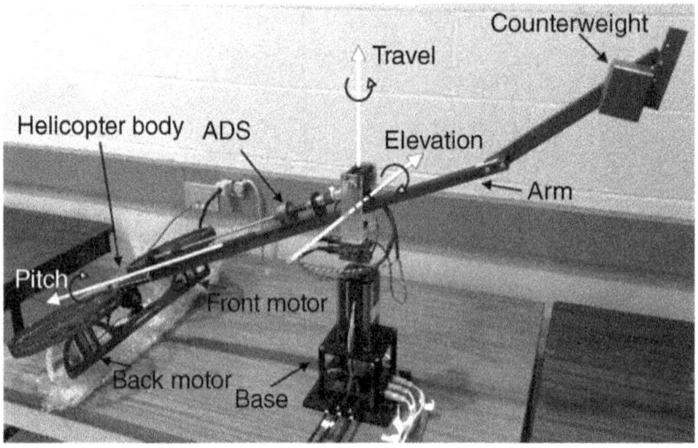

Figure 9.3 The 3DOF-Heli components.

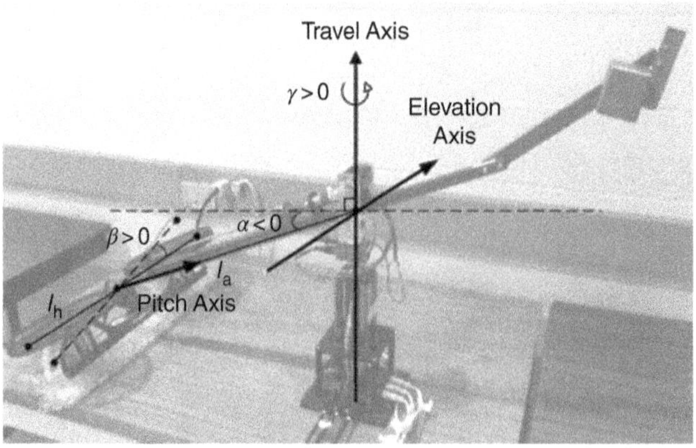

Figure 9.4 3DOF-Heli body frame.

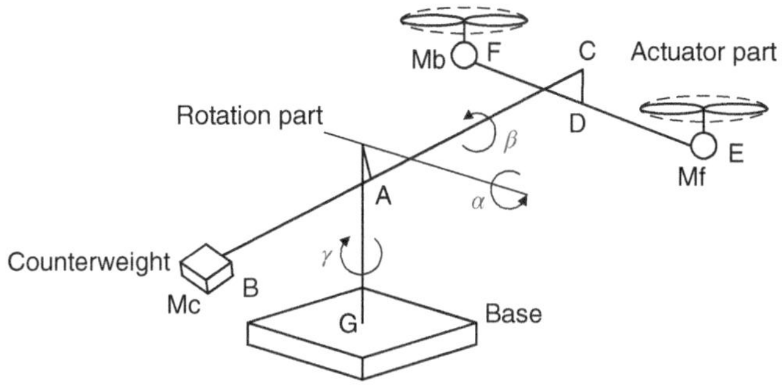

Figure 9.5 Hardware configuration of 3DOF-Heli body.

greater voltage to the front motor. The travel motion is the result of tilted thrust vectors when the body pitches. The three attitude angles are measured by encoders mounted on the instrumented joints. Therefore, pitch, elevation, and travel are the three degrees of freedom from which the 3DOF-Heli gets its name.

A convenient way to describe the motions of the 3DOF-Heli is through a body-frame defined along the axes of the equipment, centralized around the baseline rotational pivot, as shown in Figure 9.4. For a fleet of n 3DOF-Helis, the nomenclature associated with the ith 3DOF-Heli, $i \in \mathcal{I} =: \{1, \cdots, N\}$, is as follows:

- α_i is the elevation angle;
- $\dot{\alpha}_i$ is the elevation rate;
- β_i is the pitch angle;
- $\dot{\beta}_i$ is the pitch rate;
- γ_i is the travel angle;
- $\dot{\gamma}_i$ is the travel rate;
- $J_{ei} > 0$, $J_{pi} > 0$, and $J_{\gamma i} > 0$ are the moments of inertia about the elevation axis, pitch axis, and travel axis, respectively;
- K_{fi} is the force constant of the motor–propeller combination;
- l_{ai} is the distance from the elevation axis to the centre of the helicopter body;
- l_{hi} is the distance from the pitch axis to either motor;
- m_i is the effective mass of the helicopter body;
- g is the gravitational acceleration constant;
- $f_{\alpha i}, f_{\beta i}$, and $f_{\gamma i}$ denote disturbances acting on the elevation channel, pitch channel and travel axis, respectively;
- V_{fi} and V_{bi} are the voltages applied to the front motor and back motor, respectively;
- V_{si} and V_{di} denote the sum and the difference of V_{fi} and V_{bi}, respectively.

The differential equation of motion for the $i$the laboratory experimental setup, which can be derived using the Lagranges formalism, is written in the following well-known form (see examples in the literature [2, 3] and Chapter 6):

$$D(\boldsymbol{q}_i)\ddot{\boldsymbol{q}}_i + C(\boldsymbol{q}_i, \dot{\boldsymbol{q}}_i)\dot{\boldsymbol{q}}_i + g(\boldsymbol{q}_i) + \boldsymbol{F}_i^d = Q_i, i = 1, \dots, N, \tag{9.1}$$

where the inertia matrix $D(\boldsymbol{q}_i) \in \mathbb{R}^{3\times3}$, the Coriolis matrix $C(\boldsymbol{q}_i, \dot{\boldsymbol{q}}_i) \in \mathbb{R}^{3\times3}$, the gravity vector $g(\boldsymbol{q}_i) \in \mathbb{R}^3$, the disturbance input $F_i^d \in \mathbb{R}^3$, the generalized forces $Q_i \in \mathbb{R}^3$, and the state variable vector $q = (\alpha_i, \beta_i, \gamma_i) \in \mathbb{R}^3$.

For simplicity, the generalized inertia matrix $D(q)$ is often reduced to a diagonal matrix with constant entries. To be specific, the mathematical model of the form (9.1) for the ith 3DOF-Heli reads as follows:

$$J_{ei}\ddot{\alpha}_i = K_{fi}l_{ai}V_{si}\cos\beta_i - m_i g l_{ai}\cos\alpha_i + k_{1i}\sin\alpha_i - k_{2i}\dot{\alpha}_i, \tag{9.2}$$

$$J_{pi}\ddot{\beta}_i = K_{fi}l_{hi}V_{di} - k_{3i}\cos\alpha_i\sin\beta_i - k_{4i}\dot{\beta}_i, \tag{9.3}$$

$$J_{\gamma i}\ddot{\gamma}_i = K_{fi}l_{ai}V_{si}\cos\alpha_i\sin\beta_i - k_{5i}\dot{\gamma}_i, \tag{9.4}$$

where

$$V_{si} = V_{fi} + V_{bi}, \tag{9.5}$$

$$V_{di} = V_{fi} - V_{bi}, \tag{9.6}$$

and:

- *Equation (9.2) describes the elevation dynamics*: Due to the presence of the counterweight, the second-order equation of motion accounts for the pendulum-like oscillatory characteristic, where $K_{fi}l_{ai}V_{si}\cos\beta_i$ is the vertical lift thrust component, $m_i g l_{ai}\cos\alpha_i$ is the restorative spring torque, $k_{1i}\sin\alpha_i$ is the torque generated because the pivot points of the elevation and pitch motions are not exactly in the helicopter's body, and $-k_{2i}\dot{\alpha}_i$ is the unknown aerodynamic and mechanical damping torque.
- *Equation (9.3) describes the pitch dynamics*: Here, the second-order equation of motion accounts for the lightly-damped oscillatory characteristics: $K_{fi}l_{hi}V_{di}$ is the rotor torque, $k_{3i}\cos\alpha_i\sin\beta_i$ the restorative spring torque, and $-k_{4i}\dot{\beta}_i$ is the rotor damping.
- *Equation (9.4) describes the travel dynamics*: Here, $K_{fi}l_{ai}V_{si}\cos\alpha_i\sin\beta_i$ is the propulsive thrust component resulting from the pitch motion with a nonzero β_i; $-k_{5i}\dot{\gamma}_i$ is the unknown aerodynamic drag as well as some amount of mechanical friction at the hinge.

Consider the equations of motion given in (9.10)–(9.12) with relationship equations (9.6). The electric voltage V_{si} controls collectively the speed of the two propellers, and a positive V_{si} produces a positive lifting moment. The electric voltage V_{di} results in differential changes in the two rotor speeds, and a positive V_{di} produces a positive pitch moment.

Let

$$f_{\alpha i} = k_{1i}\sin\alpha_i - k_{2i}\dot{\alpha}_i, \tag{9.7}$$

$$f_{\beta i} = -k_{3i}\cos\alpha_i\sin\beta_i - k_{4i}\dot{\beta}_i, \tag{9.8}$$

$$f_{\gamma i} = -k_{5i}\dot{\gamma}_i. \tag{9.9}$$

Then Equations (9.2)–(9.4) are reduced to:

$$J_{ei}\ddot{\alpha}_i = K_{fi}l_{ai}V_{si}\cos\beta_i - m_i g l_{ai}\cos\alpha_i + f_{\alpha i}, \tag{9.10}$$

$$J_{pi}\ddot{\beta}_i = K_{fi}l_{hi}V_{di} + f_{\beta i}, \tag{9.11}$$

$$J_{\gamma i}\ddot{\gamma}_i = K_{fi}l_{ai}V_{si}\cos\alpha_i\sin\beta_i + f_{\gamma i}. \tag{9.12}$$

In Equations (9.10)–(9.12), the aerodynamic drag, the mechanical friction, and the rotor damping are not explicitly considered. Instead, we take into account both model uncertainties and friction effects, by lumping them together as the additive input disturbances $f_{\alpha i}$, $f_{\beta i}$, and $f_{\gamma i}$.

9.2.2 2DOF Model for Elevation and Pitch Control

Assuming travel motion can be achieved by highly precise pitch tracking, we simplify the 3-DOF attitude dynamics to 2-DOFs, namely elevation and pitch. The simplified model describing the elevation and pitch motions of the ith helicopter ($i \in \mathcal{I}$) is then:

$$J_{ei}\ddot{\alpha}_i = K_{fi}l_{ai}V_{si}\cos\beta_i - m_i g l_{ai}\cos\alpha_i + f_{\alpha i}, \tag{9.13}$$

$$J_{pi}\ddot{\beta}_i = K_{fi}l_{hi}V_{di} + f_{\beta i}, \tag{9.14}$$

where the pitch angle, β_i, is limited to the range $(-\pi/2, \pi/2)$ mechanically, and the nominal values of the involved parameters are as presented in Table 9.1.

Remark 9.1 We note that if the ith helicopter is equipped with an ADS, the ADS also contributes to $f_{\alpha i}(t)$ and $f_{\beta i}(t)$, regardless of whether it is static or moving along the arm. For the four helicopters in Figure 9.1, two are equipped with ADSs and the other pair are not, indicating that the four helicopters are subject to different disturbances. This is consistent with the experimental results obtained by Zhu *et al.* [4], who showed that the disturbance estimates of helicopters equipped with ADSs are different from those without ADSs.

Equations (9.13)–(9.14) can be written in the following compact form:

$$\mathbf{J}_i\ddot{\Theta}_i + \mathbf{N}_i(\Theta_i, m_i, K_{fi}) + \boldsymbol{F}_i^d = \mathbf{V}_i, i = 1, \cdots, N \tag{9.15}$$

where

$$\mathbf{J}_i = \begin{bmatrix} \dfrac{J_{ei}}{K_{fi}l_{ai}\cos\beta_i} & 0 \\ 0 & \dfrac{J_{pi}}{K_{fi}l_{hi}} \end{bmatrix} \in \mathbb{R}^{2\times2}, \Theta_i = \begin{bmatrix} \alpha_i & \beta_i \end{bmatrix}^T \in \mathbb{R}^2, \tag{9.16}$$

$$\mathbf{N}_i = \begin{bmatrix} \dfrac{m_i g}{K_{fi}\cos\beta_i} \\ 0 \end{bmatrix} \in \mathbb{R}^2, \boldsymbol{F}_i^d = \begin{bmatrix} f_{\alpha_i}^d & f_{\beta_i}^d \end{bmatrix}^T \in \mathbb{R}^2, \tag{9.17}$$

$$\mathbf{V}_i = \begin{bmatrix} V_{si} & V_{di} \end{bmatrix}^T \in \mathbb{R}^2. \tag{9.18}$$

Since pitch angle β_i is mechanically limited to the range $\left(-\frac{\pi}{2}, \frac{\pi}{2}\right)$ in all experiments, the inertia matrix $\mathbf{J}_i$ for each $i = 1, \cdots, N$ is a positive-definite matrix.

Table 9.1 Nominal parameters of the helicopters ($i = 1, 2, 3, 4$).

Parameter	Value	Parameter	Value
K_{fi}	0.1188 N/V	J_{ei}	1.034 kg·m^2
l_{ai}	0.660 m	J_{pi}	0.045 kg·m^2
l_{hi}	0.178 m	g	9.81 m/s^2
m_i	0.094 kg		

Equation (9.15) can be formulated in matrix format as:

$$\mathbf{I}\ddot{\boldsymbol{\Theta}} + \widetilde{\mathbf{N}}(\boldsymbol{\Theta}, \mathbf{m}, \mathbf{K}_f) + \boldsymbol{F}^d = \mathbf{V} \tag{9.19}$$

where $\boldsymbol{\Theta} \in \mathbb{R}^{2N}$, $\widetilde{\mathbf{N}} \in \mathbb{R}^{2N}$, $\mathbf{V} \in \mathbb{R}^{2N}$, $\boldsymbol{F}^d \in \mathbb{R}^{2N}$, $\mathbf{m} \in \mathbb{R}^N$, and $\mathbf{K}_f \in \mathbb{R}^N$ are vectors, and $\mathbf{I} \in \mathbb{R}^{2N\times 2N}$ is a diagonal inertia matrix. These vectors have the following expressions:

$$\begin{aligned}
\mathbf{I} &= \text{diag}\ [\mathbf{J}_1\ \cdots\ \mathbf{J}_N] \\
\boldsymbol{\Theta} &= [\Theta_1^T \cdots \Theta_N^T]^T = [\alpha_1\ \ \beta_1\ \ \cdots\ \ \alpha_N\ \ \beta_N]^T \\
\mathbf{m} &= [m_1\ \cdots\ m_N]^T \\
\mathbf{K}_f &= [K_{f1}\ \cdots\ K_{fN}]^T \\
\widetilde{\mathbf{N}} &= [\mathbf{N}_1^T\ \cdots\ \mathbf{N}_N^T]^T \\
\boldsymbol{F}^d &= [f_{\alpha 1}^d\ f_{\beta 1}^d\ \cdots\ f_{\alpha N}^d\ f_{\beta N}^d]^T \\
\mathbf{V} &= [\mathbf{V}_1^T\ \cdots\ \mathbf{V}_N^T]^T = [V_{s1}\ \ V_{d1}\ \ \cdots\ \ V_{sN}\ \ V_{dN}]^T.
\end{aligned}$$

Since $\beta_i(t) \neq 0$, $i = 1, \cdots, N$, for any $t \geq 0$, Equation 9.19 is equivalent to:

$$\ddot{\boldsymbol{\Theta}} + \mathbf{I}^{-1}\widetilde{\mathbf{N}} + \mathbf{I}^{-1}\boldsymbol{F}^d = \boldsymbol{\tau}, \tag{9.20}$$

where

$$\boldsymbol{\tau}_i = \mathbf{J}_i^{-1}\mathbf{V}_i, i = 1, \cdots, N, \tag{9.21}$$

$$\boldsymbol{\tau} = [\boldsymbol{\tau}_1^T\ \cdots\ \boldsymbol{\tau}_N^T]^T = [\tau_{s1}\ \tau_{d1}\ \cdots\ \tau_{sN}\ \tau_{dN}]^T. \tag{9.22}$$

We use $\Theta_d = [\alpha_d\ \ \beta_d]^T \in \mathbb{R}^2$ to denote the desired angular-position trajectories for the four helicopters, and $\dot{\Theta}_d = [\dot{\alpha}_d\ \ \dot{\beta}_d]^T \in \mathbb{R}^2$ the desired angular-velocity trajectories. For N helicopters, 2-D mutual angular-position synchronization means that:

$$\begin{aligned}
\alpha_1(t) &= \alpha_2(t) = \cdots = \alpha_N(t), \\
\beta_1(t) &= \beta_2(t) = \cdots = \beta_N(t),
\end{aligned} \tag{9.23}$$

and 2-D zero-error trajectory tracking in angular position means

$$\begin{aligned}
\tilde{\alpha}_i(t) &\triangleq \alpha_i(t) - \alpha_d(t) = 0, i = 1, \cdots, N, \\
\tilde{\beta}_i(t) &\triangleq \beta_i(t) - \beta_d(t) = 0, i = 1, \cdots, N.
\end{aligned} \tag{9.24}$$

Correspondingly, the 2-D mutual angular-velocity synchronization means

$$\begin{aligned}
\dot{\alpha}_1(t) &= \dot{\alpha}_2(t) = \cdots = \dot{\alpha}_N(t), \\
\dot{\beta}_1(t) &= \dot{\beta}_2(t) = \cdots = \dot{\beta}_N(t),
\end{aligned} \tag{9.25}$$

and the 2-D zero-error trajectory tracking in angular velocity means

$$\begin{aligned}
\dot{\tilde{\alpha}}_i(t) &\triangleq \dot{\alpha}_i(t) - \dot{\alpha}_d(t) = 0, i = 1, \cdots, N, \\
\dot{\tilde{\beta}}_i(t) &\triangleq \dot{\beta}_i(t) - \dot{\beta}_d(t) = 0, i = 1, \cdots, N.
\end{aligned} \tag{9.26}$$

In the context of this application, synchronization error is used to identify how the attitude trajectories of each 3-DOF helicopter converge with respect to each other.

Let:

$$\mathbf{V}_i = \mathbf{J}_i \boldsymbol{u}_i + \mathbf{N}_i(\Theta_i, m_i, K_{fi}), i = 1, \cdots, N, \tag{9.27}$$

$$\boldsymbol{d}_i = \mathbf{J}_i^{-1} \boldsymbol{F}_i^d, i = 1, \cdots, N, \tag{9.28}$$

$$\boldsymbol{u} = [\boldsymbol{u}_1^T \ \cdots \ \boldsymbol{u}_N^T] \in \mathbb{R}^{2N}. \tag{9.29}$$

Then, MVS (9.15) can be modelled by the vector equation (5.4) or the following two decoupled scalar equations:

$$\begin{cases} \ddot{\alpha}_i = u_{\alpha i} + d_{\alpha i}, \\ \ddot{\beta}_i = u_{\beta i} + d_{\beta i}, i \in \mathcal{I}, \end{cases} \tag{9.30}$$

where $\boldsymbol{u} = [\boldsymbol{u}_1^T, \cdots, \boldsymbol{u}_N^T]^T$ with $\boldsymbol{u}_i = [u_{\alpha i}, u_{\beta i}]^T$ $(i = 1, \cdots, N)$ is the control input vector to be determined, and $\boldsymbol{d} = [\boldsymbol{d}_1^T, \cdots, \boldsymbol{d}_N^T]^T$ with $\boldsymbol{d}_i = [d_{\alpha i}, d_{\beta i}]^T$ $(i = 1, \cdots, N)$ is the normalized disturbance input vector. This further implies that the control concepts and approaches introduced in Chapter 6 are applicable in designing $\boldsymbol{u}$ for the MVS considered.

Once $\boldsymbol{u}$ or $\boldsymbol{u}_i$ has been determined, $\mathbf{V}_i$ are computed according to Equation (9.27). Therefore, in each of the experiments below, the design of $\boldsymbol{u}$ or $\boldsymbol{u}_i$ is the main focus: we will introduce several control experiments where objectives (9.23)–(9.26) are achieved in an asymptotic manner or an approximately asymptotic manner. We start by introducing an experiment in which the GSE-based synchronized tracking controller developed in Section 5.6 is applied and implemented.

9.3 Experiment 1: GSE-based Synchronized Tracking

9.3.1 Objective

The experiments are conducted on a platform consisting of three 3-DOF helicopters, which are labelled Heli. I, Heli. II and Heli. III, respectively, as shown in Figure 9.6. The model parameters for control design are given in Table 9.2.

The objectives of this experiment are:

- to design a controller using the approach proposed in Section 5.6 for the elevation axis, to achieve synchronized trajectory tracking asymptotically; that is, to achieve (9.23)–(9.26) asymptotically;
- to implement the controller and compare its performance to that of a controller without GSE feedback (obtained by setting $\overline{b} = 0$ and $K_{Si} = 0$, $i = 1, \cdots, N$, in controller (5.142));
- to verify the robustness of the proposed controller with respect to the disturbances resulting from activated ADSs.

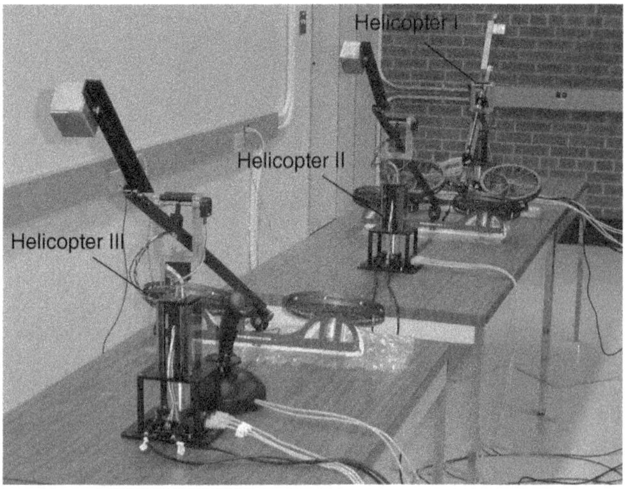

Figure 9.6 Experimental setup with three 3DOF-Helis.

Table 9.2 Parameters and control gains.

Parameter/gain	Heli. I	Heli. II	Heli. III
J_{ei}, kg·m^2	1.044	1.030	1.017
J_{pi}, kg·m^2	0.0455	0.0455	0.0455
Mass m_i, kg	0.142[a)]	0.128	0.113
K_{fi}, N/Volt	0.625	0.625	0.625
l_{ai}, m	0.648	0.648	0.648
l_{hi}, m	0.178	0.178	0.178
ADS system	Yes	No	No
Feedback gains, $[K_P, K_D]$	[30.0 15.0]	[30.0 15.0]	[30.0 15.0]
Sync. feedback gains, K_{si}	1.0	1.0	1.0
Sync. coupling gains, B_i	1.0	1.0	1.0

a) Value measured when ADS is at the farthest position from propellers, 0.170 for the middle position in the slide bar, 0.213 for the nearest position from propellers.

9.3.2 Initial Conditions and Desired Trajectories

A quintic polynomial trajectory in (9.31) is designed for the attitude motion of the elevation:

$$\alpha_d(t) = C_0 + C_1 t + C_2 t^2 + C_3 t^3 + C_4 t^4 + C_5 t^5 = \sum_{i=0}^{5} C_i t^i,$$

$$\dot{\alpha}_d(t) = C_1 + 2C_2 t + 3C_3 t^2 + 4C_4 t^3 + 5C_5 t^4 = \sum_{i=1}^{5} i \cdot C_i t^{i-1}, \tag{9.31}$$

$$\ddot{\alpha}_d(t) = 2C_2 + 6C_3t + 12C_4t^2 + 20C_5t^3 = \sum_{i=2}^{5} i(i-1) \cdot C_i t^{i-2},$$

where the coefficients $C_0 \sim C_5$ can be determined from the position, velocity, and acceleration requirements at both boundaries. If the desired maneuver is from 0 to 30° in 8 s, the coefficients are $C_0 = C_1 = C_2 = 0$, $C_3 = 0.5859$, $C_4 = -0.1099$ and $C_5 = 0.0055$. This trajectory guarantees that $\alpha(t) \in \mathcal{L}_\infty$, $\dot{\alpha}(t) \in \mathcal{L}_\infty$, and $\ddot{\alpha}(t) \in \mathcal{L}_\infty$.

9.3.3 Control Strategies

As an initial-stage experimental investigation, we applied the proposed synchronization controller to only one axis of the laboratory helicopters, i.e. the elevation axis, and use the Quanser-provided LQR controller for the other axes (pitch and travel). The synchronization process takes place along the same axis (attitude motion), but not among different axes of a single vehicle. From the dynamic equations (9.13) and (9.14), it can be seen that, despite the fact that the elevation motion and pitch motion are uncoupled, the elevation motion does not influence the pitch motion, and is completely controllable with respect to input V_{si} since β_i is mechanically limited to the range $\left(-\frac{\pi}{2}, \frac{\pi}{2}\right)$. Thus it is reasonable to apply the proposed synchronization controller to the elevation axis only and a different controller to the pitch axis.

The control law (5.142) is applied to control the elevation motion, with control parameters as given in Table 9.2. For comparison purposes between the different synchronization errors, we implement and verify the following three control strategies.

- *Control Strategy I*: The synchronization error and the coupled attitude error are chosen with L_N given as in (3.44).
- *Control Strategy II*: The synchronization error and the coupled attitude error are chosen with L_N as given in (3.45);
- *Control Strategy III*: Apply a simple variation of controller (5.142) without synchronization strategy, obtained by setting $K_{Si} = 0$ and $B_i = 0$ in (5.142).

For all the above strategies, the synchronization error defined with L_N given in (3.44) is employed as the uniform synchronization error for performance evaluation.

9.3.4 Disturbance Condition

The ADS on Heli. I is activated by a square wave command (see Figure 9.7) so as to initiate a performance disturbance to the control system. Here, 0 denotes the farthest position from the propellers, and 0.26 the nearest position). In this way, the effective mass of Heli. I will vary between 0.142 kg and 0.213 kg. If there is a parametric disturbance, such as the mass disturbance in our case, asymptotic convergence cannot be guaranteed, although the system may still be ISS.

The aim of introducing the mass disturbance in the experiments is to verify the robustness of the control strategies. We will inspect how these 3DOF-Helis respond without and with synchronization strategies, and compare the synchronization performance of different control strategies under parametric disturbance.

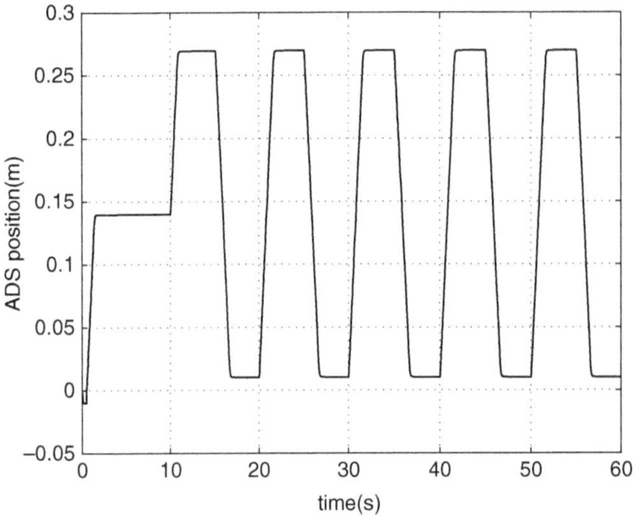

Figure 9.7 The position trajectory of the ADS.

9.3.5 Experimental Results

Figure 9.10 shows the experimental results of elevation tracking control of three 3DOF-Helis without a synchronization strategy (by setting $K_{si} = 0$ and $B_i = 0$). Figures 9.8–9.9 are the experimental results using synchronization strategy I and II, respectively. The maximal elevation tracking and synchronization errors are listed in Table 9.3. For each experiment, the three values are the maximal elevation trajectory tracking errors during maneuver (from 0 to 20 s), after maneuver (maneuver completed, after 20 s in our experiments), and the maximal synchronization error during the whole experiment, respectively.

It can be seen from Figure 9.10 that the elevation trajectory tracking error of Heli. I is large because of the ADS. Table 9.3 shows that the maximal elevation trajectory

Table 9.3 Maximal tracking and synchronization errors.

Strategy	Heli. I (°)	Heli. II (°)	Heli. III (°)
	3.136	1.968	1.978
No	1.257	0.290	0.378
	1.319	0.264	1.319
	2.466	2.217	2.206
I	0.677	0.466	0.466
	0.527	0.264	0.440
	2.385	2.261	2.259
II	0.641	0.641	0.641
	0.506	0.440	0.352

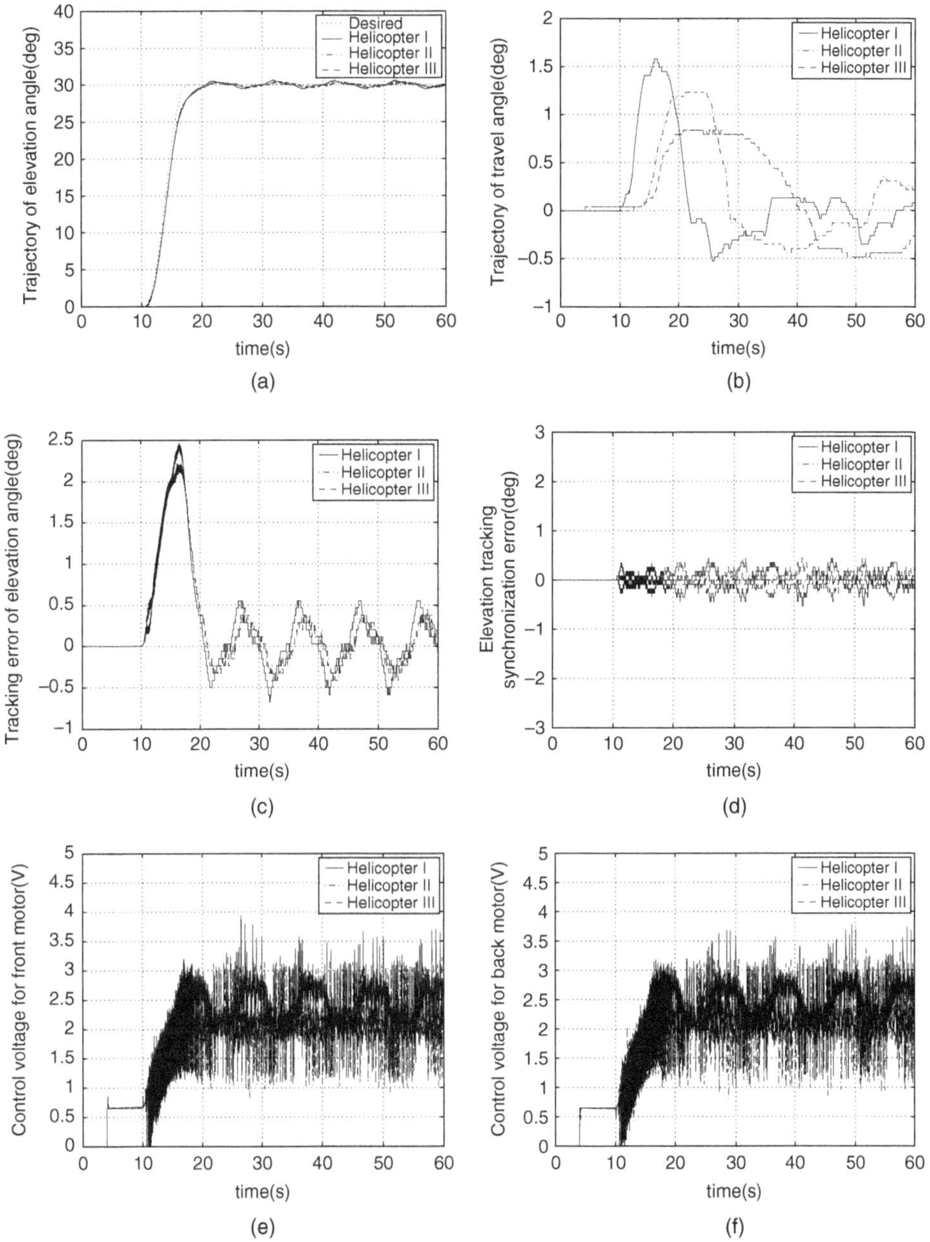

Figure 9.8 The experimental results for Control Strategy I. (a) Trajectories of elevation angle; (b) Trajectories of travel angle; (c) Tracking errors of elevation angle; (d) Synchronization errors of elevation angle; (e) Control voltages for front motors; (f) Control voltages for back motors.

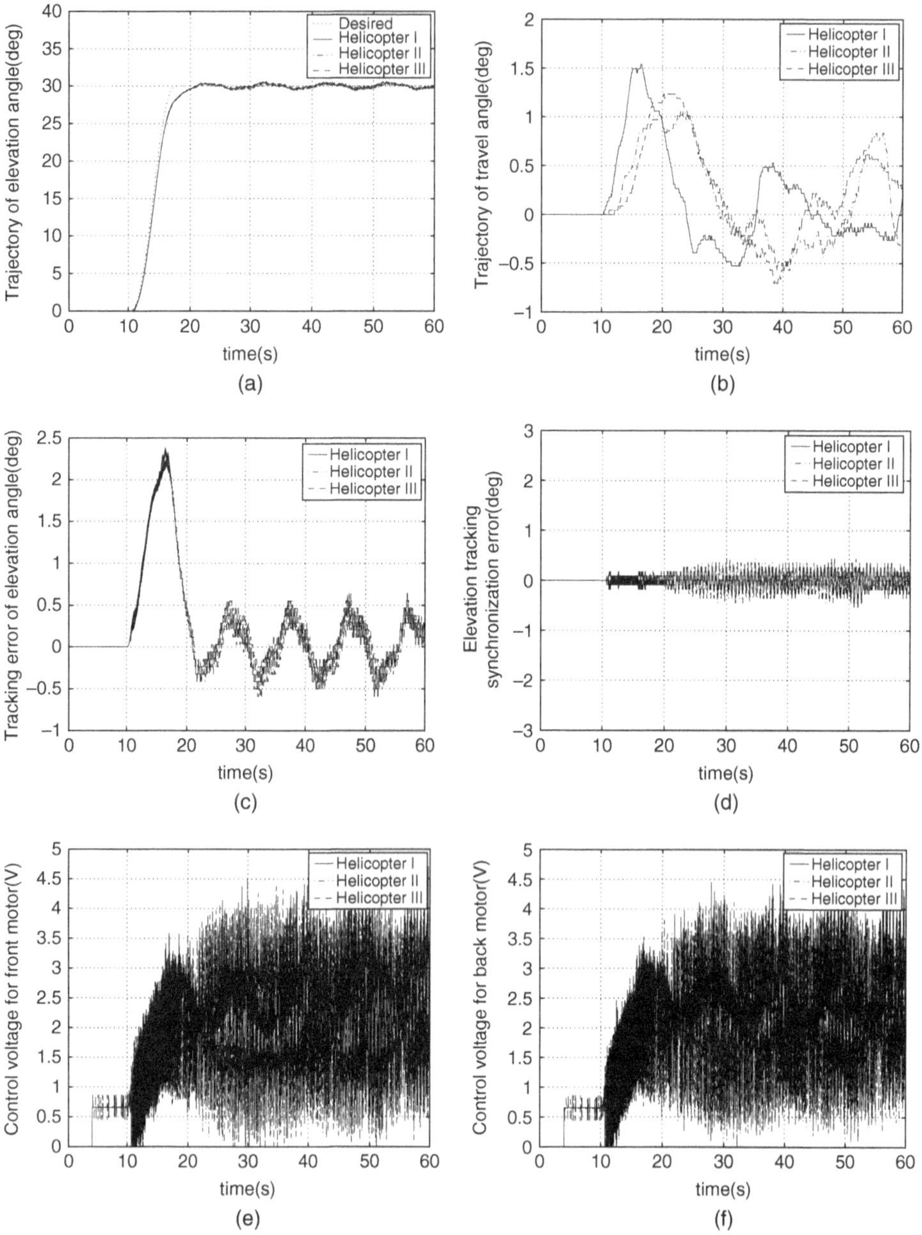

Figure 9.9 The experimental results for Control Strategy II. (a) Trajectories of elevation angle; (b) Trajectories of travel angle; (c) Tracking errors of elevation angle; (d) Synchronization errors of elevation angle; (e) Control voltages for front motors; (f) Control voltages for back motors.

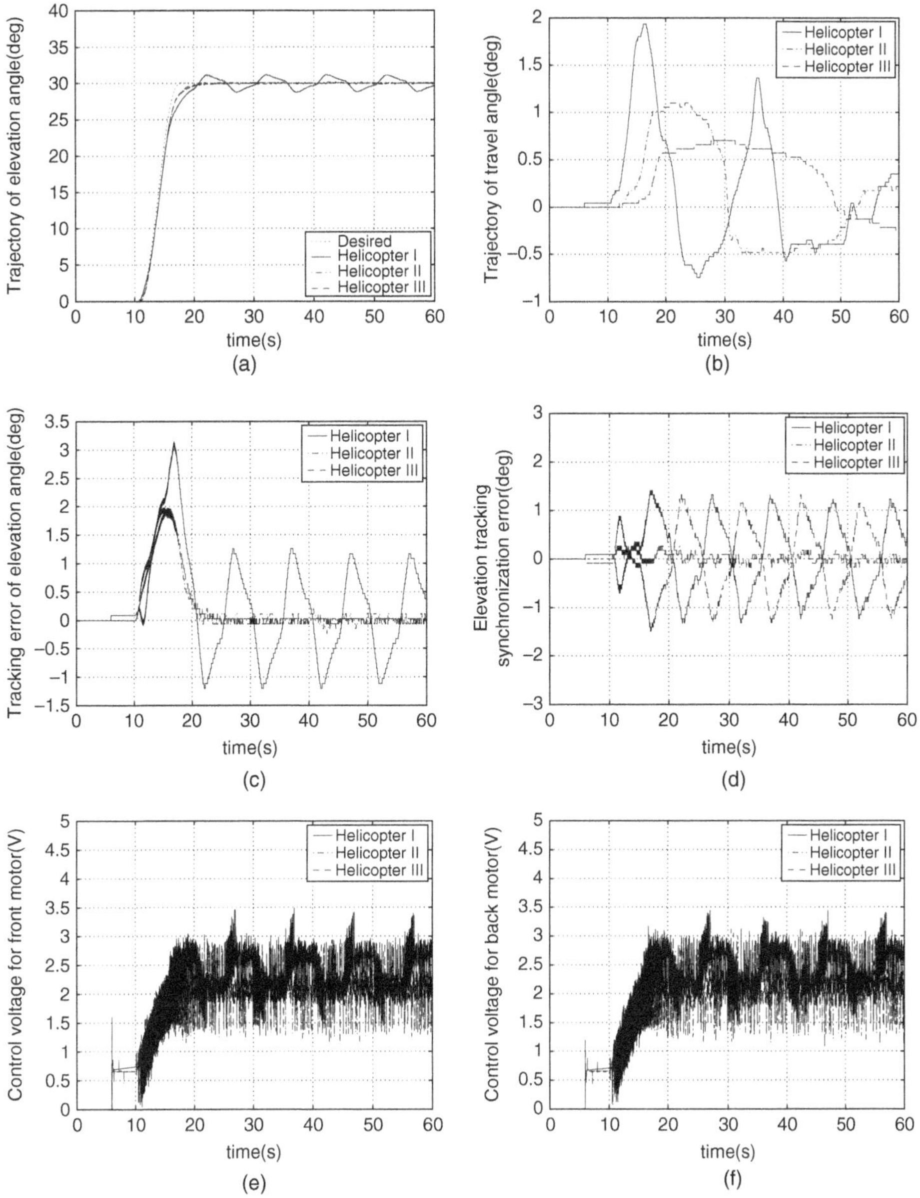

Figure 9.10 The experimental results for Control Strategy III. (a) Trajectories of elevation angle; (b) Trajectories of travel angle; (c) Tracking errors of elevation angle; (d) Synchronization errors of elevation angle; (e) Control voltages for front motors (f) Control voltages for back motors.

tracking errors during and after the maneuver are 3.136° and 1.257°, respectively. As there is no interconnection between these 3-DOF helicopters, the elevation tracking motion of Helis II and III do not react to the tracking motion of Heli. I due to ADS. Thus, asymptotic convergence of Helis II and III has been achieved. For this reason, the synchronization errors between them are large and the maximal value is 1.319°.

However, the elevation trajectory tracking and synchronization errors can be markedly reduced using the proposed synchronization strategies. For example, the maximal synchronization error is reduced to 0.527° using Control Strategy I and to 0.506° using Control Strategy II.

Strategy II is expected to produce better performance than Control Strategies I and III because it uses more neighbor information in each individual controller. However, the better performance is obtained at the cost of a greater online computational burden, less reliability, and more control effort, as can be seen from Figures 9.8–9.10 for control voltages for all the above cases.

9.3.6 Summary

In this experiment, a model-based synchronized trajectory tracking control strategy for multiple 3-DOF helicopters is presented and verified. With the proposed synchronization controller, both the attitude trajectory tracking errors and the attitude synchronization errors can achieve asymptotic convergence. The introduction of the generalized synchronization concept allows more space for designing different synchronization control strategies.

Experimental results conducted on the three 3DOF-Heli setup demonstrate the effectiveness of the proposed synchronization controller. The investigation indicates that better performance can be realized, but at the cost of computational burden, poorer reliability, and control effort. A tradeoff between synchronization performance and implementation costs should be made before choosing the synchronization strategy for a real system.

Current and future work in this area includes development of:

- an adaptive synchronization controller for systems with parametric variation
- new synchronization strategies.

9.4 Experiment 2: UDE-based Robust Synchronized Tracking

9.4.1 Objective

The objectives of the experiment are:

- to design a robust controller by applying the UDE-based approach proposed in Section 5.2 for both elevation and pitch channels;
- to implement the controller, compare its performance under different UDE parameters, and evaluate the effect of UDE parameter T_p on synchronization and tracking errors;
- to verify the robustness of the controller to disturbances from the activated ADS.

9.4.2 Initial Conditions and Desired Trajectories

The initial states of the four helicopters are specified as follows:

$$(\alpha_1(0), \dot{\alpha}_1(0), \beta_1(0), \dot{\beta}_1(0)) = (-25.7, 0, 0, 0),$$
$$(\alpha_2(0), \dot{\alpha}_2(0), \beta_2(0), \dot{\beta}_2(0)) = (-27.5, 0, 0, 0),$$

$$(\alpha_3(0), \dot{\alpha}_3(0), \beta_3(0), \dot{\beta}_3(0)) = (-18.9, 0, 0, 0),$$
$$(\alpha_4(0), \dot{\alpha}_4(0), \beta_4(0), \dot{\beta}_4(0)) = (-22.0, 0, 0, 0). \tag{9.32}$$

Without loss of generality, the following non-constant desired trajectories are adopted for experiments:

$$\alpha_d(t) = 5.73(\sin(0.25t) + \sin(0.5t) - 0.33),$$
$$\beta_d(t) = 15\sin(0.63t), \tag{9.33}$$

where all angles are given in degrees.

9.4.3 Control Strategies

Consider the directed communication topology graph shown in Figure 9.11. It is seen that only H1 has access to the desired trajectories, and Condition 3 is satisfied with $b_1 = a_{21} = a_{31} = a_{42} = 0.5$, and all other entries of $\overline{B}$ and A_4 are 0. Then, it follows that:

$$L_4 = \begin{bmatrix} 0 & 0 & 0 & 0 \\ -0.5 & 0.5 & 0 & 0 \\ -0.5 & 0 & 0.5 & 0 \\ 0 & -0.5 & 0 & 0.5 \end{bmatrix}, \quad B_4 = \begin{bmatrix} 0.5 & 0 & 0 & 0 \\ 0 & 0 & 0 & 0 \\ 0 & 0 & 0 & 0 \\ 0 & 0 & 0 & 0 \end{bmatrix}, \tag{9.34}$$

$$L_4 + B_4 = \begin{bmatrix} 0.5 & 0 & 0 & 0 \\ -0.5 & 0.5 & 0 & 0 \\ -0.5 & 0 & 0.5 & 0 \\ 0 & -0.5 & 0 & 0.5 \end{bmatrix}. \tag{9.35}$$

Consider the controller (5.19) with (5.20) and (5.32) for the motion synchronization of both elevation and pitch channels. Use $u^0_{\alpha i}$ and $u^0_{\beta i}$ to denote the two entries of controller component $\boldsymbol{u}^0_i$, $\hat{d}_{\alpha i}$ and $\hat{d}_{\beta i}$ to denote the two entries of disturbance estimate vector $\hat{\boldsymbol{d}}_i$, $\tilde{d}_{\alpha i}$ and $\tilde{d}_{\beta i}$ to denote the two entries of estimation error vector $\tilde{\boldsymbol{d}}_i$; that is, $\boldsymbol{u}^0_i = [u^0_{\alpha i}, u^0_{\beta i}]^T$, $\hat{\boldsymbol{d}}_i = [\hat{d}_{\alpha i}, \hat{d}_{\beta i}]^T$, and $\tilde{\boldsymbol{d}}_i = [\tilde{d}_{\alpha i}, \tilde{d}_{\beta i}]^T$, where $i = 1, \cdots, N$.

For the parameters of $u^0_{\alpha i}$ and $u^0_{\beta i}$, we choose

$$k^1_{\alpha i} = 2, k^2_{\alpha i} = 1.5, i = 1, 2, 3, 4,$$
$$k^1_{\beta i} = 3, k^2_{\beta i} = 1.2, i = 1, 2, 3, 4, \tag{9.36}$$

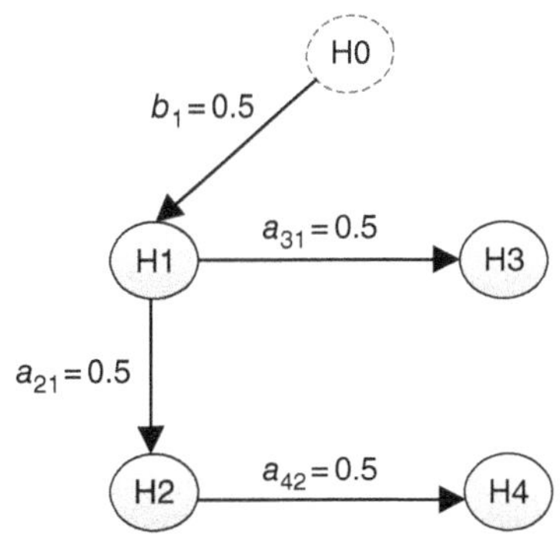

Figure 9.11 The communication topology for Experiments 2 and 3.

which results in

$$A_{\alpha i} = \begin{bmatrix} 0 & 1 \\ -2 & -1.5 \end{bmatrix}, P_{\alpha i} = \begin{bmatrix} 1.3750 & 0.2500 \\ 0.2500 & 0.5000 \end{bmatrix}, \tag{9.37}$$

$$A_{\beta i} = \begin{bmatrix} 0 & 1 \\ -3 & -1.2 \end{bmatrix}, P_{\beta i} = \begin{bmatrix} 1.8667 & 0.1667 \\ 0.1667 & 0.5556 \end{bmatrix}, \tag{9.38}$$

$$\lambda_{\max}(P_{\alpha i}) = 1.4414, \lambda_{\min}(P_{\alpha i}) = 0.4336, \tag{9.39}$$

$$\lambda_{\max}(P_{\beta i}) = 1.8875, \lambda_{\min}(P_{\beta i}) = 0.5347, \tag{9.40}$$

where the subscripts α and β are added to the symbols introduced in Section 6.2, to distinguish the two channels.

For comparison purposes, four experimental cases are studied. The same nominal controllers $u_{\alpha i}^0$ and $u_{\beta i}^0$ with parameters as in (9.36) are applied for each case. The differences between these cases are:

- For Case 1, only $u_{\alpha i}^0$ and $u_{\beta i}^0$ are applied, which corresponds to the situation in Section 4.4 and is equivalent to applying the controller (5.19) with $\hat{d}_{\rho i} \equiv 0$, $\rho = \alpha, \beta$ and $i = 1, 2, 3, 4$, to the four helicopters.
- For Case 2, UDEs (5.32) with parameters $T_\alpha = 0.5$ and $T_\beta = 0.2$ are additionally applied for each helicopter.
- For Case 3, UDEs (5.32) with *smaller* parameters of $T_\alpha = 0.1$ and $T_\beta = 0.05$ are applied for each helicopter, and none of the equipped ADSs are activated.
- For Case 4, the same controllers as for Case 3 are applied, but the ADSs of helicopters 2 and 4 are activated.

In addition, for the controller design of each helicopter, the model parameters given in Table 9.1 are also used in this experiment, which is different from the situation considered in the above Experiment 1 (where the differences among the inertial parameters of the involved helicopters are explicitly considered). This implies that larger modeling errors are neglected in constructing the controller for Experiment 2, since, as shown in Table 9.2, ADSs on the helicopters lead to obvious uncertainties in the inertial parameters J_{ei} and m_i.

9.4.4 Experimental Results and Discussions

Case 1 ***Control without UDE***

The experimental results for this case are presented in Figure 9.12. Because $d_{\alpha i}$ and $d_{\beta i}$ (which indeed exist) are neglected in the control design, neither tracking error nor synchronization error converges to close to zero. More specifically, we have the following observations:

- The magnitude of the tracking error in the elevation axis is greater than 5° for each helicopter and greater than 15° at $t = 30$ s; when t increases, the magnitude of the synchronization error between H1 and H2 (or H3 and H4) is greater than 0.5°; the synchronization error between H2 and H4 (or H1 and H3) has a smaller magnitude than that between any other two helicopters, which is an immediate consequence of the fact that both H2 and H4 are equipped with ADSs, whereas H1 and H3 are not.

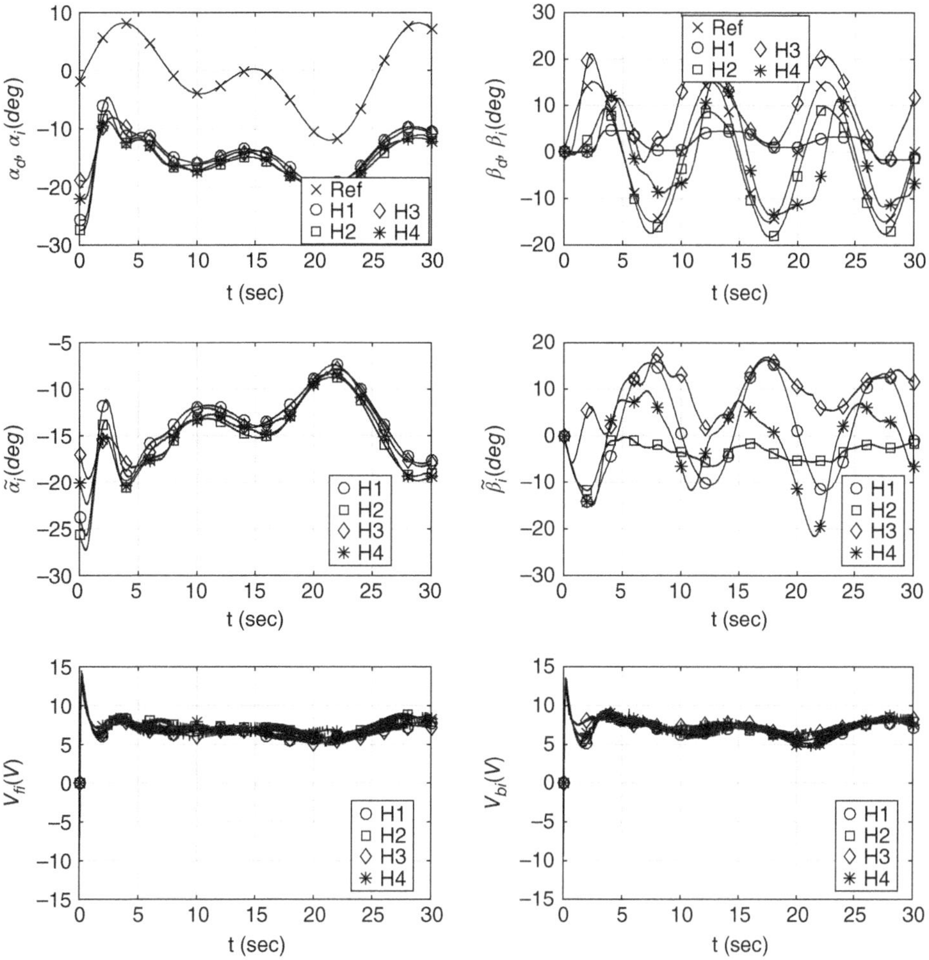

Figure 9.12 Experimental results for Case 1.

- The magnitude of the tracking error in the pitch axis is greater than 2° for each helicopter and greater than 10° at $t = 30$ s for H3; the synchronization error between any pair of helicopters is much greater than that in the elevation axis, and the magnitude of the the synchronization error between H2 and H3 is greater than 10°.
- The voltage applied to either motor has a magnitude smaller than 15 V for all $t > 0$ s and less than 10 V for $t > 1$ s.

Case 2 *using UDE with $T_\alpha = 0.5$ and $T_\beta = 0.2$*

In this case, $d_{\alpha i}$ and $d_{\beta i}$ are explicitly considered for the controller design. The corresponding experimental results are presented in Figure 9.13, which demonstrates that both tracking and synchronization errors are dramatically smaller that in Case 1. More precisely,

- The magnitude of the tracking error in the elevation axis is smaller than 2° for each helicopter, and the synchronization error between any pair of helicopters has a

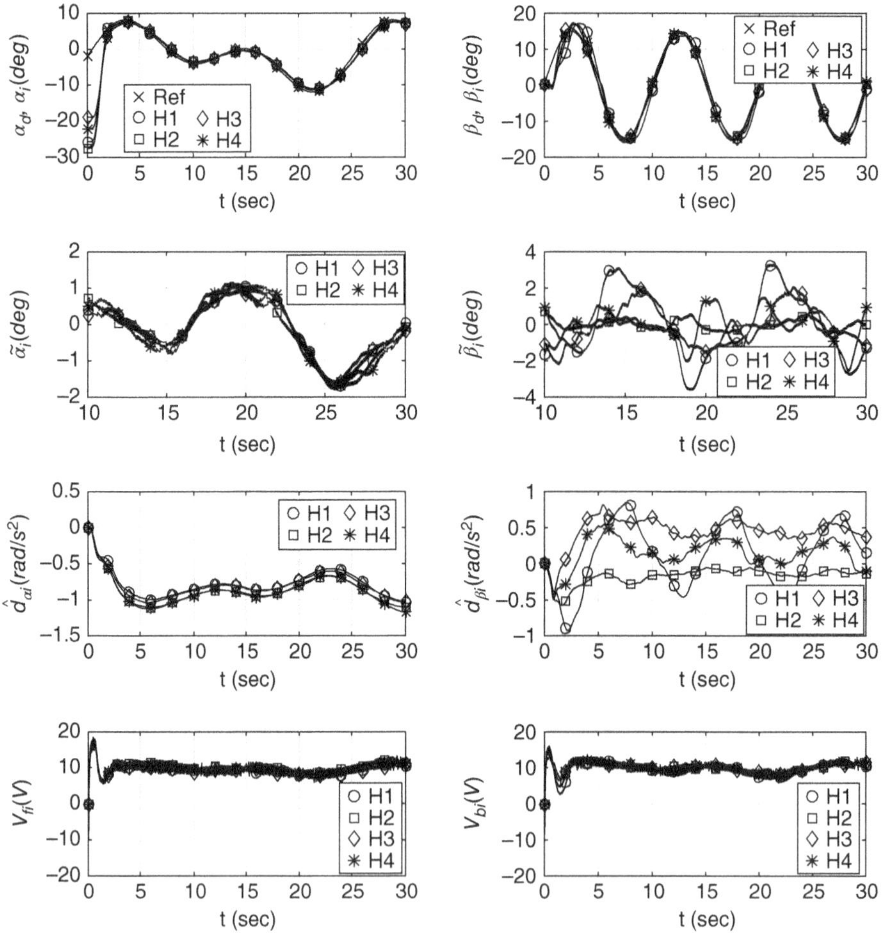

Figure 9.13 Experimental results for Case 2.

magnitude smaller than 1°; the UDE output for each helicopter is negative throughout, with a magnitude greater than 1 rad/s^2; because of the effects of static ADS on the elevation motion, the UDE outputs for H1 and H3 (which are approximately equal) are smaller than those for H2 and H4 (which are also approximately equal).

- The magnitude of the tracking error in the pitch axis is smaller than 4.5° for each helicopter, and the synchronization error between any two helicopters is less than 4°; the UDE output for H2 is negative for all $t > 0$ s, and is less than 0.2 rad/s^2 for all $t > 10$ s, whereas the UDE output for all other helicopters may be positive with a magnitude greater than 0.4 rad/s^2 (in particular, the magnitude is greater than 0.5 rad/s^2 for H1 and H3).
- The voltage applied to either the front or back motor has a magnitude near 10 V for all $t > 3$ s, with a peak greater than 15 V at the beginning.

Case 3 *using UDEs with $T_\alpha = 0.1$ and $T_\beta = 0.05$*

Compared with Case 2, UDEs with smaller parameters T_α and T_β are used in this case to give smaller tracking and synchronization errors. The experimental results are presented in Figure 9.14. It can be seen that both tracking and synchronization performances are further improved with higher voltages applied to the motors. More specifically, we have the following observations:

- The magnitude of tracking errors and synchronization errors in the elevation axis is smaller than 0.5° for all $t > 5$ s; the UDE output for each helicopter is negative (for all $t > 0$ s) with a magnitude greater than 1 rad/s^2; the UDE outputs for H1 and H3 also are smaller than those for H2 and H4 because of the effects of static ADS on the elevation motion.

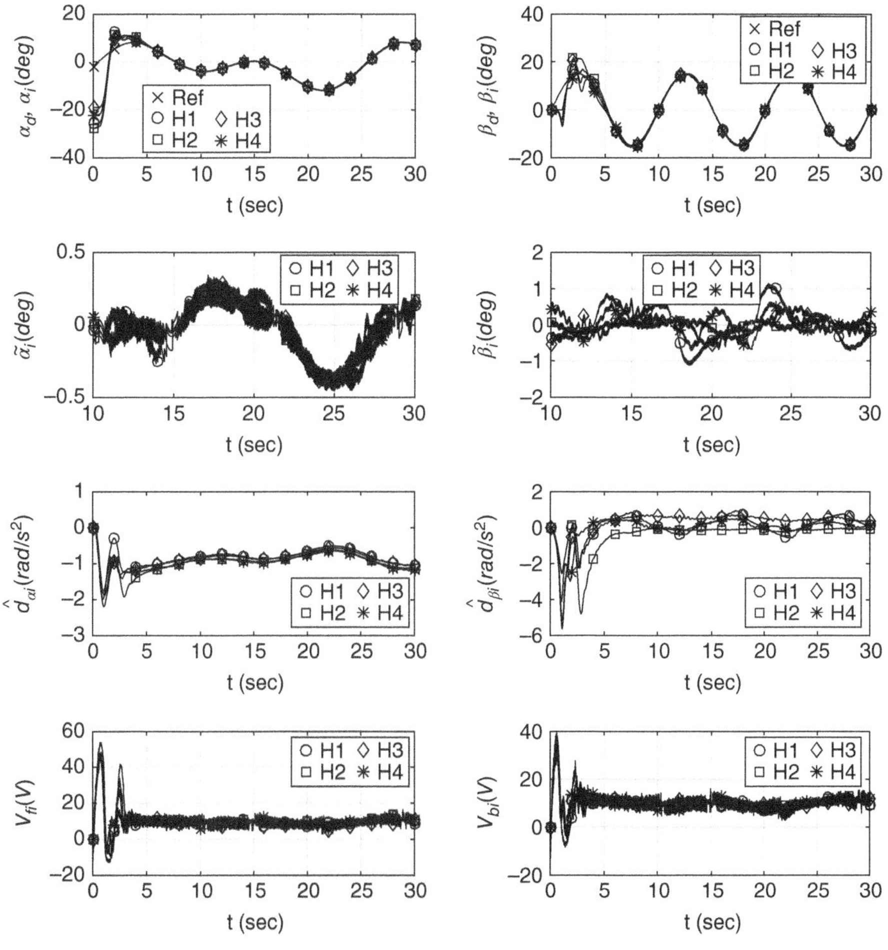

Figure 9.14 Experimental results for Case 3.

- The magnitude of tracking errors and synchronization errors in the pitch axis is smaller than 1.5° for all $t > 10$ s; the UDE output for H2 is negative for all $t > 3$ s, with a magnitude of approximately 0.1 rad/s^2 for all $t > 10$ s; the UDE output for all others may be positive or negative with a magnitude greater than 0.2 rad/s^2 (in particular, the UDE outputs for H1 and H3 are greater than 0.4 rad/s^2).
- The voltage applied to either the front or back motor has a magnitude of approximately 15 V for all $t > 5$ s, with a peak greater than (or near) 40 V at the beginning.

Case 4 *As for Case 3 but with Activated ADSs*

In this case, the ADSs on H2 and H4 are activated. Their dynamic positions along their arms are shown in Figure 9.15, with the same initial position on both, namely at −0.14 m (the farthest possible position from the propellers). The positions of ADSs are time-varying in this case, and thus different from those in the previous three cases (where the positions are fixed at −0.14 m) The experimental results for this case are presented in Figure 9.16. It can be seen that the motion of the ADSs does not dramatically decrease the steady-state performance of the tracking and synchronization errors compared with the performance achieved in Case 3. A detailed comparative analysis of the results is provided as follows:

- The magnitude of the tracking and synchronization errors in the elevation axis are nearly identical to what was achieved in Case 3; the transient performance is slightly worse (this degradation is relatively obvious for H2). The difference between the UDE outputs for H1 and H2 is greater than in Case 3 because of the motion of the ADS on H2; the difference between the UDE outputs for H2 and H4 is also slightly greater than that obtained in Case 3 because the effect of the motion of the ADS on H2 is delivered to H4, as shown in Figure 9.11.
- The magnitude of the tracking and synchronization errors in the pitch axis is also nearly identical to Case 3; the transient performance of the tracking error (or synchronization error) decreases dramatically for H2 and H4; the peaks of the UDE outputs for H2 and H4 are approximately −8 rad/s^2 and −5 rad/s^2, respectively (they are approximately −5.5 rad/s^2 and −4 rad/s^2, respectively, in Case 3).
- Both voltages applied to the front and back motors have a steady-state magnitude of approximately 15 V, but also have peaks greater than 50 V and 40 V, respectively.

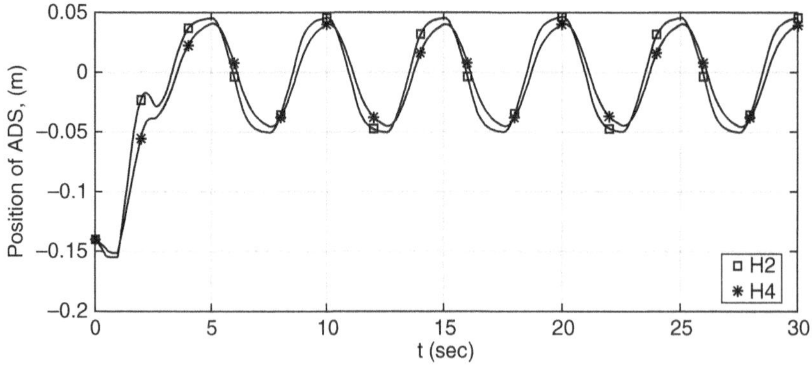

Figure 9.15 Positions of ADSs of H2 (blue) and H4 (magenta).

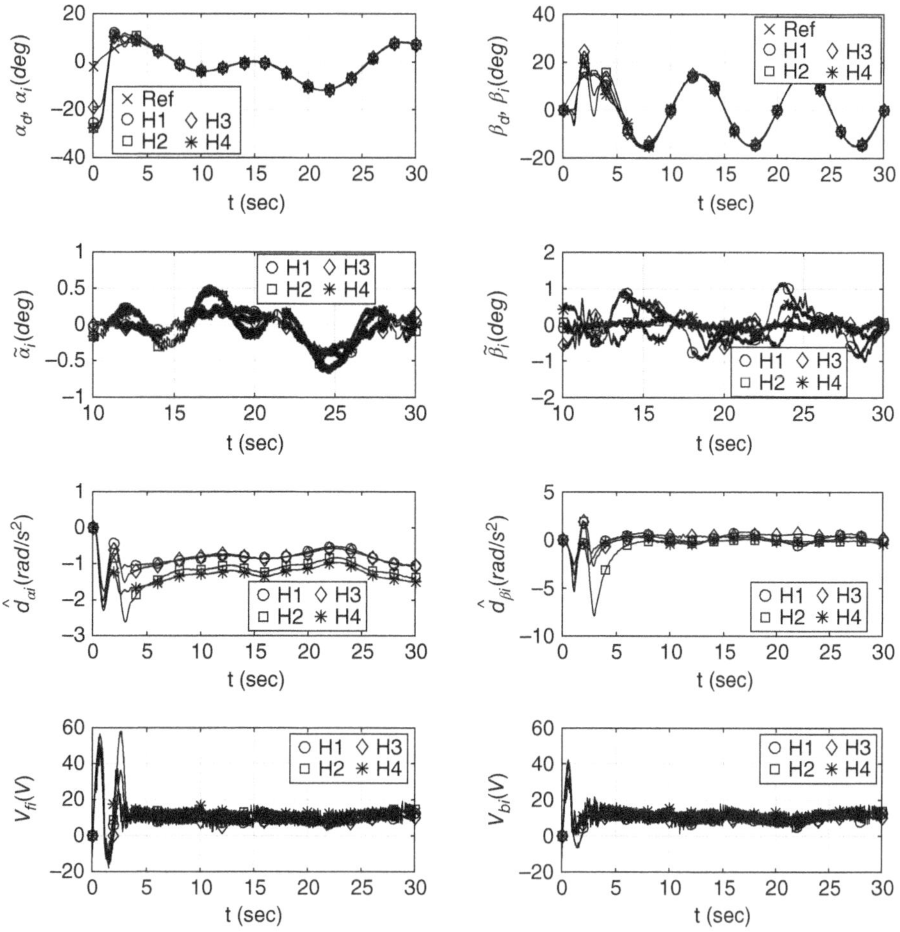

Figure 9.16 Experimental results for Case 4.

9.4.5 Summary

To show the performance differences more explicitly, the maximum magnitudes of the tracking errors for the four cases are summarized and compared in Tables 9.4 and 9.5; the responses over the same time interval $t \in [10, 30]$ are considered in each case. From these tables, it is easy to see that both the tracking and synchronization accuracy improves if UDEs with smaller UDE parameters are used.

In this experiment, the UDE-based robust synchronized tracking control of four 3-DOF helicopters has been verified under the condition that only a small subset of the helicopters has access to the desired attitude information. This robust controller is obtained by augmenting a distributed tracker with a continuous disturbance estimator that produces a signal to compensate for the effect of disturbances involved in the actual dynamics. Experimental results demonstrate that both the tracking and synchronization accuracy are greatly improved by UDEs with suitable parameters, and that the smaller the UDE parameter is, the smaller the tracking errors and synchronization errors are.

Table 9.4 Maximum tracking errors in elevation axis for $t > 10$ s.

Case	H1 (°)	H2 (°)	H3 (°)	H4 (°)
1	> 17	> 17	> 17	> 17
2	< 2	< 2	< 2	< 2
3	< 0.5	< 0.5	< 0.5	< 0.5
4	< 0.5	< 0.65	< 0.5	< 0.65

Table 9.5 Maximum tracking errors in pitch channel for $t > 10$ s.

Case	H1 (°)	H2 (°)	H3 (°)	H4 (°)
1	> 15	> 5	> 17	> 20
2	< 4	< 1	< 2	< 3
3	< 1.5	< 0.5	< 0.7	< 0.7
4	< 1.5	< 0.5	< 1	< 1

9.5 Experiment 3: Output-feedback-based Sliding-mode Control

9.5.1 Objective

The objectives of the experiment are:

- to design a robust controller by applying the sliding-mode control approach proposed in Section 5.5 for both elevation and pitch channels;
- to implement the controllers with or without disturbance compensation, and compare the performance;
- to check the control performance with respect to exogenous disturbances.

9.5.2 Initial Conditions and Desired Trajectories

The initial angular positions of the involved four helicopters are:

$$\boldsymbol{q}_1(0) = [\alpha_1(0),\ \beta_1(0)]^T = [-25.7^\circ,\ 0^\circ]^T, \tag{9.41}$$
$$\boldsymbol{q}_2(0) = [\alpha_2(0),\ \beta_2(0)]^T = [-27.5^\circ,\ 0^\circ]^T, \tag{9.42}$$
$$\boldsymbol{q}_3(0) = [\alpha_3(0),\ \beta_3(0)]^T = [-18.9^\circ,\ 0^\circ]^T, \tag{9.43}$$
$$\boldsymbol{q}_4(0) = [\alpha_4(0),\ \beta_4(0)]^T = [-22.0^\circ,\ 0^\circ]^T. \tag{9.44}$$

The initial angular velocities of all the helicopters is zero; that is, $\dot{\boldsymbol{q}}_i(0) = [0,\ 0]^T$ for each $i = 1, 2, 3, 4$. The reference trajectory of angular position $\boldsymbol{q}_d = [\alpha_d,\ \beta_d]^T$ is set to:

$$\alpha_d(t) = 10 \cdot \frac{\pi}{180} \sin\left(0.3\pi t - \frac{\pi}{2}\right), \tag{9.45}$$
$$\beta_d(t) = 15 \cdot \frac{\pi}{180} \sin(0.2\pi t), \tag{9.46}$$

so all the helicopters are expected to move sinusoidally in both channels, with amplitudes of $\pm 10°$ and $\pm 15°$ and frequencies of 0.15 Hz and 0.1 Hz for the elevation channel and pitch channel, respectively.

9.5.3 Control Strategies

Consider the same information-exchange graph as shown in Figure 9.11 for this experiment. We apply and implement the controllers (5.79), (5.94), and (5.96) for the motion synchronization of both elevation and pitch channels.

The same nominal inertial parameters used in Experiment 2 are also used to construct the controller for the helicopters in this experiment, despite the fact that the differences among the parameters of the four helicopters are obvious. This in turn implies that the results obtained in this experiments also demonstrate the system's robustness with respect to the uncertainties in inertial parameters. Furthermore, by the definition of Δ_i given by (5.18), one can see that the disturbance compensation term $\hat{\Delta}_i$ reflects not only the effect of the disturbances and uncertainties of vehicle i itself, but also that of its neighbors.

The controller gains for each helicopter are the same and given by:

$$k_{Ii} = 1, \; k_{0i} = 0.3, \; k_{1i} = 0.2, \; k_{2i} = 16, \; i = 1, 2, 3, 4. \tag{9.47}$$

These are tuned manually according to the response of the experiment, and the other parameters w_{1i}, h_{2i}, h_{3i}, h_{4i} are accordingly determined using (5.116)–(5.119).

In addition, in order to attenuate chattering caused by the discontinuous signum function, we use standard saturation functions in the control law. The slope of the saturation functions is chosen as 1000.

9.5.4 Experimental Results and Discussions

For comparison purposes, three experimental cases are designed and implemented in this experiment, corresponding to different disturbance conditions or disturbance compensation strategies.

- Case 1: the ADSs on helicopters 2 and 4 are turned off, and the control component $\hat{\Delta}_i$ is used for disturbance compensation for each helicopter.
- Case 2: the ADSs on helicopters 2 and 4 are turned on, and the control component $\hat{\Delta}_i$ is used for disturbance compensation for each helicopter.
- Case 3: the ADSs on helicopters 2 and 4 are turned off (as in Case 1), and the control component $\hat{\Delta}_i$ is removed from the implemented controller for each helicopter; in other words, the disturbance is not compensated for any helicopter.

Case 1 *ADSs on Helicopters 2 and 4 are Turned Off*

The experimental results are shown in Figures 9.17–9.19. We can see that the four helicopters reach second-order synchronized tracking; that is, both the angle and angular velocity of elevation and pitch channels converge to the desired trajectories for every helicopter as time increases. To be specific, we have the following observations:

- For the elevation axis, the relative angles among the four helicopters converge to less than 0.15°, and the tracking errors are less than 0.7°. The angular velocity tracking response does not give specific error values because the differentiator used to

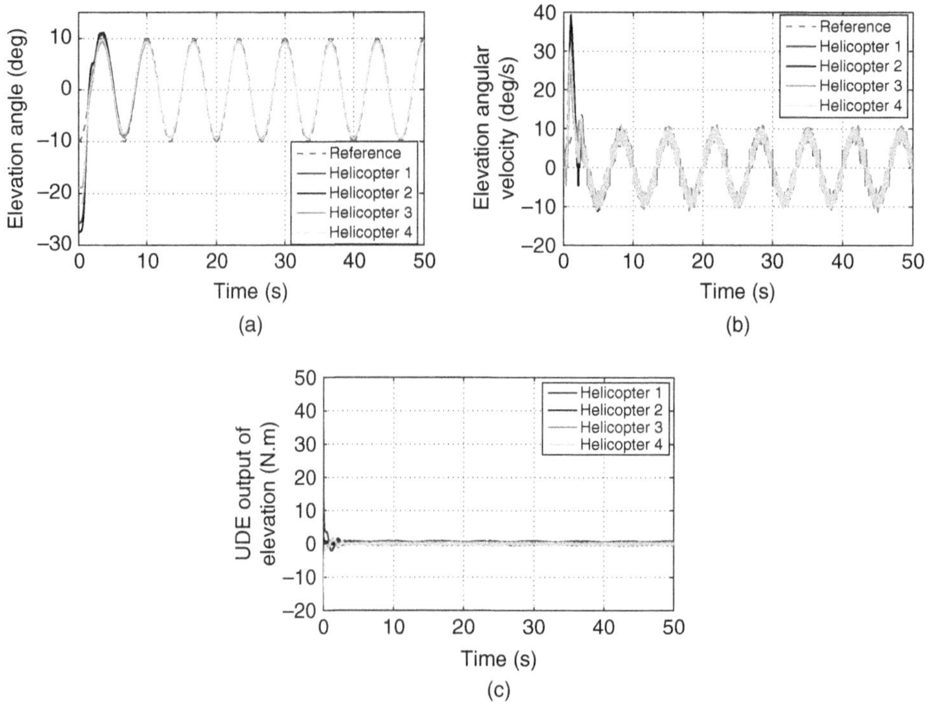

Figure 9.17 Responses of elevation axis for Case 1: (a) angle; (b) angular velocity; (c) UDE output.

obtain angular velocity brings in high-frequency noise from the discontinuous measurements of the encoders. However, it is still obvious from the plot that the angular velocities of all helicopters converge to the desired reference signal, which demonstrates the second-order consensus tracking.

- For the pitch axis, the maximum relative error among the pitch angles of the four helicopters is 1.9° and the maximum tracking error is 1.3°. The angular velocities of four helicopters also converge to the reference very well.

Case 2 ***ADSs on Helicopters 2 and 4 are Activated***

Since the ADSs are activated on helicopters 2 and 4, the movement of the ADSs introduced additional exogenous disturbances on the elevation motion of the two helicopters. The value of the additional moments added is associated with the position of ADSs and the elevation angle of the helicopter. Figure 9.20 shows the positions of ADSs moving weights on helicopters 2 and 4. We observe that the weights move toward the helicopter body by about 15 cm as the experiment begins, and then move back and forth periodically with an amplitude of 5 cm and frequency of 0.2 Hz.

The system responses are shown in Figures 9.21–9.23. It can be seen that the four helicopters achieve synchronized tracking in both axes. More precisely:

- although the ADSs disturb the elevation motion of helicopters 2 and 4, the relative elevation angles among the four helicopters are ultimately less than 0.5°, and the elevation angle tracking error is less than 1.1°;

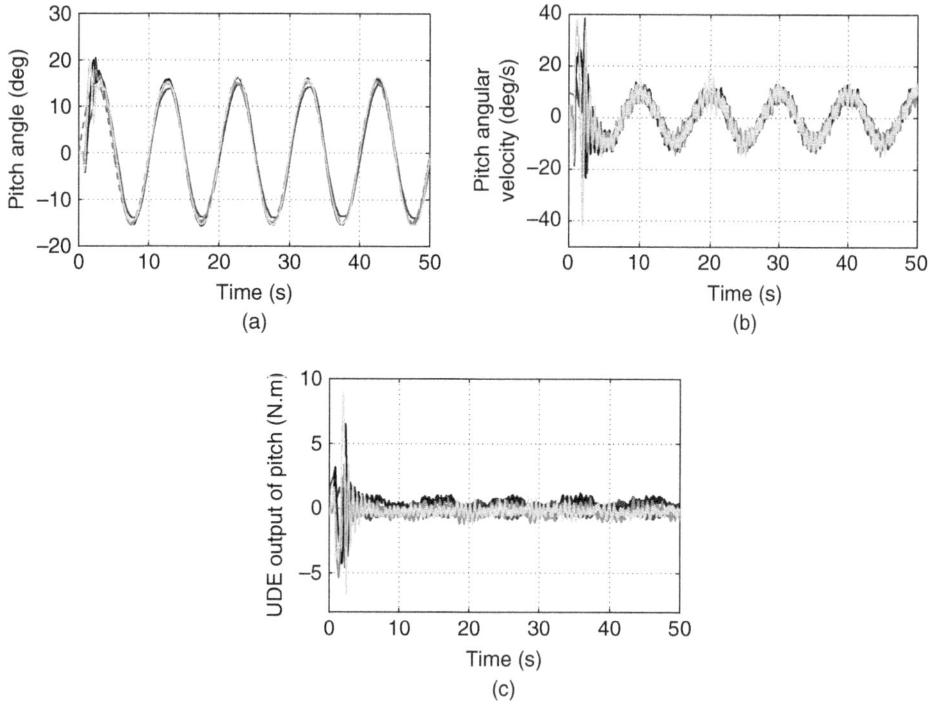

Figure 9.18 Responses of pitch axis for Case 1: (a) angle; (b) angular velocity; (c) UDE output.

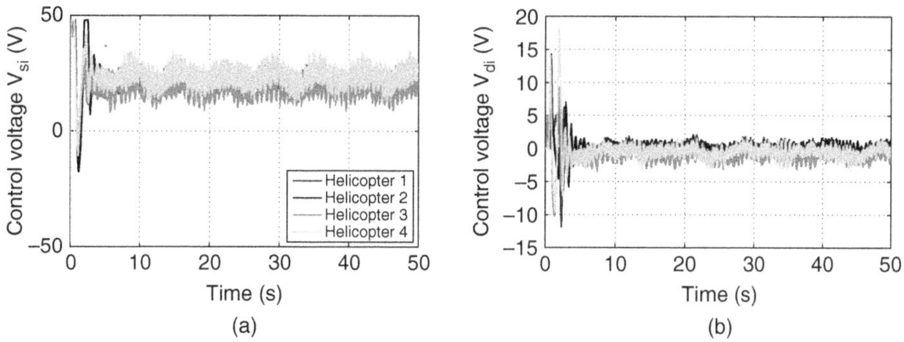

Figure 9.19 Control voltages for Case 1: (a) V_{si}; (b) V_{di}.

- responses of pitch channels are almost identical to Case 1, which is reasonable since the effect of ADSs on pitch motion is negligible.

From Figure 9.21c, we can make the following observations:

- a much larger transient peak exists for the disturbance compensation signal of each helicopter;
- the transient characteristics of the four disturbance compensation signals are obviously different;

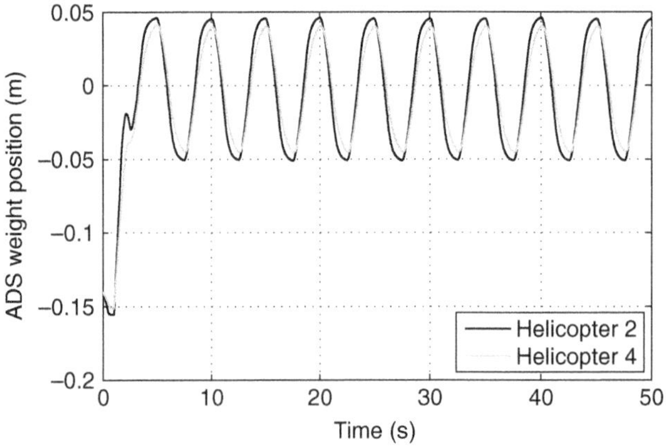

Figure 9.20 Positions of the ADSs of H2 and H4.

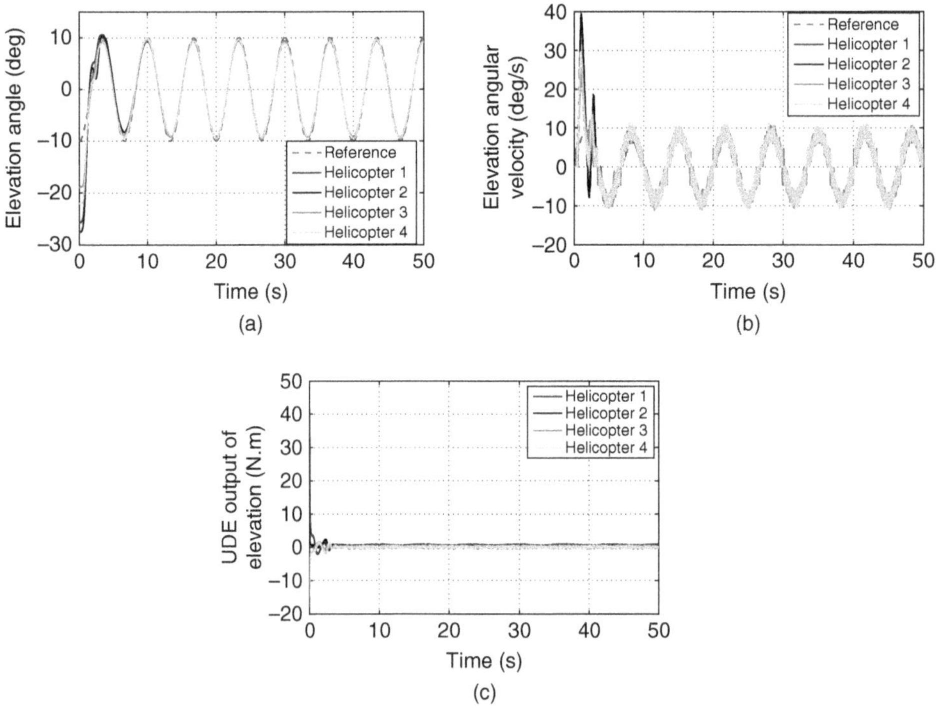

Figure 9.21 Responses of elevation axis for Case 2: (a) angle; (b) angular velocity; (c) UDE output.

- the differences among the steady-state characteristics of the four disturbance compensation signals are relatively slight.

Case 3 *Without using $\hat{\Delta}_i$ for Disturbance Compensation*

In this case, the only difference with respect to Case 1 is that the disturbance compensation component is removed by setting $\hat{F}_i(t) = 0$ for $i = 1, 2, 3, 4$. All the other settings

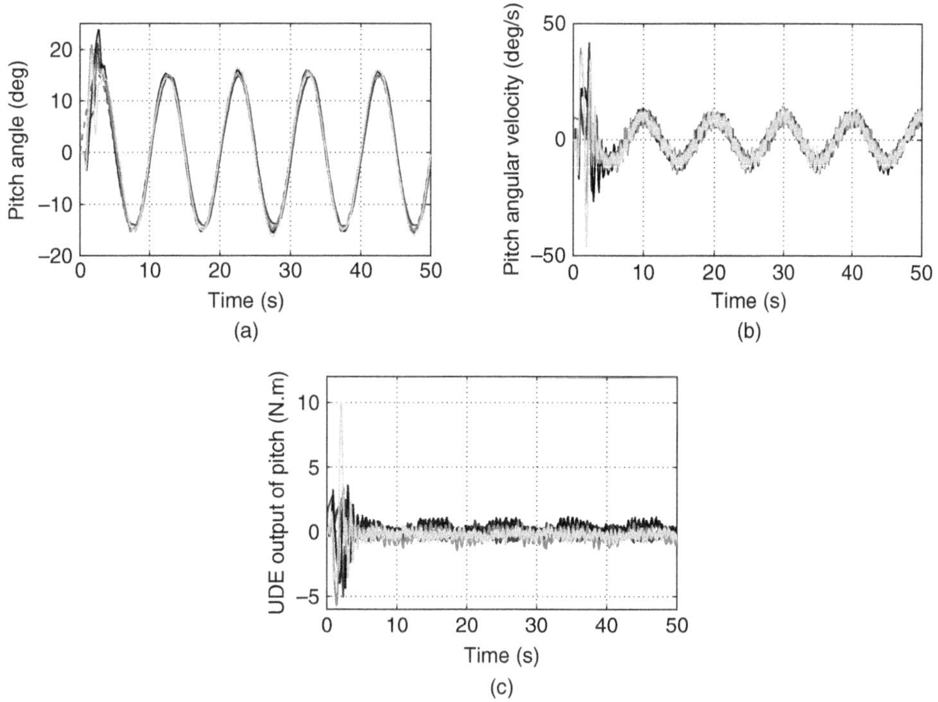

Figure 9.22 Responses of pitch axis for Case 2: (a) angle; (b) angular velocity; (c) UDE output.

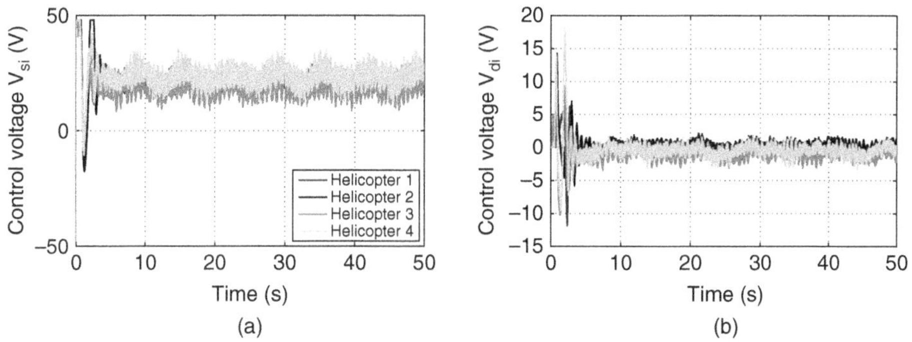

Figure 9.23 Control voltages for Case 2: (a) V_{si}; (b) V_{di}.

and conditions are the same as those in Case 1 or Case 2. The results corresponding to stationary ADSs (as the ADSs in Case 1) are shown in Figure 9.24, and the results corresponding to activated ADSs (as with the ADSs in Case 2) are shown in Figure 9.25.

It is seen that whether the ADSs are activated or not, the helicopters are not able to reach the same level of synchronization and tracking accuracy as achieved in Case 1 or 2. To be specific, the mutual synchronization errors in elevation axis between two helicopters are are nearly about 0.7 degree, and the tracking errors are nearly about 2.5 degree; the mutual synchronization error in pitch axis between helicopters 1 and 2 are

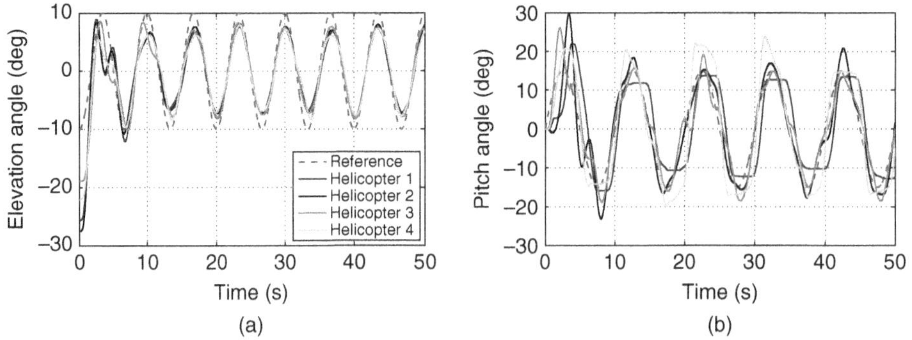

Figure 9.24 Responses for Case 3 with ADS off: (a) elevation; (b) pitch.

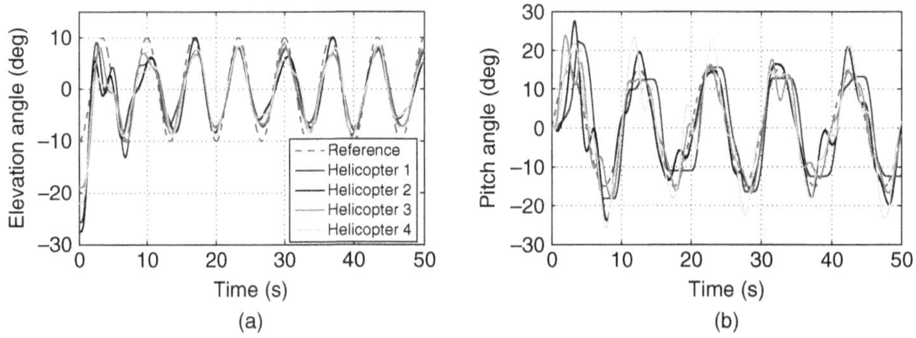

Figure 9.25 Responses for Case 3 with ADS on: (a) elevation; (b) pitch.

are nearly about 5 degree, and the tracking error in pitch axis of helicopter 2 is nearly about 5 degree.

9.5.5 Summary

In these experiments, the decentralized output-feedback synchronized tracking control strategy, as developed in Section 5.5, was implemented and verified on a platform consisting of four 3-DOF helicopters with only angular position measurements available. In particular, the effect of the disturbance compensation term $\hat{\Delta}_i$ included in controller (5.79) was evaluated.

Bibliography

1 J. Apkarian, "3-DOF helicopter reference manual," tech. rep., Quanser Consulting Inc, Canada, 2006.

2 T. Kiefer, A. Kugi, and W. Kemmetmüller, "Modeling and flatness-based control of a 3DOF helicopter laboratory experiment," in *6th IFAC Symposium on Nonlinear Control Systems*, vol. 9, 2004.

3 Z. Li, H. H. T. Liu, B. Zhu, and H. Gao, "Robust second-order consensus tracking of multiple 3-DOF laboratory helicopters via output-feedback," *IEEE Transactions on Mechatronics*, vol. 20, no. 5, pp. 2538–2549, 2015.

4 B. Zhu, H. H.-T. Liu, and Z. Li, "Robust distributed attitude synchronization of multiple three-DOF experimental helicopters," *Control Engineering Practice*, vol. 36, pp. 87–99, 2015.

Formation Control of Multiple Autonomous Vehicle Systems, First Edition. Hugh H.T. Liu and Bo Zhu.

Part V

Appendix

A

Appendix

A.1 Algebra and Matrix Theory

We now present some inequalities on vector and matrix norms. Consider vector $X \in R^{n\times 1}$ and matrix $C = [c_{ij}] \in \mathbb{R}^{m\times n}$. Define $\|X\|_2 = \left(\sum_{i=1}^{n} |x_i|^2\right)^{\frac{1}{2}}$, $\|X\|_\infty = \max_{1\leq i\leq n} |x_i|$, $\|C\|_\infty = \max_{1\leq i\leq m} \sum_{j=1}^{n} |c_{ij}|$, and $\|C\|_2 = (\lambda_{max}(C^TC))^{\frac{1}{2}}$, where $\lambda_{max}(\cdot) = \max_i |\lambda_i|$ with λ_i matrix eigenvalues.

Since the matrix norm is induced by the corresponding vector norm, we have

$$\|CX\|_\infty \leq \|C\|_\infty \|X\|_\infty. \tag{A.1}$$

For matrix $D \in \mathbb{R}^{n\times q}$, we have the following inequality

$$\|CD\| \leq \|C\|\|D\|. \tag{A.2}$$

Hints: Inequalities (A.1) and (A.2) are frequently used. In Section 5.2.4, for instance, they are used to derive equations (5.44) and (5.47).

Lemma A.1 Define the Kronecker product of two matrices $C = [c_{ij}] \in \mathbb{R}^{m\times n}$ and $D = [d_{ij}] \in \mathbb{R}^{p\times q}$ as

$$C \otimes D = \begin{bmatrix} c_{11}D & \cdots & c_{1n}D \\ \vdots & \ddots & \vdots \\ c_{m1}D & \cdots & c_{mn}D \end{bmatrix} \in R^{mp\times nq}$$

Then,

$$rank(C \otimes D) = rank(C)\, rank(D). \tag{A.3}$$

The following properties of the Kronecker product [1], will be used in this book.

Lemma A.2 Suppose that $U \in \mathbb{R}^{p\times p}$, $V \in \mathbb{R}^{q\times q}$, $X \in \mathbb{R}^{p\times p}$ and $Y \in \mathbb{R}^{q\times q}$. The following statements are true.

- $(U + X) \otimes V = U \otimes V + X \otimes V$.
- $(U \otimes V)(X \otimes Y) = UX \otimes VY$.
- $(U \otimes V)^T = U^T \otimes V^T$.
- Suppose that U and V are invertible. Then $(U \otimes V)^{-1} = U^{-1} \otimes V^{-1}$.
- If U and V are symmetric, so is $(U \otimes V)$.

Formation Control of Multiple Autonomous Vehicle Systems, First Edition. Hugh H.T. Liu and Bo Zhu.

- If U and V are symmetric positive definite (respectively, positive semidefinite), so is $(U \otimes V)$.
- Suppose that U has eigenvalues β_i with associated eigenvectors $f_i \in \mathbb{C}_p$, $i = 1, \cdots, p$, and V has eigenvalues ρ_j with associated eigenvectors $g_j \in \mathbb{C}_q$, $j = 1, \cdots, q$. Then the pq eigenvalues of $U \otimes V$ are $\beta_i \rho_j$ with associated eigenvectors $f_i \otimes g_j$, $i = 1, \cdots, p$, $j = 1, \cdots, q$.

Lemma A.3 Let $A \stackrel{\Delta}{=} [a_{ij}] \in R^{n \times n}$, and use

$$R_i(A) \stackrel{\Delta}{=} \sum_{j=1, j \neq i}^{n} |a_{ij}|, i = 1, \cdots, n, \tag{A.4}$$

to denote the deleted absolute row sums of A. Then all eigenvalues of A are located in the union of n discs

$$\bigcup_{i=1}^{n} \{z \in \mathbb{C} : |z - a_{ii}| \leq R_i(A)\}. \tag{A.5}$$

Furthermore, if a union of k of these n discs forms a connected region that is disjoint from all of the remaining $n - k$ discs, then there are precisely k eigenvalues of A in this region. (This is the Gershgorin disc theorem [2].)

A.2 Systems and Control Theory

A.2.1 Definitions of Lipschitz Condition

Consider the state $x(t)$ of the linear time-invariant equation

$$\dot{x} = Ax + Bu, \tag{A.6}$$

where $x \in \mathbb{R}^n$ is the state, $u \in \mathbb{R}^m$ is the control input, and matrices $A \in R^{n \times n}$ and $B \in R^{n \times m}$.

The solution to (A.6) is given by

$$x(t) = \exp(A(t - t_0))x(0) + \int_{t_0}^{t} \exp(-A(t - \tau))u(\tau)d\tau. \tag{A.7}$$

Lemma A.4 Suppose that $f(t, x)$ is piecewise continuous in t and satisfies the Lipschitz condition

$$\| f(t, x) - f(t, y) \| \leq L \| x - y \| \tag{A.8}$$

$\forall x, y \in \mathbb{B} = \{x \in \mathbb{R}^n | \| x - x_0 \| \leq r\}$, $\forall t \in [t_0, t_1]$, where L is the Lipschitz constant. Then, there exists some $\delta > 0$ such that the state equation $x = f(t, x)$ with $x(t_0) = x_0$ has a unique solution over $[t_0, t_0 + \delta]$. (This is Theorem 3.1 on p. 88 of Khalil *et al.* [3].)

Lemma A.5 Suppose that $f(t, x)$ is piecewise continuous in t and satisfies

$$\| f(t, x) - f(t, y) \| \leq L \| x - y \| \tag{A.9}$$

$\forall x, y \in \mathbb{R}^n, \forall t \in [t_0, t_1]$, where L is the Lipschitz constant. Then, the state equation $x = f(t, x)$ with $x(t_0) = x_0$ has a unique solution over $[t_0, t_1]$. (This is Theorem 3.2 on p. 93 of Khalil *et al.* [3].)

A.2.2 Definitions of Asymptotically Stable

Consider the autonomous system

$$\dot{x} = f(x), \tag{A.10}$$

where $f : D \mapsto \mathbb{R}^n$ is a locally Lipschitz map from a domain $D \in \mathbb{R}^n$ into $\mathbb{R}^n$. Let $x = \mathbf{0}_n$ be an equilibrium point for (A.10). Then

- the equilibrium point $x = \mathbf{0}_n$ is said to be stable if, for any $\epsilon_1 > 0$, there exists $\epsilon_2 > 0$ such that if $\| x(0) \| < \epsilon_2$, then $\| x(t) \| < \epsilon_1$ for all $t \geq 0$;
- the equilibrium point $x = \mathbf{0}_n$ is asymptotically stable if it is stable, and if there additionally exists some $\epsilon_3 > 0$ such that $\| x(0) \| < \epsilon_3$ implies that $x(t) \to \mathbf{0}_n$ as $t \to \infty$;
- the equilibrium point $x = \mathbf{0}_n$ is exponentially stable if there exist two strictly positive numbers, $\epsilon_4 > 0$ and $\epsilon_5 > 0$, such that

$$\| x(t) \| \leq \epsilon_4 \| x(0) \| \exp(-\epsilon_5 t), \forall t \geq 0, \tag{A.11}$$

in some ball around the origin $x = \mathbf{0}_n$. If the stability of the equilibrium point $x = \mathbf{0}_n$ holds for all initial states, $x = \mathbf{0}_n$ is said to be globally stable. The equilibrium point $x = \mathbf{0}_n$ is globally asymptotically stable (respectively, globally exponentially stable) when $x = \mathbf{0}_n$ is asymptotically stable (respectively, exponentially stable) for all initial states.

Lemma A.6 Let $x = \mathbf{0}_n$ be an equilibrium point for (A.10). Let $V : \mathbb{R}^n \to \mathbb{R}$ be a continuously differentiable function such that:

- $V(\mathbf{0}_n) = 0$ and $V(x) > 0, \forall x \neq \mathbf{0}_n$;
- $\| x \| \to \infty \Rightarrow V(x) \to \infty$ (a function satisfying this condition is said to be radially unbounded);
- $\dot{V}(x) < 0, \forall x \neq \mathbf{0}_n$.

Then the equilibrium point $x = \mathbf{0}_n$ is globally asymptotically stable. (see Theorem 4.2 on p. 124 of Khalil *et al.* [3]).

Definition A.1 A set M is said to be an invariant set with respect to (A.10) if $x(0) \in M$ implies $x(t) \in M, \forall t \in \mathbb{R}$. A set M is said to be a positively invariant set if $x(0) \in M$ implies $x(t) \in M, \forall t \geq 0$ (see p. 127 of Khalil *et al.* [3]).

Lemma A.7 Consider the autonomous system (A.10). Let $V : \mathbb{R}^n \to \mathbb{R}$ be a continuously differentiable function. Assume that:

- for some $c > 0$, the region $\Omega_c \stackrel{\Delta}{=} \{\in \mathbb{R}^n : V(x) < c\}$ is bounded;
- $\dot{V}(x) \leq 0, \forall x \in \Omega_c$.

Suppose E are the set of all points in Ω_c where $\dot{V}(x) = 0$, and let M be the largest invariant set in E. Then every solution $x(t)$ starting in Ω_c approaches M as $t \to \infty$ (see Theorem 3.4 on p. 272 of Slotine and Li [4]).

Lemma A.8 Consider the autonomous system (A.10). Let $V : \mathbb{R}^n \to \mathbb{R}$ be a continuously differentiable function. Assume that

- $\dot{V}(x) \leq 0, \forall x \in \mathbb{R}^n$;

- $V(x) \to \infty$ as $\| x \| \to \infty$.

Suppose E be the set of all points where $\dot{V}(x) = 0$, and let M be the largest invariant set in E. Then all solutions converge to M globally asymptotically as $t \to \infty$ (see Theorem 3.5 on p. 272 of Slotine and Li [4]).

Definition A.2 A function g is said to be uniformly continuous on a S if, given $\varepsilon > 0$, there is $\delta > 0$ (dependent only on S) such that the inequality

$$|x - y| < \varepsilon \Rightarrow |g(x) - g(y)| < \delta.$$

holds for all $x, y \in S$. For instance, a function g defined on $[0, \infty)$ is uniformly continuous if

$$\forall R > 0, \exists \eta(R) > 0, \forall t_1 > 0, t > 0, |t - t_1| < \eta \Rightarrow |g(t) - g(t_1)| < R$$

(see p. 123 of Slotine and Li [4]).

Lemma A.9 A first-order differentiable function $f(t)$ is uniformly continuous on $[0, \infty)$ if $\dot{f}(t)$ is bounded on $[0, \infty)$ (see p. 123 of Slotine and Li [4]).

Lemma A.10 If the differentiable function $f(t)$ has a finite limit as $t \to \infty$, and if $\ddot{f}(t)$ is bounded, then $\dot{f}(t) \to 0$ as $t \to \infty$ (see Lemma 4.2 on p. 123 of Slotine and Li [4]).

Definition A.3 A continuous function $\alpha : [0, a) \mapsto [0, \infty)$ is said to belong to class $\mathcal{K}$ if it is strictly increasing and $\alpha(0) = 0$. It is said to belong to class $\mathcal{K}_\infty$ if $a = \infty$ and $\alpha(r) \to \infty$ as $r \to \infty$ (see Definition 4.2 on p. 144 of Khalil *et al.* [3]).

Definition A.4 A continuous function $\beta : [0, a) \times [0, a) \mapsto [0, \infty)$ is said to belong to class $\mathcal{KL}$ if, for each fixed s, the mapping $\beta(r, s)$ belongs to class $\mathcal{K}$ with respect to r and, for each fixed r, the mapping $\beta(r, s)$ is decreasing with respect to s and $\beta(r, s) \to 0$ as $s \to \infty$ (see Definition 4.3 on p. 144 of Khalil *et al.* [3]).

A.2.3 Definitions of Input-to-state Stability

Definition A.5 Consider the system

$$\dot{x} = f(t, x, u) \tag{A.12}$$

where $f : [0, \infty) \times \mathbb{R}^n \times \mathbb{R}^m \to R^n$ is piecewise continuous in t and locally Lipschitz in x and u. The input $u(t)$ is a piecewise continuous, bounded function of t for all $t \geq 0$. System (A.12) is said to be *globally input-to-state stable* if there exist a class $\mathcal{KL}$ function β and a class $\mathcal{K}$ function γ such that for any initial state $x(t_0)$ and any bounded input $u(t)$, the solution $x(t)$ exists for all $t \geq t_0$ and satisfies

$$\|x(t)\| \leq \beta(\|x(t_0)\|, t - t_0) + \gamma\left(\sup_{t_0 \leq \tau \leq t} \|u(\tau)\|\right) \tag{A.13}$$

(see Definition 4.7 on p. 175 of Khalil *et al.* [3]).

Definition A.6 The system (A.12) is said to be *locally input-to-state stable* if there exist positive constants k_1 and k_2 such that inequality (A.13) is satisfied for $\|x(t_0)\| < k_1$ and $\sup_{t \geq t_0} \|u(t)\| < k_2$ (see Exercise 4.60 on p. 192 of Khalil *et al.* [3]).

A.2.4 Bounds of Solutions of Linear Systems

Lyapunov analysis will be used to show the boundedness and ultimate bound of solutions of some disturbed state equations. A useful lemma is as follows.

Lemma A.11 Consider the state solution $x(t)$ of the linear time-invariant equation

$$\dot{x} = Ax + Bu, \tag{A.14}$$

where $x \in \mathbb{R}^m$ is the state, $u \in \mathbb{R}$ is the continuously differentiable input, and matrices $A \in R^{m\times m}$ and $B \in R^{m\times 1}$. If A is Hurwitz, then

1. system (A.14) is globally input-to-state stable (GISS) – that is, if u is bounded for all t – then $x(t)$ with any initial state $x(t_0)$ is also bounded for all t;
2. there exist a class $\mathcal{KI}$ function β and a time $T \geq 0$ (whose value depends on $x(t_0)$ and $\|u\|_{\mathcal{L}_\infty} =: \sup_{t\geq 0}|u(t)|$) such that $x(t)$ with any $x(t_0)$ satisfies

$$\|x(t)\|_2 \leq \beta(\|x(0)\|_2, t), \forall t_0 \leq t \leq t_0 + T \tag{A.15}$$

$$\|x(t)\|_2 \leq \frac{2\lambda_{\max}(P)\|B\|_2\|u\|_{\mathcal{L}_\infty}}{\theta}\sqrt{\frac{\lambda_{\max}(P)}{\lambda_{\min}(P)}}, \forall t \geq t_0 + T \tag{A.16}$$

where $0 < \theta < 1$, $\lambda_{\max}(P)$ and $\lambda_{\min}(P)$ are the maximum and minimum eigenvalues of the symmetric positive-definite matrix $P \in \mathbb{R}^{m\times m}$ which is the solution of the Lyapunov equation

$$PA + A^T P = -I_m. \tag{A.17}$$

Proof: View system (A.14) as a perturbation of the unforced system $\dot{x} = Ax$. Point 1 can be obtained by directly applying Lemma 4.6 from Khalil *et al.* [3]. In fact, it is easy to check that all conditions of this lemma are trivially satisfied by system (A.14) since:

- the term $Ax + Bu$ is continuously differentiable and globally Lipschitz in (x, u);
- $\dot{x} = Ax$ with Hurwitz marix A has a globally exponentially stable equilibrium point at the origin $x = 0$.

To show point 2, consider the Lyapunov function

$$V(x) = x^T P x, \tag{A.18}$$

which obviously satisfies the inequalities

$$\lambda_{\min}(P)\|x\|_2^2 \leq V(x) \leq \lambda_{\max}(P)\|x\|_2^2. \tag{A.19}$$

The first-order time derivative of $V(x)$ along the trajectories of (A.14) is given by

$$\dot{V}(x) = -\|x\|_2^2 + 2x^T PBu. \tag{A.20}$$

Clearly, with any $\theta \in (0, 1)$,

$$\dot{V}(x) \leq -(1-\theta)\|x\|_2^2 - (\theta\|x\|_2 - 2\|P\|_2\|B\|_2\|u\|_\infty)\|x\|_2. \tag{A.21}$$

Furthermore,

$$\dot{V}(x) \leq -(1-\theta)\|x\|_2^2, \forall \|x\|_2 \geq \mu \tag{A.22}$$

where $\mu = \frac{2\lambda_{\max}(P)\|B\|_2\|u\|_\infty}{\theta}$ with $\lambda_{\max}(P) = \|P\|_2$.

Take $\alpha_1(\|x\|_2) = \lambda_{\min}(P)\|x\|_2^2$ and $\alpha_2(\|x\|_2) = \lambda_{\max}(P)\|x\|_2^2$ (with small abuse of the symbols α_1 and α_2). Because α_1 and α_2 are $\mathcal{K}_\infty$ functions, all conditions of Theorem 4.18 in Khalil *et al.* [3] (i.e., inequalities (4.39) and (4.40) therein) are globally satisfied. Applying this theorem directly to system (A.14) gives point 2. □

By applying the above lemma to some lower-order linear system, it is easy to derive some important properties concerning the bounds of the solution of these systems. For instance, we first consider the following first-order linear equation:

$$\dot{x} = -\frac{1}{T}x + u. \tag{A.23}$$

where T is a positive constant, and $x \in \mathbb{R}, u \in \mathbb{R}$.

Lemma A.12 The solutions of (A.23) have the following three properties.

- $x(t)$ is uniformly ultimately bounded by the bound Tu_∞, that is,

$$\lim_{t\to+\infty} |x(t)| \leq Tu_\infty. \tag{A.24}$$

- $x(t)$ is bounded by $\max(|x_0|, Tu_\infty)$ for all $t \geq 0$, that is,

$$|x(t)| \leq \max(|x_0|, Tu_\infty), \forall t \geq 0. \tag{A.25}$$

- if $\lim_{t\to+\infty} u(t) = 0$, then

$$\lim_{t\to+\infty} x(t) = 0, \tag{A.26}$$

where $x_0 = x(0)$ denotes the initial value of $x(t)$.

Proof: The third statement is true by noting that (A.23) is GISS with $\dot{d}_i$ as the input and $\tilde{d}_i$ as the state. To prove the first two statements, we first solve (A.23) and obtain:

$$x(t) = \int_0^t \exp\left(-\frac{t-\tau}{T}\right)u(\tau)d\tau + x_0\exp\left(-\frac{t}{T}\right). \tag{A.27}$$

It follows that

$$\begin{aligned} |x(t)| &\leq \int_0^t \exp\left(-\frac{t-\tau}{T}\right)|u(\tau)|d\tau + |x_0|\exp\left(-\frac{t}{T}\right) \\ &< u_\infty \int_0^t \exp\left(-\frac{t-\tau}{T}\right)d\tau + |x_0|\exp\left(-\frac{t}{T}\right) \\ &= u_\infty T + c\exp\left(-\frac{t}{T}\right), \end{aligned} \tag{A.28}$$

where $c = |x_0| - Tu_\infty$. Then, (A.24) is an immediate result of the foregoing inequality (A.28). It can also be seen that if $c \geq 0$ (i.e., $|x_0| \geq Tu_\infty$), $Tu_\infty + c\exp\left(-\frac{t}{T}\right) \leq Tu_\infty + c = |x_0|$ and thus $|x(t)| \leq |x_0|$, $\forall t \geq 0$; if $c < 0$ (i.e., $|x_0| < Tu_\infty$), then $|x(t)| < Tu_\infty$, $\forall t \geq 0$. Hence (A.25) holds. □

A.2.5 Results for Small-signal L_∞ Stability

Consider the system

$$\dot{x} = f(t, x, u), x(0) = x_0 \tag{A.29}$$

$$y = h(t, x, u) \tag{A.30}$$

where $x \in R^n$, $u \in R^m$, $y \in R^q$, $f : [0, \infty) \times D \times D_u \to R^n$ is piecewise continuous in t and locally Lipschitz in (x, u), $h : [0, \infty) \times D \times D_u \to R^q$ is piecewise continuous in t and continuous in (x, u), $D \subset R^n$ is a domain that contains $x = 0$, and $D_u \subset R^m$ is a domain containing the point $u = 0$. For each fixed $x_0 \in D$, the state model given by (A.29) and (A.30) defines an operator H that assigns to each input signal $u(t)$ the corresponding output signal $y(t)$. Suppose $x = 0$ is an equilibrium point of the unforced system

$$\dot{x} = f(t, x, 0). \tag{A.31}$$

Lemma A.13 Consider the system (A.29)–(A.30) and take $r > 0$ such that $\{\| x \| \leq r\} \subset D$. Suppose that:

- $x = 0$ is a uniformly asymptotically stable equilibrium point of (A.31), and there is a Lyapunov function $V(t, x)$ that satisfies

$$\alpha_1(\| x \|) \leq V(t, x) \leq \alpha_2(\| x \|) \tag{A.32}$$

$$\frac{\partial V}{\partial t} + \frac{\partial V}{\partial x} f(t, x, 0) \leq -\alpha_3(\| x \|) \tag{A.33}$$

$$\| \frac{\partial V}{\partial x} \| \leq \alpha_4(\| x \|) \tag{A.34}$$

  for all $(t, x) \in [0, \infty) \times D$ for some class $\mathcal{K}$ functions α_1 to α_4
- f and h satisfy the inequalities

$$\| f(t, x, u) - f(t, x, 0) \| \leq \alpha_5(\| u \|) \tag{A.35}$$

$$\| h(t, x, u) \| \leq \alpha_6(\| x \|) + \alpha_7(\| u \|) + \eta \tag{A.36}$$

  for all $(t, x, u) \in [0, \infty) \times D \times D_u$ for some class $\mathcal{K}$ functions α_5 to α_7, and a nonnegative constant η.

Then, for each x_0 with $x_0 \leq \alpha_2^{-1}(\alpha_1(r))$, the system (A.29)–(A.30) is small-signal L_∞ stable. Moreover, the system trajectory satisfies the inequality (i.e., inequality (5.21) in the proof of this theorem on p. 207 of Khalil *et al.* [3]):

$$\|x(t)\| \leq \beta(\|x_0\|, t) + \gamma\left(\sup_{t_0 \leq t \leq \tau} \|u(t)\|\right), t_0 \leq t \leq \tau \tag{A.37}$$

for each x_0 with $\|x_0\| \leq \alpha_2^{-1}(\alpha_1(r))$ and $\sup_{t_0 \leq t \leq \tau}\|u(t)\| \leq r_u$, where $r_u > 0$ is small enough such that $\{\| u \| \leq r_u\} \subset D_u$ and $\alpha_3^{-1}\left(\frac{\alpha_4(r)\alpha_5(\sup t_0 \leq t \leq \tau \|u(t)\|)}{\theta}\right) < \alpha_2^{-1}(\alpha_1(r))$.

Lemma A.14 Suppose that, in some neighborhood of the point $(x, u) = (0, 0)$, the function $f(t, x, u)$ is continuously differentiable, the Jacobian matrices $[\partial f/\partial x]$ and $[\partial f/\partial u]$ are bounded, uniformly in t, and $h(t, x, u)$ satisfies (A.36). If the unforced system (A.31) has a uniformly asymptotically stable equilibrium at the point $x = 0$, then the system (A.29)-(A.30) is small-signal L_∞ stable (see Corollary 5.3 on p. 208 of Khalil *et al.* [3]).

Generalized saturation functions are used in control design. They are defined as follows.

Lemma A.15 A function $\sigma : R \to R$ is said to be a generalized saturation function (GSF) with bound $\overline{\sigma}$, if it is locally Lipschitz, nondecreasing, and satisfies the following:

P1: $x\sigma(x) > 0, \forall x \neq 0$;
P2: $|\sigma(x)| \leq \overline{\sigma}, \forall x \in R$

(see Aguiñaga-Ruiz *et al.* [5]). A strictly increasing continuously differentiable GSF, $\sigma(x)$, has the following three properties (as proven in Aguiñaga-Ruiz *et al.*'s Lemma 1):

P3: the derivative of σ with respect to its argument (i.e., $\sigma'(x) = \frac{d\sigma}{dx}(x)$) is positive and bounded, i.e., there exist a constant $\sigma'_M \in (0, \infty)$ such that $0 < \sigma'(x) \leq \sigma'_M, \forall x \in R$;
P4: $\sigma(x)$ is globally Lipschitz, i.e.,

$$|\sigma(x_1) - \sigma(x_2)| \leq \sigma'_M |x_1 - x_2|, \forall x_1, x_2 \in R; \tag{A.38}$$

P5:

$$\int_0^x \sigma(\tau) d\tau \to +\infty \quad as \quad |x| \to +\infty. \tag{A.39}$$

It is easy to check that $\tanh(x)$ and $\frac{x}{\sqrt{1+x^2}}$ are two GSFs satisfying P1–P5 with $\overline{\sigma} = 1$.

A.3 Proofs

A.3.1 Proof of Theorem 5.6

Proof: With (5.44), the following inequalities hold:

$$\begin{aligned} |\Delta_i(t)| \leq \|\boldsymbol{\Delta}(t)\|_\infty &\leq \|L + \overline{B}\|_\infty \|\tilde{\boldsymbol{d}}(t)\|_\infty \\ &= \|L + \overline{B}\|_\infty \max_{1 \leq i \leq N} \left(\tilde{\boldsymbol{d}}_i(t)\right), \forall t \geq 0, i \in \mathcal{I}. \end{aligned} \tag{A.40}$$

Combining (5.36) and (A.40) results in (5.45). Because statement 2 of the theorem is identical to that of Theorem 5.5, it can be proven in the same manner.

Apply Lemma A.1 to systems (5.39) with $t_0 = t_d$. In particular, by analogy with inequality (A.16), we conclude that there is $t_q > t_d$ such that

$$\|[\boldsymbol{e}_i(t), \dot{\boldsymbol{e}}_i(t)]\|_2 \leq \frac{2b_\Delta \lambda_{\max}(P_i)}{\theta} \sqrt{\frac{\lambda_{\max}(P_{qi})}{\lambda_{\min}(P_i)}}, \forall t \geq t, i \in \mathcal{I}. \tag{A.41}$$

This result, along with the fact that $|\boldsymbol{e}_i(t)| \leq \|[\boldsymbol{e}_i(t), \dot{\boldsymbol{e}}_i(t)]\|_2, \forall t \geq 0$, indicates that

$$|\boldsymbol{e}_i(t)| \leq \frac{2b_{\Delta q} \lambda_{\max}(P_i)}{\theta} \sqrt{\frac{\lambda_{\max}(P_i)}{\lambda_{\min}(P_i)}}, \forall t \geq t_q, i \in \mathcal{I}. \tag{A.42}$$

We here choose θ that additionally satisfies $\frac{b_\Delta}{\theta} \leq \frac{\|L+\bar{B}\|_\infty T\bar{d}_d}{\theta}$ for (A.42). Note that this condition can always be satisfied because θ can be any number in the set $\{\theta \mid 0 < \theta < 1\}$.

It is then easy to obtain (5.46) by combining (A.42) and the inequality $\|\boldsymbol{e}_q(t)\|_2 \leq \sqrt{N}\max_{1\leq i\leq N}(|\boldsymbol{e}_{qi}(t)|)$, $\forall t \geq 0$. Inequality (5.47) follows from (5.46) by noting the property (3.54) of the type III LSEs. This ends the proof of this theorem. □

A.3.2 Proof of Lemma 5.10

Proof: To prove the first statement, consider the Lyapunov function candidate

$$V_i = k_i^1 \int_0^{\boldsymbol{e}_{qi}} \sigma_1(\tau)d\tau + \frac{1}{2}\dot{\boldsymbol{e}}_{qi}^2, \tag{A.43}$$

for each $i \in \mathcal{I}$, which is positive definite and radially unbounded with respect to $\boldsymbol{e}_{qi}$ and $\dot{\boldsymbol{e}}_{qi}$ due to (A.39). Differentiating V_i along the trajectories of (5.57), gives

$$\dot{V}_i = k_i^1\sigma_1(\boldsymbol{e}_{qi})\dot{\boldsymbol{e}}_{qi} + \dot{\boldsymbol{e}}_{qi}\left(-k_i^1\sigma_1(\boldsymbol{e}_{qi}) - k_i^2\sigma_2(\dot{\boldsymbol{e}}_{qi})\right) = -\dot{\boldsymbol{e}}_{qi}k_i^2\sigma_2(\dot{\boldsymbol{e}}_{qi}) \leq 0$$

where the odd property of $\sigma_2(\cdot)$ is used. Thus, $\boldsymbol{e}_i$ and $\dot{\boldsymbol{e}}_i$ are bounded for all $t \geq 0$ under any initial condition.

Noting $\dot{V}_i \equiv 0$ implies $\dot{\boldsymbol{e}}_{qi} \equiv 0$, we obtain $\lim_{t\to+\infty} \dot{\boldsymbol{e}}_{qi}(t) = 0$ by LaSalle's invariance principle. Since the condition of the Barbalat lemma is satisfied ($\dddot{\boldsymbol{e}}_{qi}(t)$ is obviously bounded), we obtain $\lim_{t\to+\infty} \ddot{\boldsymbol{e}}_{qi}(t) = 0$. As a result, $\lim_{t\to+\infty} \boldsymbol{e}_{qi}(t) = 0$ due to (5.57). This completes the proof of the first statement.

The second statement is an immediate result of the first statement due to the properties of the Type III LSEs. Let $w_i = -k_i^1\sigma_1(\boldsymbol{e}_{qi}) - k_i^2\sigma_2(\dot{\boldsymbol{e}}_{qi})$, $W = [w_1, \cdots, w_N]^T$, $B = [b_1, \cdots, b_N]^T$ and $U = [\boldsymbol{u}_1^0, \cdots, \boldsymbol{u}_N^0]^T$. Then, the nominal control vector satisfies $HU = B\ddot{\boldsymbol{q}}_d + W$ and

$$U = H^{-1}(B\ddot{\boldsymbol{q}}_d + W) \tag{A.44}$$

which further shows that U is unique and bounded since both W and $\ddot{\boldsymbol{q}}_d$ are bounded under the considered conditions. Applying (A.1) to (A.44) and considering the parameter condition (5.59), yields

$$\begin{aligned} \|U\|_\infty &\leq \|H^{-1}\|_\infty \|B\ddot{\boldsymbol{q}}_d + W\|_\infty \\ &\leq \|H^{-1}\|_\infty \max_{1\leq i\leq N}\left(b_i\bar{q}_{dd} + k_i^1\bar{\sigma}_1 + k_i^2\bar{\sigma}_2\right) \\ &\leq \bar{\mu} \end{aligned} \tag{A.45}$$

Now, the third statement is clear since $\|U\|_\infty = \max_{1\leq i\leq N}|\boldsymbol{u}_i^0|$. □

A.3.3 Proof of Lemma 5.13

Proof: We aim to show the small-signal L_∞ stability of system (5.60) by applying Lemma A.14. The proof is then reduced to checking that all the conditions of the lemma are satisfied. For each $i = 1, \ldots, n$, we use $x_{ei} =: (x_{i1}, x_{i2})^T = (\boldsymbol{e}_{qi}, \dot{\boldsymbol{e}}_{qi})^T$ and

$y_{ei} = x_{ei}$ to denote the system state and output, respectively. Then, the second-order equation included in (5.60) can be rewritten into

$$\dot{x}_{ei} = f(x_{ei}, \tilde{\Delta}_i), y_{ei} = h(x_{ei}, \tilde{\Delta}_i), i \in \mathcal{I}, \tag{A.46}$$

where

$$f\left(x_{ei}, \tilde{\Delta}_i\right) = \begin{bmatrix} x_{i2} \\ -k_i^1\sigma_1(x_{i1}) - k_i^2\sigma_2(x_{i2}) + \tilde{\Delta}_i \end{bmatrix}, h\left(x_{ei}, \tilde{\Delta}_i\right) = x_{ei}. \tag{A.47}$$

The Jacobian matrices are given by

$$\frac{\partial f(x_{ei}, \tilde{\Delta}_i)}{\partial x_{ei}} = \begin{bmatrix} 0 & 1 \\ -k_i^1 \dfrac{\partial\sigma_1(x_{i1})}{\partial x_{i1}} & -k_i^2 \dfrac{\partial\sigma_2(x_{i2})}{\partial x_{i2}} \end{bmatrix},$$

$$\frac{\partial f(x_{ei}, \tilde{\Delta}_i)}{\partial \tilde{\Delta}_i} = \begin{bmatrix} 0 \\ 1 \end{bmatrix}, i \in \mathcal{I}. \tag{A.48}$$

By property P3 of saturation functions, $f(x_{ei}, \tilde{\Delta}_i)$ is continuously differentiable and the Jacobian matrices $\frac{\partial f(x_{ei}, \tilde{\Delta}_i)}{\partial x_{ei}}$ and $\frac{\partial f(x_{ei}, \tilde{\Delta}_i)}{\partial \tilde{\Delta}_i}$ are bounded uniformly in t. Moreover,

$$\|h(x_{ei}, \tilde{\Delta}_i)\| = \|x_{ei}\|, i \in \mathcal{I}, \tag{A.49}$$

in any neighborhood of $(x_{ei} = 0, \tilde{\Delta}_i = 0)$. Note that $(\boldsymbol{e}_{qi}, \dot{\boldsymbol{e}}_{qi}) = (0, 0)$ is a globally uniformly asymptotically stable equilibrium point of the unforced system (5.57). Thus, all the conditions of Lemma A.14 are satisfied and the second-order equation included in (5.60) is small-signal L_∞ stable. This easily ensures that the third-order system (5.60) is also small-signal L_∞ stable.

We here apply Lemma A.13 to the system to show it is also LISS. The proof is reduced to checking that all the conditions of Lemma A.13 are satisfied. Since the system clearly satisfies the conditions of the converse Lyapunov theorem (i.e., Theorem 4.16 in Khalil *et al.* [3]), there exists a Lyapunov function $V(t, x_{ei})$ satisfying the inequalities (A.32)–(A.34). Moreover, all the other conditions of Lemma A.13 (i.e., (A.35)–(A.36)) also hold since $\|f(x_{ei}, \tilde{\Delta}_i) - f(x_{ei}, 0)\| = |\tilde{\Delta}_i|$ and $\|h(x_{ei}, \tilde{\Delta}_i)\| = \|x_{ei}\|$. Thus, all the results included in Lemma A.13 hold. So, similar to (A.37), the inequality

$$\|x_{ei}(t)\| \leq \beta(\|x_{ei}(0)\|, t) + \gamma\left(\sup_{0 \leq t \leq \tau} |\tilde{\Delta}_i(t)|\right), \quad t \geq 0 \tag{A.50}$$

holds for any $\|x_{ei}(0)\| < r_{x0}$ and $\sup_{0 \leq t \leq \tau} |\tilde{\Delta}_i(t)| < r_u$, where $r_{x0} > 0$ is equivalent to $\alpha_2^{-1}(\alpha_1(r))$ in Lemma A.13, and β, γ and r_u have the same meanings as in Lemma A.13.

Since the unforced system (5.57) is globally asymptotically stable, r can be arbitrarily large. Because the mapping from $\tilde{\Delta}_i$ to x_{ei} is causal, $x_{ei}(t)$ depends only on $\tilde{\Delta}_i(t^*)$ for $0 \leq t^* \leq t$. Thus, the supremum on the right-hand side of (A.50) can be taken over $[0, t]$. Thus,

$$\|x_{ei}(t)\| \leq \beta(\|x_{ei}(0)\|, t) + \gamma\left(\sup_{0 \leq \tau \leq t} |\tilde{\Delta}_i(\tau)|\right), \quad t \geq 0, \tag{A.51}$$

for $\|x_{ei}(0)\| < r_{x0}$ and $\sup_{0 \leq \tau \leq t} |\tilde{\Delta}_i(\tau)| < r_u$, which implies that the second-order equation included in (5.60) is LISS by definition. This ends the proof of Lemma 5.13. □

Bibliography

1 A. J. Laub, *Matrix Analysis for Scientists and Engineers*. Siam, 2005.

2 R. A. Horn and C. R. Johnson, *Matrix Analysis*. Cambridge University Press, 1985.

3 H. K. Khalil, *Nonlinear Systems*, Prentice Hall, 2002.

4 J.-J. E. Slotine and W. Li, *Applied Nonlinear Control*, Prentice-Hall, 1991.

5 E. Aguiñaga-Ruiz, A. Zavala-Río, V. Santibáñez, and F. Reyes, "Global trajectory tracking through static feedback for robot manipulators with bounded inputs," *IEEE Transactions on Control Systems Technology*, vol. 17, no. 4, pp. 934–944, 2009.

Formation Control of Multiple Autonomous Vehicle Systems, First Edition. Hugh H.T. Liu and Bo Zhu.
© 2018 John Wiley & Sons Ltd. Published 2018 by John Wiley & Sons Ltd.

Index